THE PHYSICAL METALLURGY OF STEELS

McGRAW-HILL SERIES IN MATERIALS SCIENCE AND ENGINEERING

Blatt: *Physics of Electronic Conduction in Solids*
Brick, Pense, and Gordon: *Structure and Properties of Engineering Materials*
Buerger: *Contemporary Crystallography*
Buerger: *Introduction to Crystal Geometry*
Darken and Gurry: *Physical Chemistry of Metals*
Dieter: *Mechanical Metallurgy*
Drauglis, Gretz, and Jaffee: *Molecular Processes on Solid Surfaces*
Elliott: *Constitution of Binary Alloys, First Supplement*
Flemings: *Solidification Processing*
Fontana and Greene: *Corrosion Engineering*
Gaskell: *Introduction to Metallurgical Thermodynamics*
Gordon: *Principles of Phase Diagrams in Materials Systems*
Guy: *Introduction to Materials Science*
Hansen: *Constitution of Binary Alloys*
Kanninen, Adler, Rosenfield, and Jaffee: *Inelastic Behavior of Solids*
Kehl: *The Principles of Metallographic Laboratory Practice*
Leslie: *The Physical Metallurgy of Steels*
Mills, Ascher, and Jaffee: *Critical Phenomena in Alloys, Magnets, and Superconductors*
Murr: *Electron Optical Applications in Materials Science*
Rhines: *Phase Diagrams in Metallurgy: Their Development and Application*
Rosenfield, Hahn, Bement, and Jaffee: *Dislocation Dynamics*
Rosenqvist: *Principles of Extractive Metallurgy*
Rozenfeld: *Corrosion Inhibitors*
Rudman, Stringer, and Jaffee: *Phase Stability in Metals and Alloys*
Shewmon: *Diffusion in Solids*
Shewmon: *Transformations in Metals*
Shunk: *Constitution of Binary Alloys, Second Supplement*
Smith: *Structure and Properties of Engineering Alloys*
Wert and Thomson: *Physics of Solids*

THE PHYSICAL METALLURGY OF STEELS

William C. Leslie
University of Michigan

Hemisphere Publishing Corporation
Washington New York London

McGraw-Hill Book Company
New York St. Louis San Francisco Auckland Bogotá
Hamburg Johannesburg London Madrid Mexico
Montreal New Delhi Panama Paris São Paulo
Singapore Sydney Tokyo Toronto

THE PHYSICAL METALLURGY OF STEELS

1 2 3 4 5 6 7 8 9 0 BCBC 8 9 8 7 6 5 4 3 2 1

This book was set in Press Roman by Communication Crafts, Ltd. The editors were Diane Heiberg, Valerie M. Ziobro, and Elizabeth Dugger. BookCrafters, Inc. was printer and binder.

Library of Congress Cataloging in Publication Data

Leslie, William C date.
The physical metallurgy of steels.

(McGraw-Hill series in materials science and engineering)
Bibliography: p.
Includes index.
1. Steel–Metallurgy. I. Title.
TN693.I7L45 669'.96142 80-28332
ISBN 0-07-037780-4

To the Memory of
ALFRED SHOU-BING KEH
who would have made of this a better text

CONTENTS

PREFACE

Formerly, it was the general practice to teach physical metallurgy at the undergraduate level in at least two separate parts, ferrous and nonferrous. Later, this separation was decried and most schools began to teach physical metallurgy as a unified subject. For introductory courses, this is the most effective method, but it does not suffice, for obvious reasons. The importance of steels as engineering materials is so overwhelming and the characteristics of body-centered cubic metals are so singular that a separate course of instruction in steels is required for the proper education of metallurgists. As the ores of metals other than iron, aluminum, magnesium, and titanium are depleted and as the less abundant metals become more costly, steels will become even more important than they are now and it will be essential to achieve optimum properties with minimum use of alloying elements.

The book is intended to be used in a course for advanced undergraduate or first-year graduate students, extending over one or two semesters. In the past, courses in the physical metallurgy of steels emphasized the characteristics of steels intended to be heat-treated by the user and largely ignored the vast majority of steels that are not so treated. One of the purposes of this text is to remedy this omission. It presupposes some knowledge of physical chemistry, mechanics of materials, dislocations and other lattice defects, and phase transformations in solids. An attempt is made to apply the findings of basic research to economical engineering materials. Although primarily intended for university students, this book should be useful to practicing engineers who wish to improve their understanding of the properties of steels.

Our knowledge of body-centered cubic metals and why they act as they do has expanded greatly in the past two decades so that we can teach their properties with greater confidence than was formerly the case. Therefore, the book follows the scheme of discussing systems of steadily increasing complexity, beginning with the properties of high-purity iron, then considering the effects of the addition of interstitial solute atoms, substitutional atoms, interactions between solutes in iron, the properties of carbon steels, high-strength low-alloy steels, heat-treated steels, thermomechanical treatments, very high strength steels and fracture prevention, surface hardening processes, steels for magnetic and electrical applications, stainless steels, and tool steels.

There are many thousands of published articles dealing with the physical metallurgy of steels and although the book makes greater use of literature references than is customary in textbooks, it is obviously impossible to provide a complete coverage of the literature. I have favored recent references in the perhaps optimistic view that we do learn from our predecessors. This decision means that many original sources are not given due credit, and for these omissions I apologize. The references included will provide readers with an entry to the pertinent literature, in which they can pursue their own interests.

The text had its genesis in a series of lectures to graduate students at The Ohio State University in 1964–65 while the author was serving as Battelle Visiting Professor of Metallurgy. These lectures were revised and expanded for classes taught at The University of Michigan in the period 1973–1978. The book was given its final form during a sabbatical leave in the Department of Mining and Metallurgy, University of Melbourne, from January through June 1979. The author is grateful to that department for the provision of office space and facilities and especially to Ms. Nola Sim for her faultless typing of the manuscript. The tedious task of typing many revisions and additions was done very competently and cheerfully by Ms. Kay Carroll of the Department of Materials and Metallurgical Engineering, The University of Michigan.

I am indebted to my former colleagues, Drs. L. J. Cuddy, H. Hu, J. T. Michalak, P. R. Mould, G. R. Speich, W. A. Spitzig, and H. A. Wriedt for reviews of portions of the text and for other assistance. Whatever merit is found here is due in large measure to their efforts; the remaining deficiencies are my own responsibility.

William C. Leslie

CHAPTER

ONE

PROPERTIES OF HIGH-PURITY IRON

I-1. IDEAL STRENGTH OF IRON CRYSTALS

Although we are herein primarily concerned with the relations between structure and properties in steels, it is obvious that an understanding of these relations must be based upon knowledge of the properties of the base material, body-centered cubic (bcc) iron. Let us begin by considering the upper limit of the strength of iron—its ideal strength—set by the magnitude of interatomic forces. This ideal strength can be measured by determining the strength of individual whisker crystals,[1] or calculated approximately by several different means. One of these[2] relates the ideal tensile strength to Young's modulus E, equilibrium surface energy γ, and the interplanar spacing a_0 of planes perpendicular to the applied stress:

$$\sigma_{\max} = \left(\frac{E\gamma}{a_0}\right)^{1/2} \tag{I-1}$$

The approximate ideal shear strength can be calculated from the relation developed by Frenkel[3]:

$$\tau_{\max} = \frac{Gb}{2\pi d} \tag{I-2}$$

where G = shear modulus
b = repeat distance in the shear direction
d = interplanar spacing of the shear planes

[1] S.S. Brenner, *J. Appl. Phys.*, 27:854 (1956).
[2] A. Kelly, *Strong Solids*, Oxford Univ. Press, New York, 1966.
[3] J. Frenkel, *Z. Phys.*, 37:572 (1926).

Table I-1 Comparison of calculated and measured ideal strength of bcc Fe crystals

	Orientation	Strain (σ_{max}/E)	Strength (GPa)
a. Uniaxial tensile strength σ_{max}			
Observed	⟨111⟩	0.049	13.1
Observed	⟨100⟩	0.033	4.8
Calculated	⟨111⟩	0.23	46.0
Calculated	⟨100⟩	0.13	30.0
b. Shear strength τ_{max}			
Observed	110 ⟨111⟩		3.58
Calculated	110 ⟨111⟩		6.6–7.8

These procedures evade the problem of ignorance of the cohesive forces in a solid. In principle, methods based on models of interatomic forces should be more precise than these approximations, but a knowledge of interatomic forces is lacking; only static lattices are considered, and vacant lattice sites are ignored. Milstein[1] represented the interatomic forces in iron by a Morse potential function, in an attempt to calculate the uniaxial ideal tensile and compressive stresses in ⟨100⟩.

The results of experiments by Brenner and simple calculations by Kelly are summarized in Table I-1. It is obvious that there is no great disparity between the measured and calculated values, even though the latter can be considered only as order-of-magnitude approximations.[2] Milstein's calculations of ideal tensile strength produced estimates less than one-tenth of the measured tensile strength of iron, which probably indicates that the Morse potential model is inadequate.

The tensile strength of our strongest iron-base engineering material, heavily drawn pearlitic steel wire, can approach 4.8 GPa. Although we hasten to add that this strength is achieved by different mechanisms than that which governs the strength of iron whiskers, nevertheless, in an engineering material we have reached a strength equal to about one-third of the ideal strength of iron.

I-2. YIELDING AND PLASTIC FLOW OF BODY-CENTERED CUBIC IRON

A larger tonnage of iron is used because it is inexpensive, comparatively soft, and easily formed than because of its strength. Therefore, we must consider closely the plastic

[1] F. Milstein, *Phys. Rev.*, B3:1130 (1971).
[2] N.H. Macmillan, *J. Mater. Sci.*, 7:239 (1972).

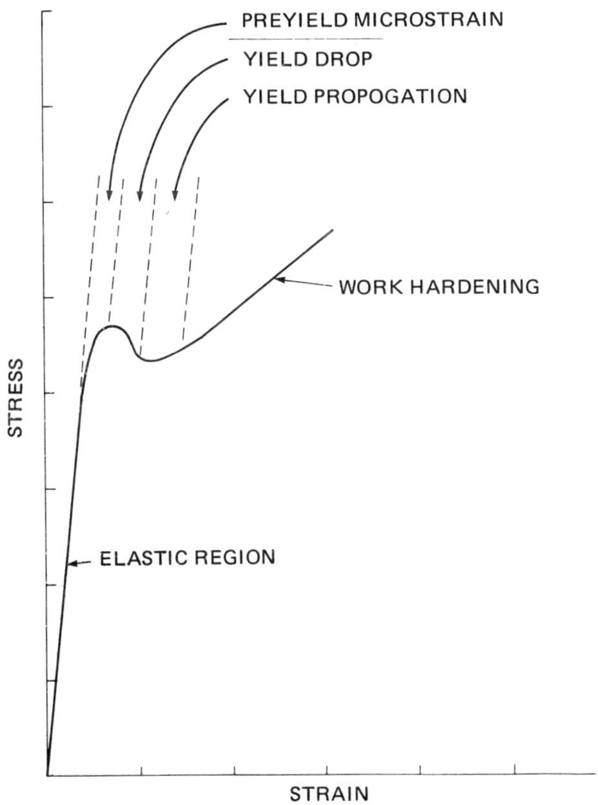

Figure I-1. Typical flow curve of polycrystalline bcc iron in uniaxial tension.

deformation of iron. Body-centered cubic iron always slips in the close-packed direction, ⟨111⟩. The shortest lattice vector extends from the atom at the corner of a unit cell to the atom at the center; thus the Burgers vector is $(a/2)\langle 111\rangle$ of magnitude 0.248 nm. The slip plane of an edge dislocation is defined by the dislocation line and its Burgers vector; therefore, edge dislocations slip only on {110} and {112} and perhaps on {123}.[1] The Burgers vector and line of a screw dislocation are parallel and do not define a plane, which means that screw dislocations can glide on any plane containing a ⟨111⟩ direction and can cross-slip. The slip plane chosen in any given instance presumably is the one on which the resolved shear stress is a maximum.

For high-purity, polycrystalline, annealed bcc iron containing essentially no interstitial solutes, the flow curve in tension appears as in Fig. I-1 at conventional testing-machine strain rates and at temperatures slightly below ambient. As load is applied in a uniaxial test the iron first yields elastically, but the extent of this elastic deformation is severely limited. It is followed by preyield microstrain, during which dislocations move in a few grains, presumably at sites of stress concentration. As straining con-

[1] J.P. Hirth and J. Lothe, *Theory of Dislocations,* McGraw-Hill, New York, 1968, p. 258.

tinues, a drop in the applied load is encountered. The testing machine is imposing an essentially constant strain rate on a specimen containing a small number of mobile dislocations. To maintain this strain rate in the specimen, the load must increase, with two effects:

1. The density of mobile dislocations is increased by dislocation multiplication.
2. The average dislocation velocity rises very rapidly.

In a restricted volume of the specimen, the localized strain rate exceeds that of the machine, and a sudden decrease occurs in the load being applied to the specimen. In the yield propagation region, the band (or bands) of highly localized deformation (Lüders bands) spread from one or several sources until they fill the entire gauge section, then uniform strain hardening begins.

It was assumed for many years that the presence of interstitial solutes (C, N) was a prerequisite for such abrupt, discontinuous yielding, and that it occurred because of the sudden unpinning of dislocations. This concept has been abandoned in favor of an explanation based on dislocation dynamics, discussed later.

Recent years have seen a marked improvement in our understanding of the details of the flow curve shown in Fig. I-1. Under a given applied stress, edge and screw dislocations in bcc iron do not move with the same facility. Edge dislocations move readily at stresses well below the macroscopic yield stress, and their rate of motion increases rapidly with increasing stress. In a high-purity Fe-3%Si alloy, Saka and Imura,[1] by direct observation in a high-voltage transmission microscope, obtained the following relation between average velocity of edge dislocations and applied shear stress:

$$\bar{\nu} = \left(\frac{\tau}{\tau_0}\right)^m \tag{I-3}$$

where τ = shear stress resolved on the slip plane
τ_0 = shear stress corresponding to unit velocity = 157 ± 5 MPa
m = 35 ± 5

Therefore, when stress is applied to an annealed, high-purity iron, the first flow to occur is *microscopic*. The stress required for microscopic yielding can be divided into two parts[2]:

σ_E stress at which linear elastic behavior ends (the elastic limit) and hysteresis begins
σ_A stress necessary for irreversible (plastic) deformation—that is, the stress at which the hysteresis loop fails to close. This is the anelastic limit

As shown in Fig. I-2, the elastic limit σ_E represents the stress required for reversible motion of edge dislocations, and the anelastic limit σ_A is the stress required for

[1] H. Saka and T. Imura, *J. Phys. Soc. Jpn.*, 32:702 (1972).
[2] N. Brown and R.A. Ekvall, *Acta Metall.*, 8:1101 (1962).

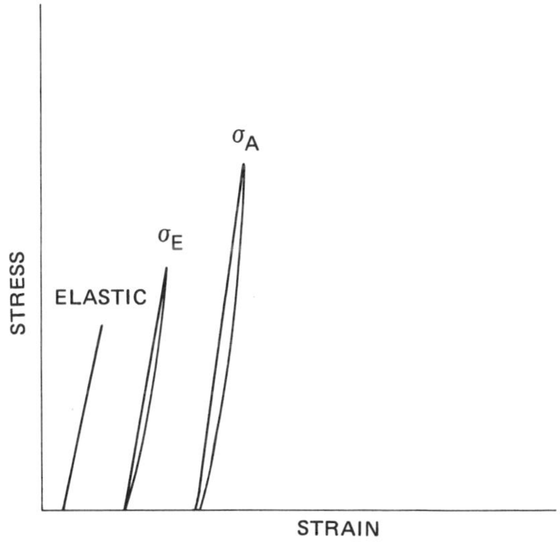

Figure I-2. Stress–strain curves of iron, in the microstrain region.

their long-range irreversible motion. These stresses are low, less than 35 MPa at room temperature, and show only a slight increase with decreasing temperature. Meakin[1] reports that σ_E for iron at room temperature is about 2 MPa. They can be measured only in the absence of dislocation pinning by interstitial impurities, and in an annealed iron the required sensitivity of strain measurement is about 10^{-6}. The extent of pre-yield microstrain can be greatly increased by prior plastic deformation, which creates large numbers of mobile dislocations. In this circumstance, microplasticity can be observed with strain sensitivities of about 10^{-5}.[2]

At stresses below the macroscopic yield stress, the motion of screw dislocations is insignificant. As the stress is increased, the available edge dislocations are exhausted. Macroscopic yielding depends upon the motion of screw dislocations, which is accompanied by dislocation multiplication via the double-cross-slip mechanism. The macroscopic yield stress—the stress at which the plastic strain rate becomes a large fraction of the strain rate imposed by the testing machine—can be considered as the stress at which abundant cross slip begins. In iron and other body-centered cubic metals, this stress is determined by the stress required to move screw dislocations.

I-3. STRAIN HARDENING OF IRON

As iron is cold-worked, the dislocation density increases, and, in the initial stages of working, the density is proportional to true strain (Fig. I-3). The dislocation density for a given strain is greater in fine-grained than in coarse-grained iron. The relation between dislocation density and flow stress in iron is shown in Fig. I-4. The straight

[1] J.D. Meakin, *Can. J. Phys.*, 45:1121 (1967).
[2] H.D. Solomon, C.J. McMahon, Jr., and W.C. Leslie, *Trans. Am. Soc. Met.*, 62:886 (1969).

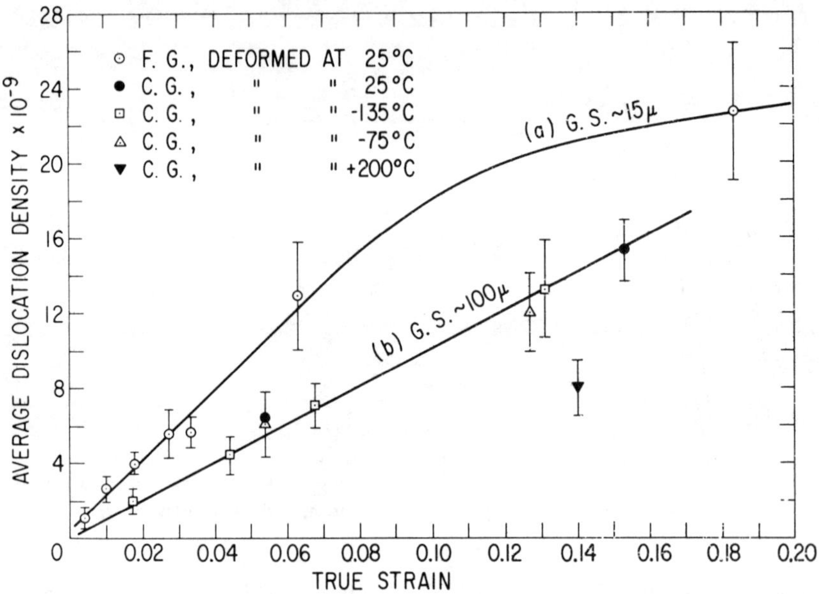

Figure I-3. Relationship between average dislocation density and plastic strain of iron. CG = coarse grain, FG = fine grain. (A.S. Keh and S. Weissmann, *Electron Microscopy and the Strength of Crystals,* G. Thomas and J. Washburn, eds., Interscience, New York, 1963, by permission.)

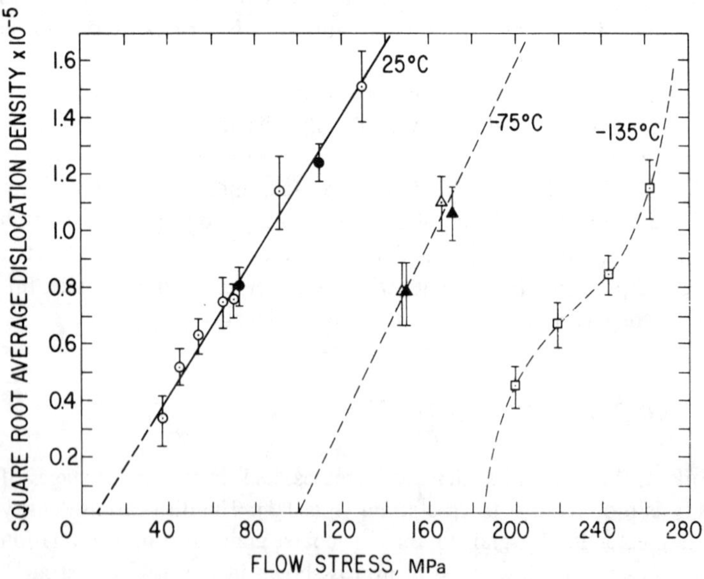

Figure I-4. Relationship between square root of average dislocation density and flow stress of iron. ○△□, G.S. ~15 μm; ●▲, G.S. ~100 μm. (A.S. Keh and S. Weissmann, *Electron Microscopy and the Strength of Crystals,* G. Thomas and J. Washburn, eds., Interscience, New York, 1963, by permission.)

line at room temperature can be represented by

$$\sigma = \sigma_0 + \alpha G b \sqrt{N} \tag{I-4}$$

where α = proportionality constant
G = shear modulus
b = magnitude of Burgers vector
N = dislocation density

The rate of work hardening of iron is greatest at about room temperature, decreases to a minimum at temperatures between −40 and −75°C, then increases at lower temperatures.

For expressing the stress–strain relation in a concise form, the empirical equation

$$\sigma = K\epsilon^n \tag{I-5}$$

where σ is true flow stress, ϵ is true strain, and K and n are constants, developed originally by Ludwik,[1] is used almost to the exclusion of all others. It is applied only in the region of uniform plastic strain. It does not allow for elastic deformation at low stresses, nor for necking. Obviously in Eq. I-5, n is the slope of a log true stress–log true strain plot, with K being the true stress at a strain of unity. As discussed later, the magnitude of n, the strain-hardening exponent, is an indication of the stretchability of low-carbon steel sheets. The limit of uniform plastic strain in a tension test is numerically equal to the strain-hardening exponent. In the manner of Dieter,[2] the change in load when necking begins is defined by

$$dP = 0 \tag{I-6}$$

The load P is equal to stress σ times area

$$P = \sigma A \tag{I-7}$$

at the onset of instability

$$dP = \sigma\, dA + A\, d\sigma = 0 \tag{I-8}$$

assuming constant volume,

$$\frac{dL}{L} = -\frac{dA}{A} \tag{I-9}$$

where L is the gauge length of the specimen.

$$-\frac{dA}{A} = \frac{d\sigma}{\sigma} = d\epsilon \tag{I-10}$$

Then $$\frac{d\sigma}{d\epsilon} = \sigma \tag{I-11}$$

and $$\sigma = K\epsilon^n$$

[1] P. Ludwik, *Elemente der technologischen Mechanik,* Springer-Verlag, Berlin, 1909, p. 32.
[2] G.E. Dieter, *Mechanical Metallurgy,* 2nd ed., McGraw-Hill, New York, 1976, p. 342.

$$\frac{d\sigma}{d\epsilon} = nK\epsilon^{n-1} = K\epsilon^{n} \tag{I-12}$$

and $$\epsilon_u = n \tag{I-13}$$

(ϵ_u, the limit of uniform strain, is equal to n.)

As mentioned before, Eq. I-5 is empirical and is often not strictly obeyed. For example, when a strain-induced phase transformation occurs, as in a metastable austenitic stainless steel, the plot of log true stress vs. log true strain is concave upward—that is, n increases with increasing strain. For reasons that are still not apparent, the same plots for low-carbon steel sheets are often concave downward; the value of n *decreases* with increasing strain. Examples of such behavior have been shown by Morrison.[1] The curved plot is often approximated by two straight-line portions, with the change in slope occurring at a strain of about 0.08.

The value of n increases with increasing grain size; Morrison[1] suggests the relation

$$n = \frac{5}{10 + d^{-1/2}} \tag{I-14}$$

where d is intercept grain diameter in millimeters. The grain size of commercially produced low-carbon steel sheet is usually in the range ASTM 6 to 8 (0.045 to 0.022 mm). From Eq. I-14, this would give a variation in n of 0.04, which is significant. Unfortunately, when grains become larger than about ASTM 6, roughening of the sheet surface ("orange-peel" effect) is encountered during drawing.

The work hardening of iron at large strains depends upon the mode of deformation. In wire drawing, in which the material is subject to severe geometrical restraints, the hardening of polycrystalline iron becomes linear with strain, at least up to a true strain of 10.[2,3] The strain-hardening rate per unit true strain is about 152 MPa. As shown in Fig. I-5, Ti-gettered iron, with an initial grain size of ASTM 5 to 6, can be cold drawn to a tensile strength of about 1650 MPa at a true strain of 10. This linear strain hardening is related to the development of ribbonlike cells having a ⟨110⟩ fiber texture. This will be considered in greater detail in Sec. I-5. Because of the great complexity of the defect structure, no reliable relation between structure and flow stress has been developed.[4]

In contrast, work hardening of iron in torsion reaches saturation.[5] At a large strain, somewhat less than 4.8, the flow stress becomes independent of strain, as illustrated in Fig. I-5. Work hardening at that point is balanced by dynamic recovery. The same situation probably exists in cold rolling, where geometrical restraints are less severe than in wire drawing, but at strains of 2.8 the flow stress was still increasing.[6] In that instance, the Ludwik equation was followed up to the maximum true strain of 2.8.

[1] W.B. Morrison, *Trans. Am. Soc. Met.*, 59:824 (1966).
[2] G. Langford and M. Cohen, *Trans. Am. Soc. Met.*, 62:623 (1969).
[3] G. Langford, P.K. Nagata, R.J. Sober, and W.C. Leslie, *Met. Trans.*, 3:1843 (1972).
[4] G. Langford and M. Cohen, *Met. Trans. A*, 6A:901 (1975).
[5] C.M. Young, L.J. Anderson, and O.D. Sherby, *Met. Trans.*, 5:519 (1974).
[6] J.S.H. Lake, *AIM Conference*, Newcastle, New South Wales, 1977, Paper 4A-2.

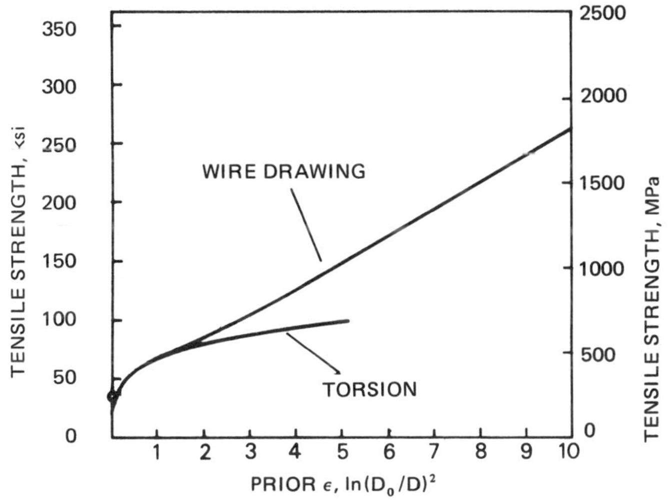

Figure I-5. Tensile strength-prior strain plots for iron gettered with 0.15 wt % Ti.

During the deformation of iron at room temperature dislocation lines are fairly straight if the strain is less than 1%. After 1% strain, the dislocations become more irregular and interactions of dislocations of different Burgers vectors occurs. Dislocations start to cluster inside individual grains, and these clusters eventually join together to form a cell structure as the amount of deformation is increased (Fig. I-6). Cells can be seen after about 3.5% strain. After more than about 9% strain, most grains are subdivided into well-defined cells whose walls consist of a tangled network of dislocations. Within each cell the dislocation density is low. With increasing amounts of deformation, the cell walls become more sharply defined and tend to align themselves in certain crystallographic directions. In most instances, the cell walls lie along {112}- and {123}-type planes after deformation at room temperature. The average size of the well-defined cells decreases slightly in the early stages of deformation, then reaches a constant value near 1.5 μm (Fig. I-7). The misorientation across a cell wall is typically about 3.5°. Decreasing the temperature at which the specimen is deformed increases the strain required for cell formation (Fig. I-8), or, conversely, it increases the number of long, straight screw dislocations by inhibiting cross slip. The size of cells increases with increasing deformation temperature, as shown in Fig. I-9.

The substructure of iron after large deformation, about 20% reduction or more, cannot be described solely on the basis of dislocation glide, dislocation cell formation, and changes in cell dimensions. It has become apparent[1-4] that in heavily deformed metals the deformation occurs by a series of different processes, each of which operates within a particular range of deformation. With increasing deformation of alpha iron, the sequence is

[1] J. Grewen, J. Huber, and M. Hatherly, *Met. Forum,* 1:115 (1978).
[2] P.S. Mathur and W.A. Backofen, *Met. Trans.,* 4:643 (1973).
[3] A.S. Malin and M. Hatherly, *Met. Sci.,* 13:463 (1979).
[4] M. Hatherly and A.S. Malin, *Met. Technol. (NY),* 6:308 (1979).

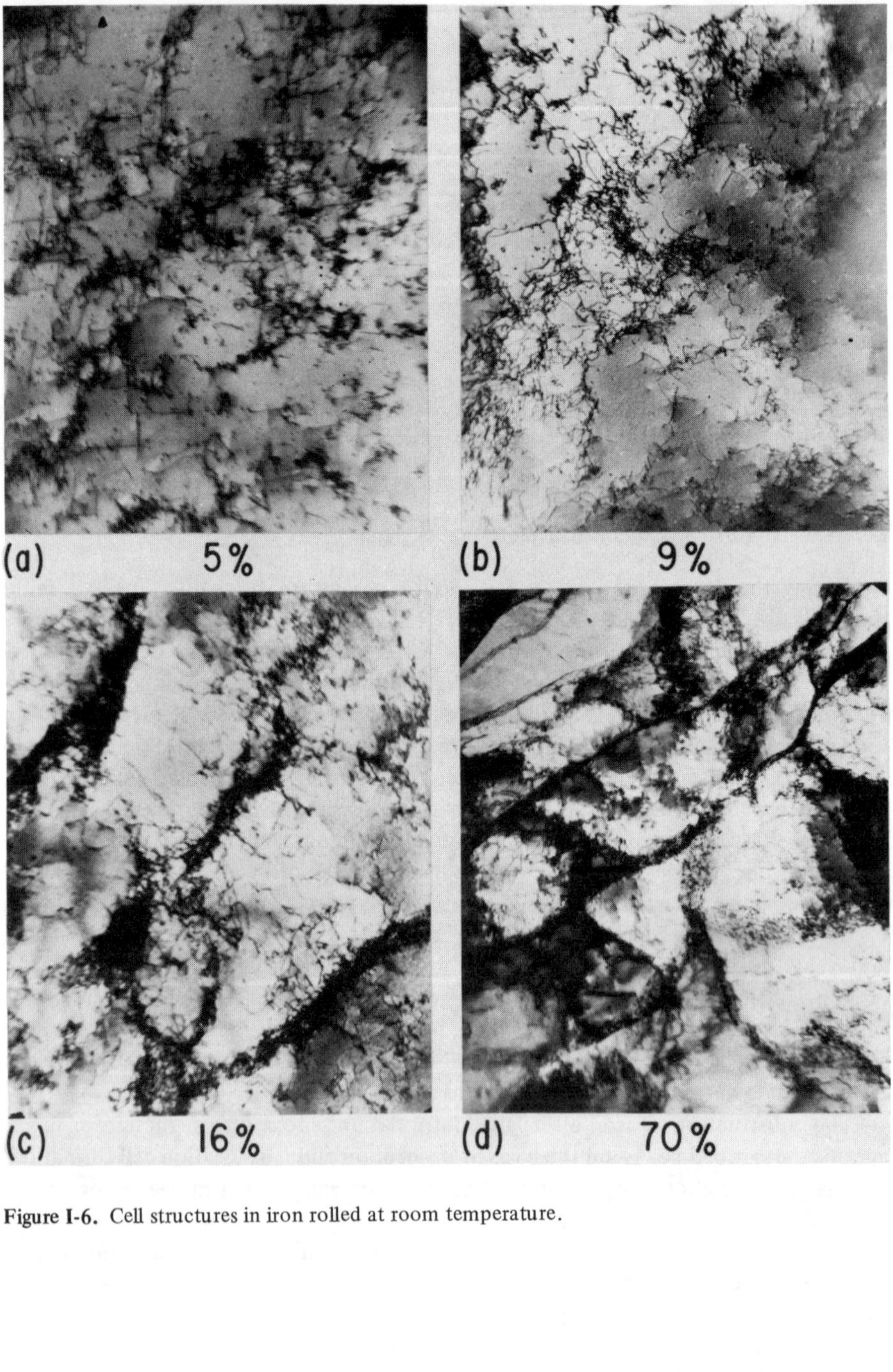

Figure I-6. Cell structures in iron rolled at room temperature.

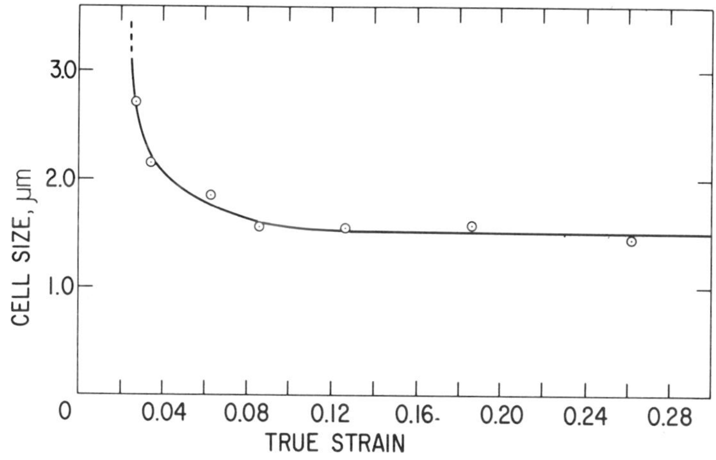

Figure I-7. Dependence of cell size of iron on the amount of deformation at 25°C. (A.S. Keh and S. Weissmann, *Electron Microscopy and the Strength of Crystals,* G. Thomas and J. Washburn, eds., Interscience, New York, 1963, by permission.)

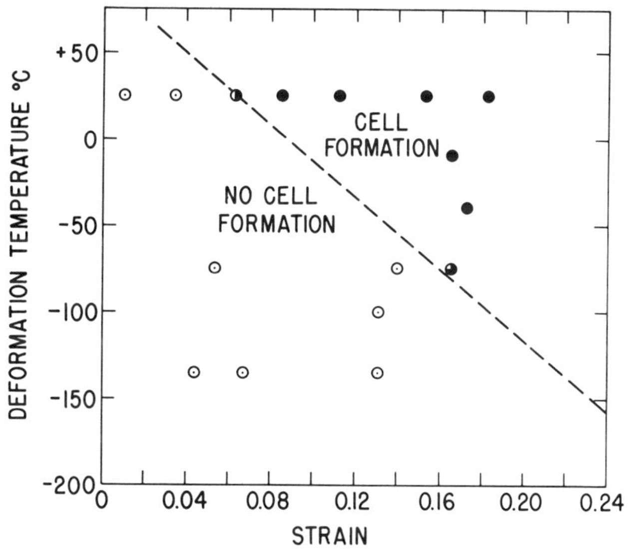

Figure I-8. Tendency for cell formation in iron as a function of strain and deformation temperature. (A.S. Keh and S. Weissmann, *Electron Microscopy and the Strength of Crystals,* G. Thomas and J. Washburn, eds., Interscience, New York, 1963, by permission.)

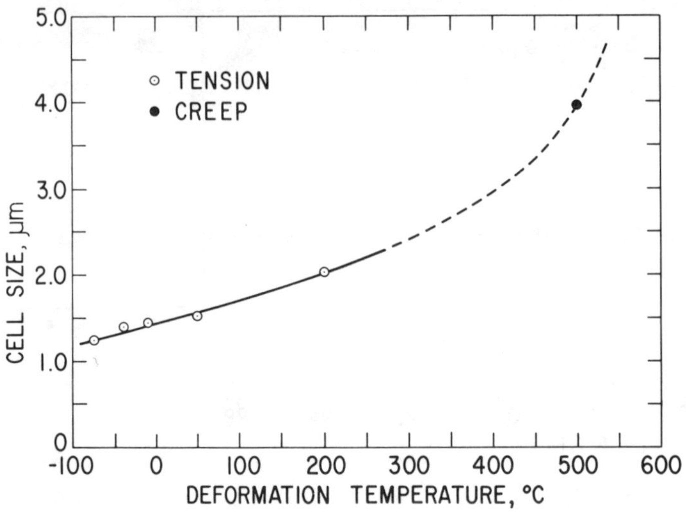

Figure I-9. Dependence of cell size of iron on the deformation temperature. (A.S. Keh and S. Weissmann, *Electron Microscopy and the Strength of Crystals.* G. Thomas and J. Washburn, eds., Interscience, New York, 1963, by permission.)

1. Slip by dislocation glide
2. Formation of microbands
3. Macroscopic shearing

Microbands in iron form on {110}. They consist of regions of highly concentrated slip, usually 0.2 to 0.3 μm thick, 20 to 30 μm wide, and up to 40 μm long. Macroscopic shear bands form at about 35° to the rolling plane, parallel to the transverse direction, when the capacity for work hardening is exhausted. The shear bands are sheets of long, thin cells.

Microbands, when grouped in clusters, can produce a continuous orientation change from one side of the cluster to another. These clusters, or transition bands,[1] form the connections between parts of grains that have rotated in different directions during deformation because of constraints imposed by neighboring grains, or by the deformation process itself.

I-4. MECHANICAL TWINS

Mechanical twins are very common in alpha iron deformed at low temperatures or at a high rate, or when the iron is alloyed with Si. They are more prevalent in coarse-grained than in fine-grained iron. In all body-centered cubic metals, the twinning shear

[1] H. Hu, in *Recovery and Recrystallization of Metals,* Interscience, New York, 1963, p. 311.

is $1/\sqrt{2}$ in a ⟨111⟩ direction on a {112} plane. This can be generated by a shear of (1/6)⟨111⟩ on successive {112} planes, as shown in Fig. I-10. Although a twin can be described formally in this manner, it should not be inferred that this is necessarily the way in which mechanical twins form and grow in a bcc lattice. Mechanical twins (deformation twins, Neumann bands) form as thin plates, often not over 5×10^{-3} mm thick, probably because of the high elastic strain energy involved.

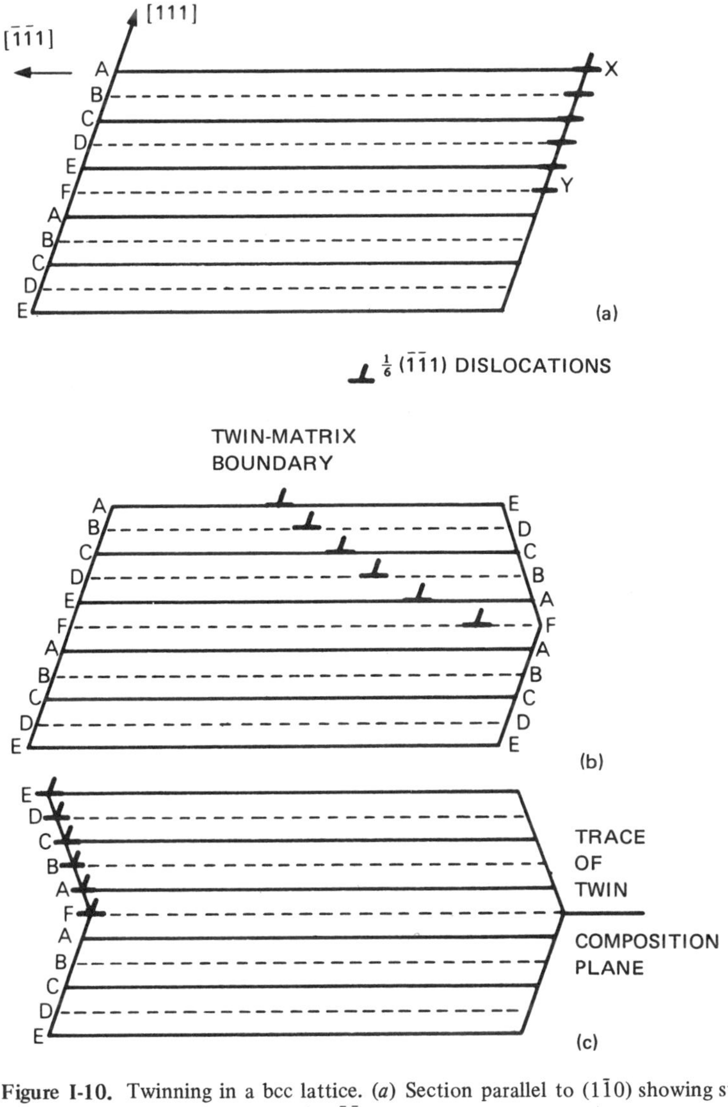

Figure I-10. Twinning in a bcc lattice. (*a*) Section parallel to $(1\bar{1}0)$ showing stacking sequence of (112) planes. *X-Y* is a row of $(1/6)(\bar{1}\bar{1}1)$ twinning dislocations. (*b*) The twinning dislocations have moved into the crystal to produce a twin-oriented region, changing the stacking sequence. (*c*) A twinned crystal. (Reprinted with permission from D. Hull, *Introduction to Dislocations,* 2d ed., Pergamon Press, Ltd., Oxford, 1975.)

These twins have several characteristics:

1. They can form and grow at very high rates, perhaps with an upper limit of the velocity of an elastic shear wave in the crystal.
2. Some plastic deformation by slip usually precedes twinning.
3. Twinning can be prevented by a preexisting dislocation structure (dislocations arranged in cell walls).
4. The stress required to produce mechanical twins in a given material seems to be nearly constant.

The high rates of twin formation and growth is indicated by their high density in explosively shock-loaded metals.[1] The observation that plastic deformation by slip precedes twinning may indicate that the prior slip is required to create stress concentrations such as at the intersection of two well-defined slip bands. However, prior deformation by slip can *prevent* subsequent twinning in iron if the deformation is done at ambient temperatures so that dislocation cells are formed. Presumably, the existence of glide dislocations on many slip systems allows the material to respond rapidly to subsequent high applied stress without the need for twinning. The apparent constancy of stress required for twinning is indicated roughly by a comparison of Figs. I-18 and I-23. When deformed slowly at constant temperatures, polycrystalline iron begins to twin profusely at about 75 K and a stress of about 650 MPa. The lower yield stress remains nearly constant as the temperature is decreased. When deformed rapidly at room temperature, the same sort of iron deforms profusely by twinning at a stress of about 550 MPa.

The mechanism of formation of twins remains obscure,[2] resembling martensite in that respect. Since the work of Cottrell and Bilby,[3] most theoretical treatments have involved a dislocation pole mechanism, but there is no proof of the generality of this process. Processes involving homogeneous shear of a small region into a twin orientation,[4] or creation of a twin by the action of a single high-speed dislocation,[5] similarly lack proof.

I-5. DEFORMATION TEXTURES

When a metal is deformed, its grains are reoriented, with certain orientations being preferred over others. This development of a preferred orientation (texture) depends upon at least four factors:

[1] W.C. Leslie, in *Metallurgical Effects at High Strain Rates*, Plenum Press, New York, 1973, p. 571.

[2] J.W. Christian, in *Physics of Strength and Plasticity*, MIT Press, Cambridge, Mass., 1969, p. 85.

[3] A.H. Cottrell and B.A. Bilby, *Phil. Mag.*, 42:573 (1951).

[4] R.L. Bell and R.W. Cahn, *Proc. Roy. Soc. London*, A239:494 (1957).

[5] I. Ishioka, *J. Appl. Phys.*, 46:4271 (1975).

1. The type of deformation
2. The amount of deformation
3. The temperature of deformation
4. The mechanics of slip in the metal

With regard to the first factor, rolling and wire drawing are the most common cold-working operations, so textures developed by these procedures have been most thoroughly studied.

The texture existing in a polycrystalline metal is usually described by pole figures,[1] obtained by X-ray transmission and reflection. The transmission technique is used in the region extending 60° inward from the periphery of the pole figure; the reflection technique can be used in the region extending 50° outward from the center of the pole figure. The area of overlap can be used to convert the transmission or reflection intensities to a common basis. Pole figures can be determined automatically, with a gain in convenience but with perhaps some loss in sensitivity. Needless to say, an understanding and appreciation of pole figures requires a familiarity with standard stereographic projections.

Pole figures can be supplemented by inverse (axial) pole figures, but both of these are attempts to portray a three-dimensional distribution of crystal orientations on a two-dimensional plot. However, in recent years two mathematical techniques have been developed for combining information from several two-dimensional pole figures to deduce a three-dimensional orientation distribution. One of these techniques[2-4] employs expansion of a series of spherical harmonics; the other[5,6] uses a trial and error method, employing iterations to produce a final orientation distribution. Both procedures require a digital computer.

Because of the large number of slip systems in bcc metals, the development of cold-rolling texture in alpha iron is obscure. It is sometimes described as (001) [110], and this is the principal component, but not the only one. A definite texture develops after about 30% reduction, and if the sheet is thin, ≅3 mm in thickness or less, there is no difference between surface and center texture after about 50% reduction. The texture becomes more pronounced with increasing deformation. The cold-rolling texture is not affected by small amounts of carbon, nor by up to several percent of substitutional solutes. A typical pole figure for heavily cold-rolled iron is shown in Fig. I-11.

The development of the texture with increasing deformation can be divided, quite arbitrarily, into three parts:

[1] C.S. Barrett and T.B. Massalski, *Structure of Metals,* 3d ed., McGraw-Hill, New York, 1966, p. 193.

[2] R.J. Roe, *J. Appl. Phys.,* 36:2024 (1965).

[3] R.J. Roe, *J. Appl. Phys.,* 37:2069 (1966).

[4] H.J. Bunge, *Z. Metallk,* 56:872 (1965).

[5] R.O. Williams, *Trans. AIME,* 242:105 (1968).

[6] R.O. Williams, *J. Appl. Phys.,* 39:4329 (1968).

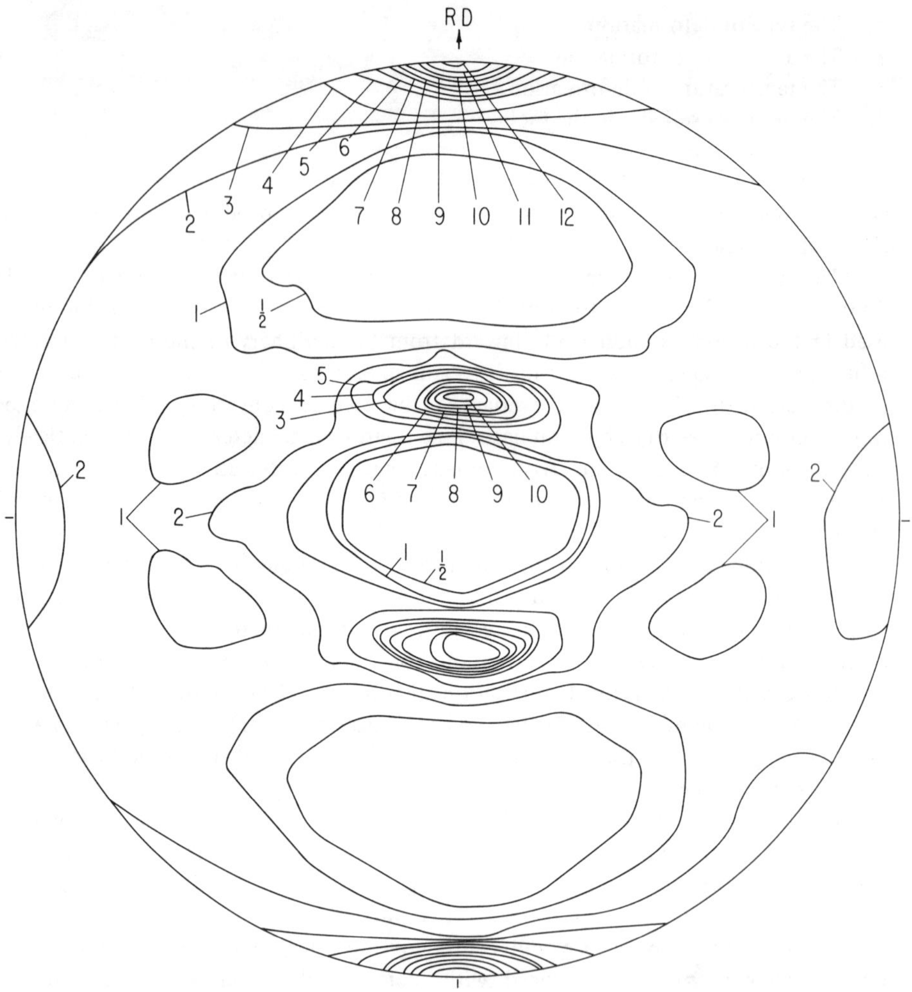

Figure I-11. (110) pole figure of iron cold-rolled to 90% reduction of thickness. A pole density of unity corresponds to randomness. (W.C. Leslie, *Trans. AIME,* 221:752 (1961).)

1. Rotation of the lattice into an incomplete fiber texture, with [110] at 60° to the rolling direction.
2. Rotation around the [110] axis of the fiber texture of part 1 to an end orientation of (112) [1$\bar{1}$0].
3. Rotation around [110] parallel to the rolling direction toward (100) [011] and (111) [1$\bar{1}$0].

These processses do not occur separately; all three may be going on simultaneously. There is no satisfactory explanation of how this particular texture is developed.

The texture of wire-drawn bcc metals is better understood. When the structure of drawn iron wires is examined by transmission electron microscopy, in two orthogonal sections, it is found to consist of ribbon-shaped cells, curled about one another, as illustrated in Fig. I-12. The bcc wire texture ⟨110⟩ prevents the cells from becoming equiaxed. They become ribbon-shaped through a reduction of their thickness but not their width. This shape change has been explained by Hosford.[1] Two of the four ⟨111⟩ slip directions lie perpendicular to the wire axis (Fig. I-13), so that slip in these two directions does not elongate the wire. (The angles between ⟨110⟩ and ⟨111⟩ are 90° and 35.26°.) Slip in the other two ⟨111⟩ directions permits extension of the wire by a thinning of each grain and cell rather than by a symmetrical reduction of grain diameter. After the ⟨110⟩ fiber texture develops, additional slip produces plane strain rather than axially symmetric flow of individual grains. If the grains are to maintain continuity, they must curl around one another.

I-6. THE EFFECTS OF GRAIN SIZE ON YIELDING

The very useful empirical relation between lower yield stress and grain size,

$$\sigma_{LYS} = \sigma_0 + Kd^{-1/2} \tag{I-15}$$

where d, average intercept distance, is in millimeters, was developed by Hall[2] and used to a greater extent by Petch.[3] The proper value of the exponent of d was in dispute for some years, but the works of Morrison[4] and Miller,[5] which together covered a range of grain sizes from 0.35 to 400 μm (ASTM 19.5 to 0), verified that the exponent should be $-\frac{1}{2}$. Their data are shown in Fig. I-14. Note particularly that the verification of this relation required working far outside the normal range of grain sizes in commercial steels, which is about ASTM 5 to 12.

Hirth[6] has reviewed the efforts made to provide a theoretical basis for the Hall-Petch relation. It does not appear to be possible with current knowledge to decide which, if any, of the proposed mechanisms is applicable.

As indicated in Fig. I-14, when the impurity or alloy content of iron is changed, the values of σ_0 and K in the Hall-Petch relation also change. Morrison,[4] working with unalloyed but impure alpha iron and low-carbon steels, found σ_0 (the intercept) to be 88.9 MPa and K (the slope) to be 18.1 MPa/mm$^{-1/2}$. As will be seen later, this variation of slope and intercept creates considerable difficulty in the determination of solid solution strengthening in polycrystalline iron.

When iron or low-carbon steel is strained in tension, the plastic strain concentrated in the Lüders band (the Lüders strain) increases with decreasing grain size and the variation is linear with $d^{-1/2}$, at least within a limited range of grain sizes (Fig. I-15). As a general rule, the mechanical properties of metals improve as their grain size

[1] W.F. Hosford, *Trans. AIME,* 230:12 (1964).
[2] E.O. Hall, *Proc. Phys. Soc.,* 64B:747 (1951).
[3] N.J. Petch, *J. Iron Steel Inst.,* 174:25 (1953).
[4] W.B. Morrison, *Trans. Am. Soc. Met.,* 59:824 (1966).
[5] R.L. Miller, *Met. Trans.,* 3:905 (1972).
[6] J.P. Hirth, *Met. Trans.,* 3:3047 (1972).

Figure I-12. Cell structure in drawn iron wire, $\epsilon = 3.8$. (*a*) Longitudinal section. (*b*) Transverse section. (G. Langford and M. Cohen, *Trans. Am. Soc. Met.*, 62:628 (1969), by permission.)

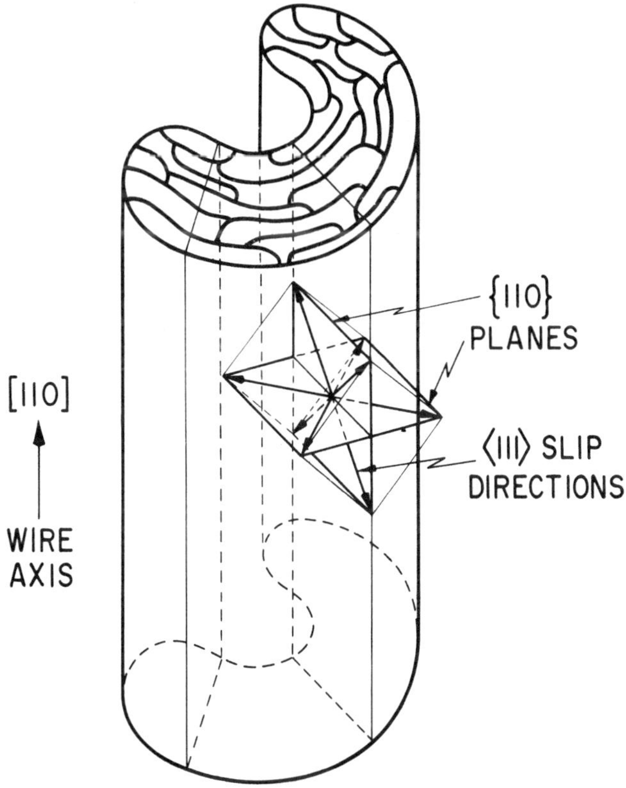

Figure I-13. Schematic representation of grain and cell shape in (110) textured bcc wire. (G. Langford and M. Cohen, *Trans. Am. Soc. Met.*, 62:628 (1969), by permission.)

is reduced, but there is at least one limit to this process which should be noted. If iron is made with grains smaller than about 0.6 μm in "diameter," there is no uniform elongation in a tension test; the specimen breaks during the Lüders strain. This effect is illustrated in Fig. I-16. This phenomenon is not often encountered because of the exceptionally small grain size required. Grains 0.6 μm in diameter are equivalent to a grain size of ASTM 18. In such metals, the Lüders band may occupy less than 15% of the gauge section. Although the local deformation is high, there is no uniform elongation. This phenomenon is not unique to iron; it also occurs in face-centered cubic (fcc) and hexagonal close-packed (hcp) alloys. With grain sizes larger than 10 μm (ASTM 10), uniform elongation is independent of grain size, as shown in Fig. I-17.

I-7. THE TEMPERATURE DEPENDENCE AND STRAIN-RATE DEPENDENCE OF THE YIELD STRESS OF IRON

The most distinctive feature of the plastic deformation of iron and other bcc metals is the large temperature dependence and strain-rate dependence of the *macroscopic*

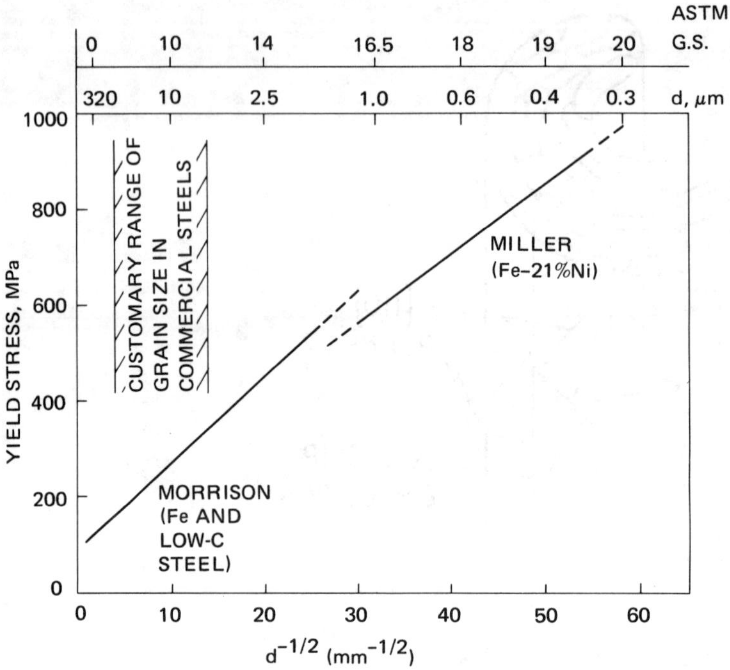

Figure I-14. The effect of grain size on the lower yield stress of alpha iron.

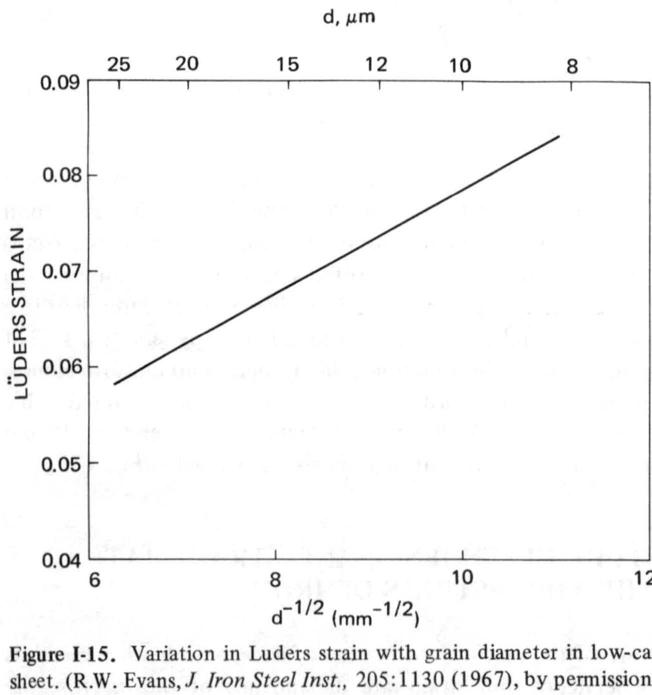

Figure I-15. Variation in Luders strain with grain diameter in low-carbon, annealed rimmed steel sheet. (R.W. Evans, *J. Iron Steel Inst.*, 205:1130 (1967), by permission.)

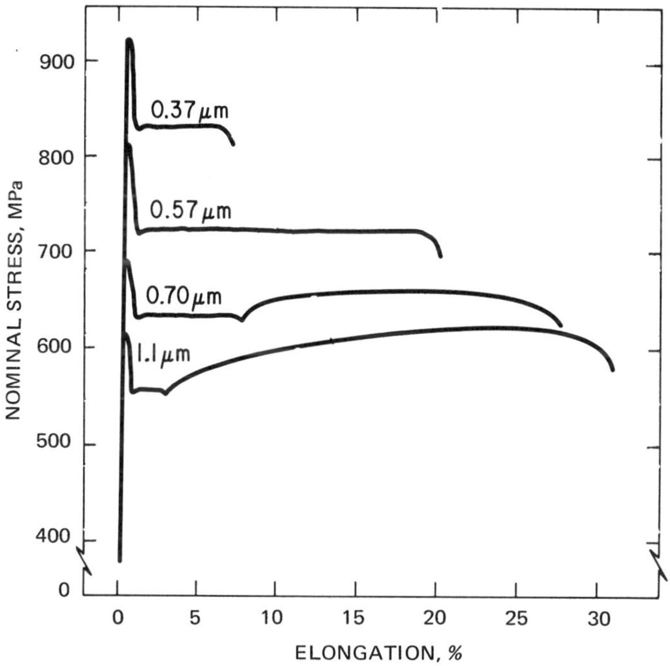

Figure I-16. The effect of grain size on elongation in a tension test. (R.L. Miller, *Met. Trans.*, 3:905 (1972), by permission.)

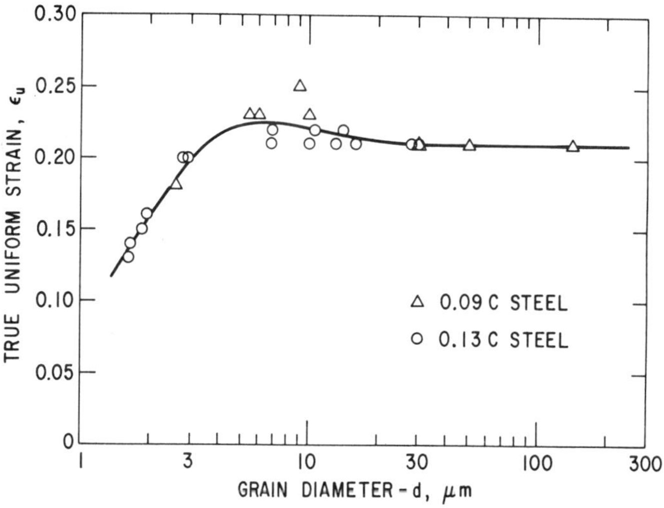

Figure I-17. Variation of uniform elongation with grain size in low-carbon steel. (W.B. Morrison and R.L. Miller, "The Ductility of Ultrafine-Grain Alloys," in *Ultrafine-Grain Metals,* (J.J. Burke and V. Weiss, eds., Syracuse Univ. Press, Syracuse, New York, 1970, p. 185, by permission.)

flow stress. It is only within the last decade or so that the important features of this temperature and strain-rate dependence have become obvious. For a long time, the situation was confused by the presence and effects of interstitial solutes in iron. These can be removed by a variety of techniques: for example, by long-time treatment of the iron in a closed system in a stream of dry H_2 gettered by ZrH_2.[1,2] The C, N, and O are removed from the Fe and picked up by the ZrH_2. When a larger quantity of interstitial-free Fe is required, without the problem of grain-size control inherent in the ZrH_2 technique, it can be obtained by the addition of a small amount of Ti, about 0.15%, to an iron that has been melted in vacuum to the minimum content of C, N, O, and S. The Ti getters these elements, especially if the resulting alloy is slowly cooled, or given an isothermal treatment near 800°C.[3,4] When these measures are taken, the temperature dependence of the yield stress becomes sharper than when interstitial solutes are present. The situation is illustrated in Fig. I-18. The curve obviously has two principal parts:

1. A portion extending from −196 to about −23°C. In this region the yield stress is markedly dependent upon temperature and strain rate.
2. A strictly linear portion extending from 100 to 600°C, in which the yield stress is nearly independent of temperature, declining only as the shear modulus decreases with increasing temperature (detailed later, in Fig. III-6). Above 600°C the yield stress declines rapidly.

These two portions are joined by a transition region, extending from −23 to 100°C, in which the temperature dependence of the yield stress changes rapidly. The elimination of the effects of interstitial solutes has served to narrow the transition region and to make it more obvious that two separate modes of deformation operate above and below ambient temperature, at the usual testing-machine strain rates.

At a strain rate of 2.5×10^{-4} s^{-1}, no discontinuous yielding occurred at room temperature or above, but, as noted previously, upper and lower yield points were obtained at all temperatures below ambient, indicating again that the presence of interstitial solutes is not essential for discontinuous yielding.

Figure I-18 also includes plots of σ_E, the elastic limit, and σ_A, the anelastic limit, vs. temperature. As stated before, these stresses depend only slightly upon temperature.

Much subsequent work[5-8] with iron of higher purity has indicated that:

[1] D.F. Stein, J.R. Low, Jr., and A.U. Seybolt, *Acta Metall.*, 11:1253 (1963).
[2] W.A. Spitzig and A.S. Keh, *Acta Metall.*, 18:611 (1970).
[3] R.P. Steijn and R.M. Brick, *Trans. Am. Soc. Met.*, 46:1406 (1954).
[4] W.C. Leslie and R.J. Sober, *Trans. Am. Soc. Met.*, 60:99 (1967).
[5] H. Matsui, S. Moriya, S. Takaki, and H. Kimura, *Trans. Jpn. Inst. Met.*, 19:163 (1978).
[6] D.J. Quesnel, A. Sato, and M. Meshii, *Mat. Sci. Engr.*, 18:199 (1975).
[7] J. Diehl, M. Schreiner, S. Staiger, and S. Zwiesele, *Scr. Metall.*, 10:949 (1976).
[8] D. Tseng and K. Tangri, *Scr. Metall.*, 11:719 (1977).

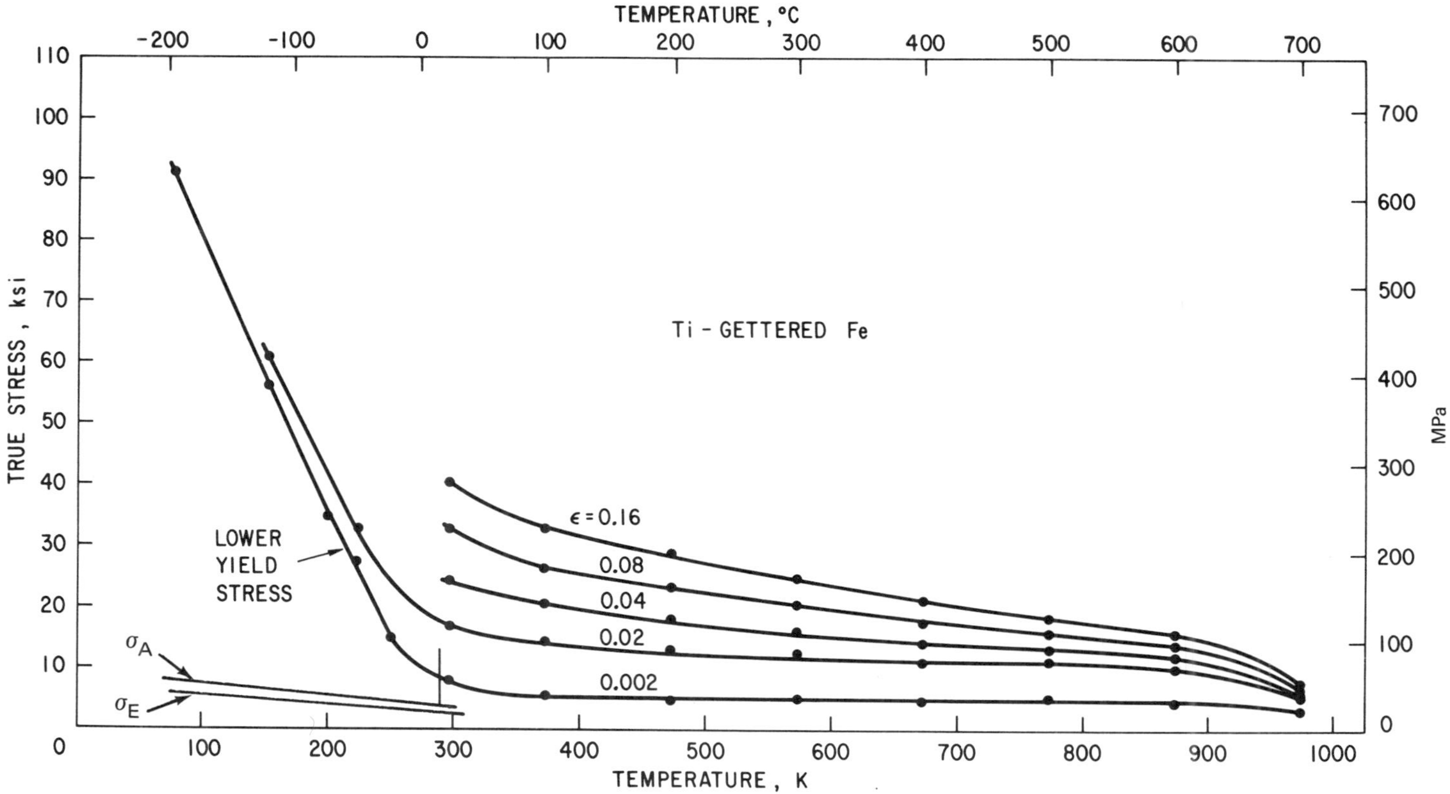

Figure I-18. Temperature dependence of yield and flow stresses in Fe–0.15%Ti. $\dot{\epsilon} \cong 2.5 \times 10^{-4}\ s^{-1}$, G.S. ASTM 5–6. σ_E is the elastic limit and σ_A the anelastic limit. (W.C. Leslie, *Met. Trans.*, 3:9 (1972).)

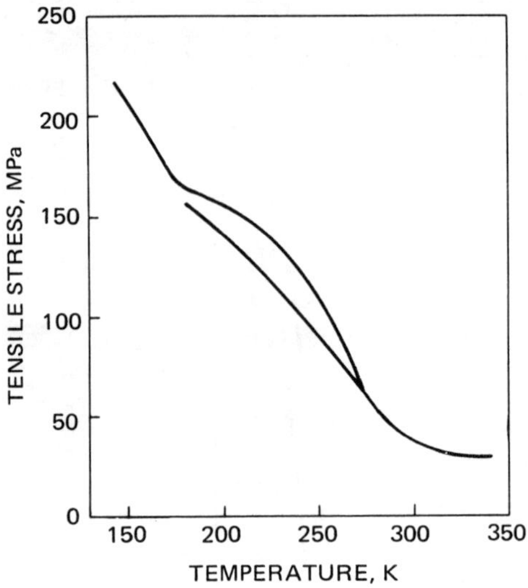

Figure I-19. Limits of the "hump" in the flow stress–temperature curve of high-purity iron. Stress at $\epsilon = 0.001$; $\dot{\epsilon}$, about $1 \times 10^{-4}\ s^{-1}$. (H. Matsui, S. Moriya, S. Takaki, and H. Kimura, *Trans. Jpn. Inst. Met.,* 19:163 (1978), by permission.)

1. Large-scale plastic deformation of iron by slip can occur at 4.2 K and the yield stress continues to rise as that temperature is approached.
2. A hump (concave downward region) appears in the yield stress–temperature plot between about 180 and 280 K, as illustrated in Fig. I-19.
3. The strain rate sensitivity $\Delta\sigma/\Delta \ln \dot{\epsilon}$ of the flow stress of iron reaches a maximum at about 250 K, coincident with the hump in the yield stress–temperature curve.
4. The rate of work hardening $d\sigma/d\epsilon$ is at a minimum in the temperature range between 175 and 230 K.[1]

These features all appear to be intrinsic to bcc iron.

The very large temperature dependence of the macroscopic yield stress of iron has been attributed to one or another of three factors:

1. Interstitial impurities
2. Periodic energy barriers (Peierls–Nabarro force)
3. Thermally activated sessile-glissile transformations of screw dislocations

There is now nearly complete agreement that the probable cause is either 2 or 3, but the available evidence does not allow us to distinguish between them.

The *Peierls–Nabarro stress* is the stress required to move a dislocation through a

[1] A.S. Keh and S. Weissmann, *Electron Microscopy and the Strength of Crystals,* Interscience, New York, 1962, p. 231.

perfect lattice. It arises because of the periodic nature of the lattice. The presence of a dislocation displaces surrounding atoms from their equilibrium positions in the perfect lattice, with the magnitude of the displacement decreasing with increasing distance from the core of the dislocation. These displacements are used to specify the *width* of a dislocation, which is defined as the distance along the slip direction, on the slip plane, at which the atomic displacement has decreased to half of the maximum displacement. The width w depends on the force–displacement relation between individual atoms. Because this is not precisely known, the width is also uncertain, but it is usually assumed to be between 1 and 10 atom diameters. The customary assumption, from Frenkel,[1] is that the shearing force required to slide one plane of atoms over the atoms on an adjacent plane is given by the sinusoidal relation

$$\tau = \frac{b}{a}\frac{G}{2\pi}\sin\frac{2\pi X}{b} \tag{I-16}$$

where b = magnitude of Burgers vector
a = spacing between atom planes
G = shear modulus
X = distance from the equilibrium (low-energy) position of the dislocation

To obtain the maximum shearing force (Peierls–Nabarro stress), X is replaced by w,

$$w = \frac{a}{1-\nu} \tag{I-17}$$

where ν = Poisson's ratio

to give the approximate relation

$$\tau_{\max} = 2G\exp\left(\frac{-2\pi w}{b}\right) \tag{I-18}$$

As stated before, the term w in the exponent is greatly dependent upon the form of the force–displacement relation.

Crystals with narrow dislocations, such as Si, Ge, or Al_2O_3, are hard; those with wide dislocations, such as Au, are soft. Body-centered cubic metals are hard at low temperatures and soft at elevated temperatures (Fig. I-18)—that is, the Peierls stress is predominant at low temperatures. At 0 K the dislocations in iron are assumed to lie in positions of minimum energy along ⟨111⟩, and to behave as rigid rods. A stress equal to the Peierls stress is required to move the dislocation to an adjacent equilibrium posi-

[1] J. Frenkel, *Z. Phys.*, 37:572 (1926).

tion. As the temperature is increased, thermal fluctuations assist the dislocation in surmounting the activation energy barrier. The frequency of such vibrations is given by the Maxwell–Boltzmann distribution law, with an upper limit of the Debye frequency (frequency of oscillation of a single atom), about $5 \times 10^{12}\ s^{-1}$. Thermal fluctuations of magnitude equal to or greater than 0.16 aJ (1 eV) can occur at room temperature. These are sufficient for the overcoming of short-range obstacles, such as the Peierls stress. Thus, at ambient temperature the dislocation line will not lie in a single valley of the potential energy surface but may include kinks which lie in an adjacent valley, as illustrated in Fig. I-20, where AB is the dislocation line and CD is an adjacent equilibrium position. Subsequent movement of the dislocation can occur by glide of the two kinks, as indicated by the arrows in Fig. I-20*b*.

A second distinctive feature of bcc metals is asymmetry of plastic flow.[1] Slip on {112} planes is easier when the {112} ⟨111⟩ slip system corresponds to a possible twinning system than when it does not. Also, the temperature dependence of the yield stress is greatest for slip in the antitwinning sense, intermediate in the twinning sense, and least for slip on {110}.[2] This led to the suspicion that screw dislocations in bcc metals could dissociate. Such dissociation must be slight, for the stacking fault energy in bcc metals is very high. When a stacking fault forms in a fcc or hcp metal, there is

[1] J.W. Christian, *2nd Int. Conf. Strength of Metals and Alloys,* Am. Soc. Met., Metals Park, Ohio, 1:31 (1970).

[2] W.A. Spitzig and A.S. Keh, *Acta Metall.,* 18:1021 (1970).

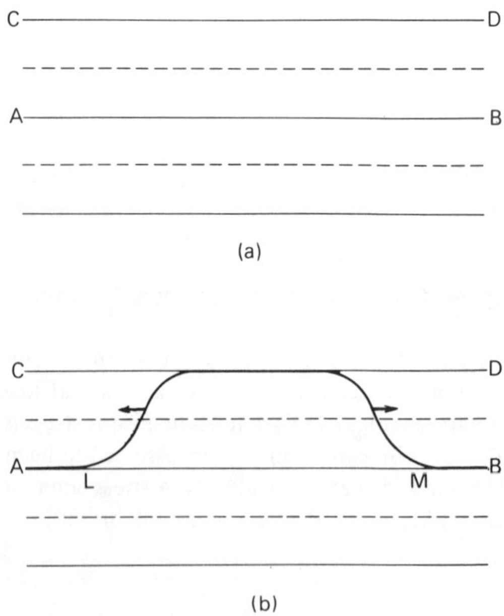

Figure I-20. Formation of a dislocation kink. (*a*) Dislocation *AB* in position of minimum potential energy. (*b*) Double kink. (A. Seeger, H. Donth, and J. Pfaff, *Discussions Faraday Soc.* 23:19 (1957), by permission.)

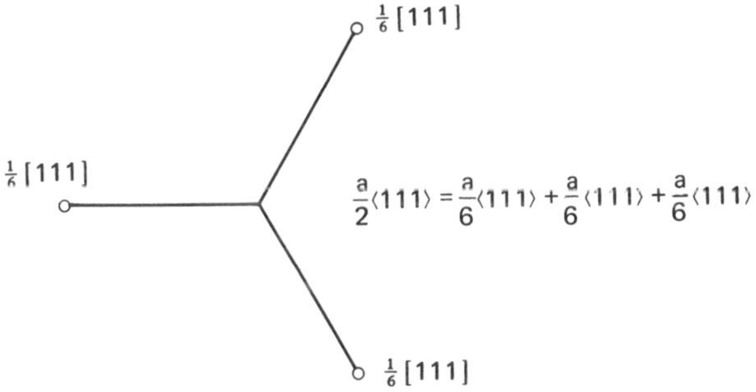

Figure I-21. Possible dissociation of screw dislocation on (112) planes in bcc iron. (P.B. Hirsch, *Proc. 1st ICSMA*, Suppl. *Trans. Jpn. Inst. Met.*, 9:XXX (1968), by permission.)

no change in nearest-neighbor atom distance, but a stacking fault in a bcc metal involves a change in the near-neighbor distance. It is doubtful that stacking faults have ever been observed in any bcc metal or alloy. It has been proposed[1,2] that the core of an $(a/2)[111]$ screw dislocation dissociates into a pattern of stacking faults, one or two Burgers vectors wide, lying on three {112} planes:

$$\tfrac{1}{2}[111] \rightarrow \tfrac{1}{6}[111] + \tfrac{1}{6}[111] + \tfrac{1}{6}[111] \tag{I-19}$$

This situation is shown in Fig. I-21, where the three partial dislocations lie in {112} in a three-fold symmetrical splitting. This configuration is sessile (immobile) and must be changed (constricted) by thermal activation before the dislocation can glide, as shown schematically in Fig. I-22.

At temperatures *above* ambient, the yield stress of an interstitial-free iron is controlled by the stress required to overcome long-range stress fields, and this stress cannot be reduced by thermal fluctuations. Therefore, at temperatures above ambient the yield stress reduces with increasing temperature only at the rate required by the reduction of shear modulus with temperature (see Fig. III-4).

If the yield stress of a metal is highly sensitive to temperature, it will also be highly sensitive to strain rate. Figure I-23 shows the effect of strain rate, over nine orders of magnitude, on the yield and flow stress of iron. The yield strength of iron at room temperature, at a strain rate of 10^5 s^{-1}, is nearly as great as the yield strength at −196°C at conventional slow testing-machine strain rates. In both instances, the mechanism of plastic deformation changed from slip to twinning, and thereafter the yield strength remained constant.

Most industrial cold-forming operations are done in the strain-rate range from 10^0 to 10^3 s^{-1}. The yield stress of iron doubles within this range, but with commercial alloys the effect is much less. All strengthening mechanisms reduce both the strain-rate dependence and temperature dependence of the yield stress.

[1] V. Vitek, *Cryst. Lattice Defects*, 5:1 (1974).
[2] P.B. Hirsch, *Proc. 1st ICSMA*, suppl. Trans. Jpn. Inst. Met., 2:XXX (1968).

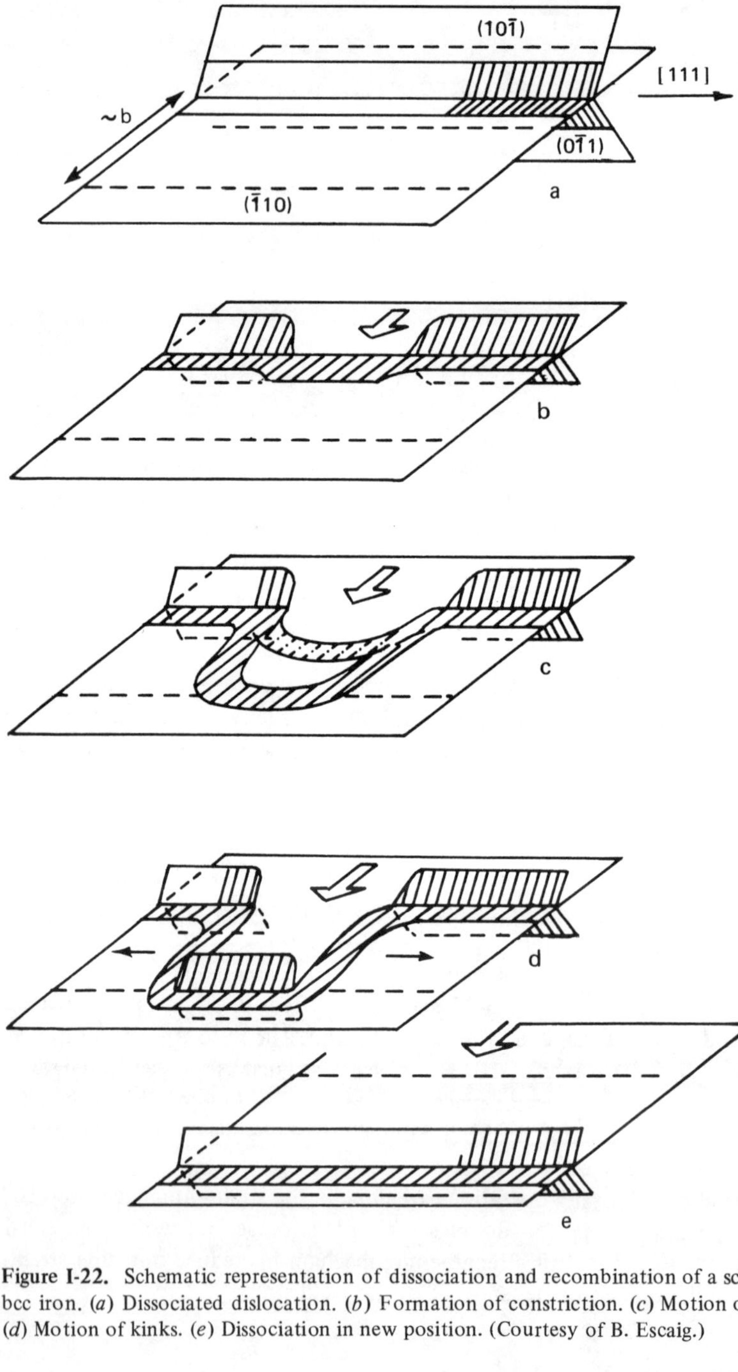

Figure I-22. Schematic representation of dissociation and recombination of a screw dislocation in bcc iron. (*a*) Dissociated dislocation. (*b*) Formation of constriction. (*c*) Motion of glissile segment. (*d*) Motion of kinks. (*e*) Dissociation in new position. (Courtesy of B. Escaig.)

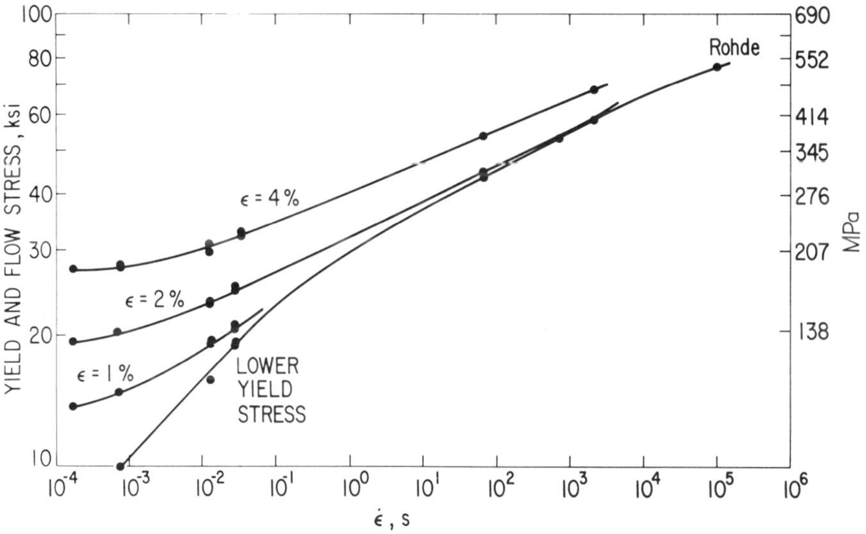

Figure I-23. The effect of strain rate on the strength of polycrystalline bcc iron at room temperature. (W.C. Leslie, R.J. Sober, S.G. Babcock, and S.J. Green, *Trans. Am. Soc. Met.*, 62:690 (1969).)

I-8. THE DISLOCATION DYNAMICS THEORY OF YIELDING

In the dislocation dynamics theory,[1] three factors are considered necessary for abrupt yielding:

1. A low density of mobile dislocations in the specimen before straining;
2. Rapid generation and multiplication of dislocations; and
3. A small to moderate dependence of dislocation velocity on stress.

In annealed, low-carbon steels, the dislocations existing before straining are firmly locked by segregated C and N atoms. These pinned dislocations are usually not freed by the applied stress; rather, new ones are generated. In annealed, high-purity, dislocation-free iron the dislocation density is low, and, as mentioned previously, only the edge components move at low stresses.

Dislocations can be created homogeneously or heterogeneously, but the latter is much more common. In homogeneous nucleation, dislocations are formed in a perfect lattice by the action of stress alone. They can be formed heterogeneously with the help of preexisting defects in the crystal. Such defects serve to concentrate stresses to create dislocations or to move them out of boundaries. Stress concentrations can be ranked in a rough order of effectiveness in generating dislocations:

[1] W.G. Johnston and J.J. Gilman, *J. Appl. Phys.*, 30:129 (1959).

1. Grain corners, where four grains meet at a point
2. Grain edges, along which three grains meet
3. Stress raisers at external surfaces
4. Matrix-second phase interfaces
5. Grain boundaries
6. Subboundaries
7. Twin boundaries

Ledges or adsorbed dislocations on high-angle boundaries or interfaces can be removed by application of stress. However, all of these actual or potential sources cannot account for the high density of dislocations seen in cold-deformed metals. Some process of multiplication must occur, involving the familiar Frank–Read source, or a variant thereof.

A Frank–Read source can be single- or double-ended. The single-ended type is illustrated in Fig. I-24. *BC* is a portion of a glide dislocation lying in its slip plane and *AB* is a part of the dislocation not on the slip plane and therefore immobile, that is, *BC* is glissile and *AB* is sessile. When a shear stress is applied to the crystal in such a manner as to make *BC* rotate clockwise, then the portion *BC* will become a spiral and the upper portion of the crystal will be displaced relative to the lower portion by one Burgers vector per revolution of *BC*.

The operation of the more familiar double-ended Frank–Read source is shown in

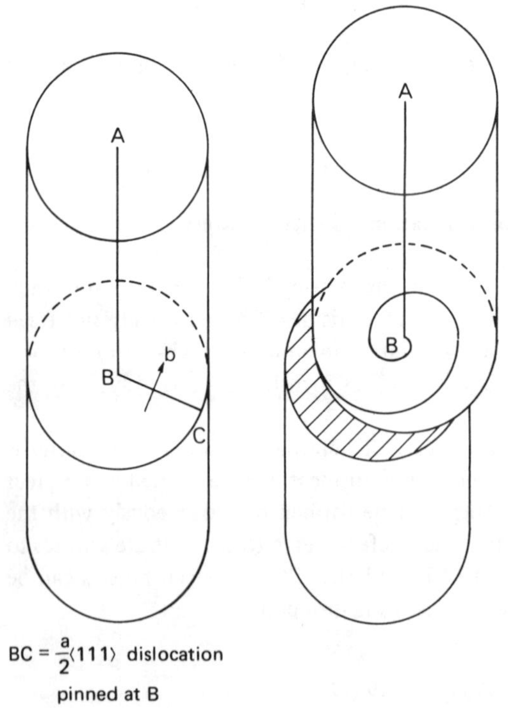

Figure I-24. Operation of a single-ended Frank–Read source. (D. Hull, *Introduction to Dislocations,* 1st ed., Pergamon Press, Ltd., Oxford, 1965, by permission.)

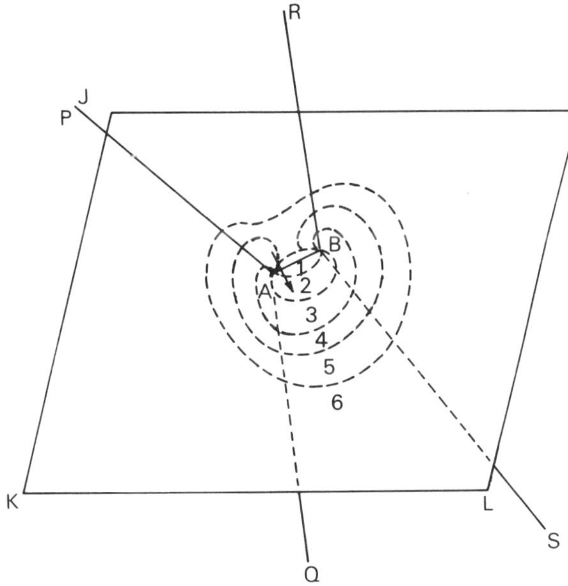

Figure I-25. Operation of a double-ended Frank–Read source. (A.H. Cottrell, *The Mechanical Properties of Matter,* John Wiley and Sons, Inc., New York, 1964, by permission.)

Fig. I-25. The mobile dislocation line *AB* on the slip plane *JKLM* is anchored at its points of intersection with two other dislocations, *PAQ* and *RBS*. Application of a shear stress τ in the indicated direction on the slip plane will cause the line *AB* to bow out into an arc of a circle. If the source dislocation *AB* is in a screw orientation, its spiraling will cause two edge segments of opposite sign to approach one another and mutually annihilate, between positions 5 and 6 of the figure. This sort of dislocation mill has been seen to operate in silicon, where the dislocations are restricted to slip planes. Such restriction is an obvious requirement for the continued operation of the Frank–Read source. However, as previously stated, in bcc iron at ambient or higher temperatures, screw dislocations can glide on any plane in the ⟨111⟩ zone. Therefore, the mobile dislocation will seldom complete a cycle in the Frank–Read mechanism before some obstacle causes it to move out of the slip plane. In iron, slip bands broaden with increasing stress because of easy cross slip of screw dislocations. Cross slip of a portion of a screw dislocation will create a jog, a small segment of edge dislocation (Fig. I-26). This segment can glide (move conservatively) only in the direction of its Burgers vector, which is parallel to the screw dislocation. The effects produced by jogs depend upon their height. Very small jogs (one atom high) can be dragged along with the screw dislocation, leaving behind a trail of either vacancies or interstitial atoms. If the jog is of intermediate height, two segments of edge dislocation of opposite sign will move toward each other until they are parallel and held in place by elastic interaction. This situation is shown in Fig. I-27. The result is the creation of a dislocation dipole. If the separation of the two edge dislocation segments is greater than about 20 nm, they can move independently, creating a Frank–Read source.

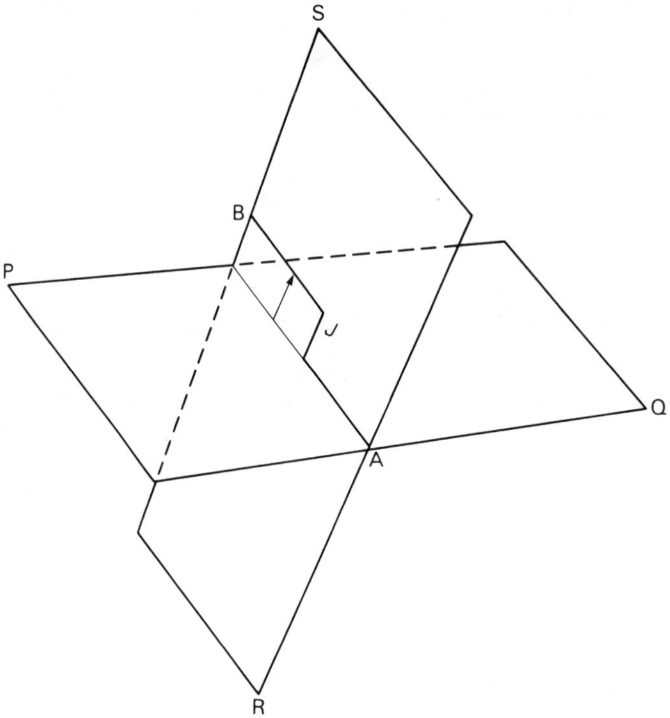

Figure I-26. Formation of a jog by cross-slip of a screw dislocation *AB* from the primary plane *PQ* to the secondary plane *SR*. (A.H. Cottrell, *The Mechanical Properties of Solids,* John Wiley and Sons, New York, 1964, by permission.)

The multiple cross-slip mechanism seems to be the principal mode of dislocation multiplication in metals such as iron in which the screw dislocations are narrow—that is, those with a high stacking-fault energy. Figure I-28 illustrates the operation. A screw dislocation lies on the (110) slip plane, along *CD*. Meeting some obstacle, it cross-slips onto the (101) plane for a short distance, then cross-slips again onto a (110) plane. The jogs *AC* and *BD* are predominantly edge in character and can glide only on the (101) plane. The result is two Frank-Read sources operating on two parallel slip planes. The process need not stop there, of course, and the outcome is a continuous dislocation line lying on many parallel slip planes connected by jogs. This produces more rapid multiplication than can be obtained from a simple Frank-Read source and wavy, poorly defined slip bands.

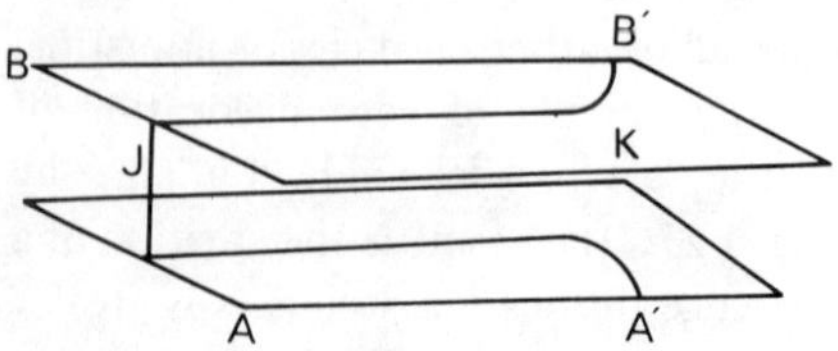

Figure I-27. Edge dislocation dipole formed by trailing of a jog, *J*, behind a gliding screw dislocation, $AJB \rightarrow A'JB'$. (A.H. Cottrell, *The Mechanical Properties of Solids,* John Wiley and Sons, New York, 1964, by permission.)

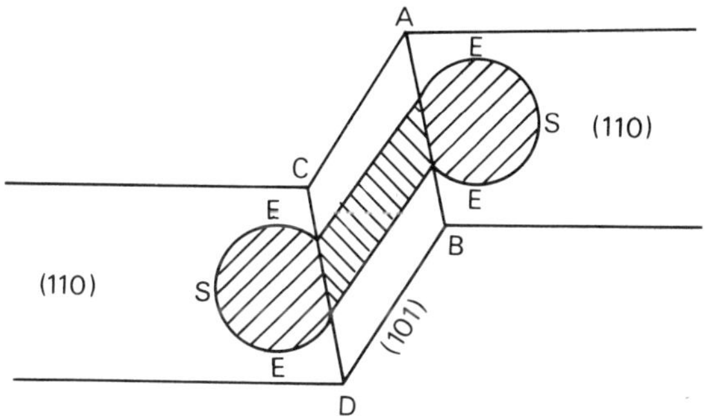

Figure I-28. Double cross-slip of a screw dislocation; S is the screw component, E the edge component. (D. Hull, *Introduction to Dislocations,* 1st ed., Pergamon Press, Ltd., Oxford, 1965, by permission.)

Of the three factors necessary for discontinuous yielding we have now discussed two: a low density of mobile dislocations, and rapid generation and multiplication of dislocations. We now consider the third factor, the dependence of dislocation velocity on stress, the goal being an expression that will yield a stress-strain curve of the proper shape and magnitude. In doing so we will follow the procedure used by Hahn.[1]

The strain rate in a crystal deforming under stress is usually represented by

$$\frac{d\epsilon}{dt} = \dot{\epsilon} = 0.5 b \rho_m \bar{v} \tag{I-20}$$

where b = magnitude of Burgers vector
ρ_m = density of mobile dislocations (total length per unit volume)
$\bar{v}$ = average velocity of dislocations

The factor of 0.5 is employed to represent the contribution of a single dislocation of unit velocity and unit length moving in a direction determined by the maximum resolved shear stress. Dislocation velocity is best measured in a high-voltage transmission electron microscope, with a specimen of the greatest thickness consistent with electron beam transmission, strained in situ. In alpha iron, as a consequence of the much higher velocity of edge components as compared to screws, most of the dislocations seen after strains of the order of 0.2% are screws, the edge components having moved out of the crystal.

The quantity ρ_m in Eq. I-20 is a source of considerable uncertainty, but it is usually assumed to be a fixed fraction of the total dislocation density,

$$\rho_m = f\rho \tag{I-21}$$

with f being about 0.1.

The multiplication of dislocations is usually represented by a parabolic function

$$\rho = C\epsilon_p^{\beta} \tag{I-22}$$

where ϵ_p is the plastic strain and β is about 0.9. The multiplication factor,

$$C = \frac{d\rho}{d\epsilon}$$

[1] G.T. Hahn, *Acta Metall.*, 10:727 (1962).

Table I-2 Typical values of the multiplication factor *C*

Material		C
Single crystal	Cu	$4 \times 10^6/\text{cm}^2$
Polycrystalline	Ta	$1.7 \times 10^{10}/\text{cm}^2$
Polycrystalline	Fe	$1.6 \times 10^9/\text{cm}^2$

can be applied in the plastic strain range $10^{-3} < \epsilon_p < 10^{-1}$. The relation is frequently written

$$\rho = \rho_0 + C\epsilon_p^\beta \tag{I-23}$$

in which ρ_0 represents the average density of mobile dislocations present before straining (see Table I-2).

As indicated previously (Eq. I-3), the average velocity of dislocations depends upon stress according to

$$\bar{v} = \left(\frac{\tau}{\tau_0}\right)^m \tag{I-3}$$

The magnitude of m varies greatly between different materials (see Table I-3).

It is necessary to introduce strain hardening, and for simplicity a linear relation is used

$$\Delta\sigma = q\epsilon_p \tag{I-24}$$

where σ = tensile stress corresponding to τ
q = work-hardening coefficient

The tensile stress is assumed to be twice the shear stress, and the change in the tensile stress is about equal to the increment required to maintain a given average velocity. Equation I-3 is then rewritten:

$$\bar{v} = \left(\frac{\sigma - q\epsilon_p}{2\tau_0}\right)^m \tag{I-25}$$

Table I-3 Magnitude of stress component of dislocation velocity

Material	Temperature	m
Si	600–900°C	1.4
Ge	420–700°C	1.4–1.9
Fe-3%Si	room temperature	35
Cu	room temperature	200
Ag	room temperature	300

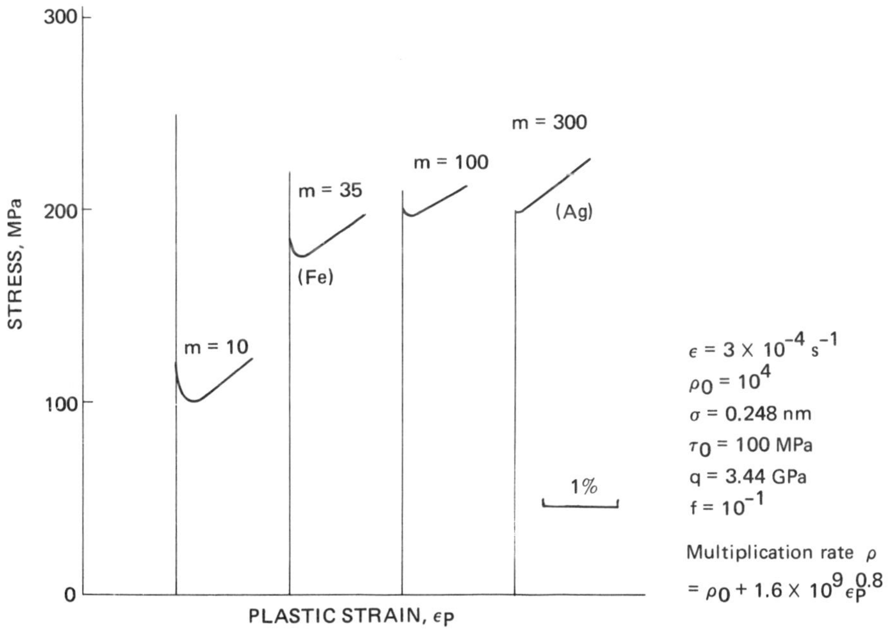

Figure I-29. Effect of m on yield drop in tension. (Reprinted with permission from G.J. Hahn, *Acta Metall.,* 10:727 (1962).)

or $$\bar{v} = (2\tau_0)^{-m}(\sigma - q\epsilon_p)^m \tag{I-26}$$

Equations I-20, 21, 25, and 26 are combined to yield an expression relating plastic strain rate, plastic strain, and flow stress in bcc iron:

$$\dot{\epsilon}_p = 0.5bf(\rho_0 + C\epsilon_p^\beta)(2\tau_0)^{-m}(\sigma - q\epsilon_p)^m \tag{I-27}$$

For simplicity, elastic strain is ignored. The flow stress σ can be expressed in terms of strain and strain rate

$$\sigma = q\epsilon_p + 2\tau_0\left[\frac{\dot{\epsilon}}{0.5bf(\rho_0 + C\epsilon_p^\beta)}\right]^{1/m} \tag{I-28}$$

From this expression, the stress–strain curve can be calculated. Very large yield drops can be obtained without recourse to the unlocking of pinned dislocations. The results are shown in Fig. I-29. Extreme yield drops are typical of materials with low values of m, whereas no yield drops occur for metals such as copper or silver, which have m values of 200 to 300. Iron occupies an intermediate position.

As concerns alpha iron, the weakness of this calculation arises from our lack of knowledge of the stress dependence of the average velocity of *screw* dislocations. The value of 35 ± 5 customarily employed applies only to *edge* dislocations. The motion of screws does not occur until macroscopic yielding, and then their motion is retarded by jogs.[1] At the yield stress of Fe–3%Si (147 MPa), the average velocity of edges and

[1] H. Saka and T. Imura, *J. Phys. Soc. Jpn.,* 32:702 (1972).

screws differs by a factor of about 5, being about 2×10^{-2} cm/s for edge and about 10^{-3} cm/s for screws. We have no knowledge of their relative velocities in unalloyed iron. Certainly, the yield stress for iron shown in Fig. I-29 is too high by a factor of about three. Nevertheless, the model gives a reasonable approximation of the yielding behavior of alpha iron at ambient temperatures or below, with a distinct upper yield stress, abrupt yielding, a lower yield stress, and homogeneous work hardening. We can also state that the macroscopic flow initiated at the upper yield stress coincides with the motion, cross slip, and multiplication of screw dislocations.

In polycrystals, the preyield microstrain is not general, but is confined to a few grains.[1] Individual grains yield by slip bands typically coming only out of one boundary. At stresses below the upper yield stress, slip bands will traverse some of the grains. In coarse-grained metals, most of the grains will be crossed by slip bands; but in fine-grained metals, as few as 1% of the grains may be crossed. Before a slip band can cross a grain, it is necessary for the applied stress to be sufficient to activate a dislocation source, generally at a grain boundary, and to propagate the dislocations across the grain, along the operative slip system.

As the stress continues to increase, slip bands propagate through several grains in a group. The length of these regions increases rapidly because, as we have seen, the dislocation velocity increases rapidly with increasing stress, and the stress concentration at the end of the slip band increases with the length of the band. Once the entire cross section has been traversed, the slipped region is called a Lüders band (in the sheet-metal trade, a stretcher strain). Because of the local reduction in cross section, the Lüders bandfront will contain many concentrations of shear stress, so the yield point strain (Lüders strain) is propagated over the gauge length of the specimen at a nearly constant stress.

I-9. THERMALLY ACTIVATED FLOW

A glance at Fig. I-18 will indicate that the yield or flow stress of iron can be reasonably assumed to be made up of two components, one of which is independent of temperature, except as the shear modulus depends upon temperature, and the other of which is strongly dependent on temperature.[2] The athermal portion has come to be called the *internal* stress τ_i. In Fig. I-18 it can be considered to be the yield stress at 100°C and above. It is the stress required to move a dislocation through the long-range stress fields of other dislocations, dislocation tangles, and cell walls. The internal stress is customarily related to the total dislocation density ρ by

$$\tau_i = \alpha G b \rho^{1/2} \tag{I-29}$$

where α is a constant and G and b are the usual symbols for shear modulus and magnitude of Burgers vector. Work hardening, then, is associated with increasing τ_i with increasing strain, because of the rise in dislocation density.

The thermal component τ^* is called the *effective* stress. It is dependent on both

[1] P.J. Worthington and E. Smith, *Acta Metall.*, 12:1277 (1964).
[2] H. Conrad, *Iron and Its Dilute Solid Solutions,* Interscience, New York, 1961, p. 315.

temperature and strain rate. This component results from short-range obstacles, such as dislocation core dissociation, Peierls stress, and impurity atoms. The shear stress τ at any point along the shear stress–shear strain curve can be expressed as

$$\tau = \tau_i + \tau^* \tag{I-30}$$

Dislocations held up at energy barriers can overcome them by applied stress and by thermal energy. This necessary activation energy, per unit length of dislocation line l, we will call ΔH. If a dislocation of Burgers vector b moves through a distance d in overcoming a barrier, then we have a quantity v^* with the dimensions of volume, called the *activation volume.*

$$v^* = bdl \tag{I-31}$$

The activation energy (activation enthalpy) is given by

$$\Delta H = \Delta H_b - v^*\tau \tag{I-32}$$

where ΔH_b is the energy of the dislocation line over length l. ΔH is less than ΔH_b by an amount equal to the work done by the applied stress.

ΔH_b can be divided into two parts:

(a) $v^*\tau_i$, the energy required to push the dislocation through long-range internal stress fields (not thermally activated), and
(b) ΔH^*, the energy required to move the dislocation through short-range obstacles (thermally activated).

Thus,

$$\Delta H_b = \Delta H^* + v^*\tau_i \tag{I-33}$$

Substituting from Eq. I-30,

$$\Delta H_b = \Delta H^* + v^*(\tau - \tau^*) \tag{I-34}$$

and from Eq. I-32

$$\Delta H = \Delta H^* - v^*(\tau - \tau_i) \tag{I-35}$$

Let N equal the number of mobile dislocations per unit volume, all arrested by energy barriers. When a dislocation moves through an area A in overcoming a barrier, the strain rate is

$$\dot{\epsilon} = bAN\nu \exp\left(-\frac{\Delta H}{kT}\right) \tag{I-36}$$

where k = Boltzmann's constant

ν = frequency with which the dislocation tries to overcome the barrier (The upper limit to ν would be the frequency of atomic vibration, $\cong 5 \times 10^{12}\,\mathrm{s}^{-1}$.)

From Eq. I-35,

$$\dot{\epsilon} = bAN\nu \exp\left[-\frac{\Delta H^* - \nu^*(\tau - \tau_i)}{kT}\right] \tag{I-37}$$

Rearranging and solving for τ,

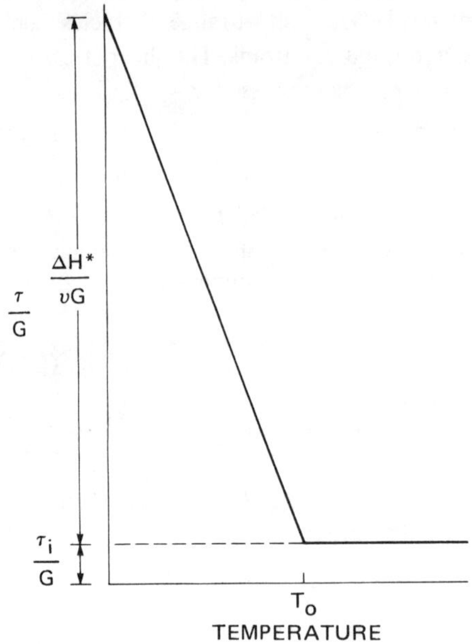

Figure I-30. Schematic representation of the flow stress of a metal with a high temperature dependence of the flow stress.

$$\tau = \tau_i + \frac{\Delta H^* - kT \ln (bAN\nu/\dot{\epsilon})}{v^*} \tag{I-38}$$

Eq. I-38 applies when $T \leqslant T_0$, where T_0 is the temperature at which the second term on the right-hand side of Eq. I-38 goes to zero. For polycrystalline iron at a low strain rate, T_0 would be about 100°C (Fig. I-18), and

$$T_0 = \frac{\Delta H^*}{k \ln (bAN\nu/\dot{\epsilon})} \tag{I-39}$$

Although this analysis certainly provides a close approximation to reality (Figs. I-18 and I-30), it must be remembered that it is based on the assumption of a single thermally activated rate-controlling process for the motion of dislocations over the entire temperature range.

I-10. RECOVERY

When cold-worked metals are heated, two processes, recovery and recrystallization, compete for the stored energy of cold work. Recrystallization is defined as reorientation by passage of a high-angle boundary. Recovery is a general term applied to all other changes that reduce the stored energy of deformation. In iron this stored energy is usually in the range 20 to 40 J/mol, or about 2 to 10% of the energy of cold work.

The processes of recovery can be followed by several techniques:

1. Calorimetry, for the direct measurement of release of stored energy;
2. Decreases in electrical resistivity;
3. Decreases in X-ray line broadening;
4. Decreases in flow stress or hardness; and

5. Measurement of decreases in dislocation density and changes in dislocation arrangement.

Recovery in cold-deformed or irradiated iron can begin at about 190 K. This recovery stage, detected by decreases in electrical resistivity, was at first thought to be due to monovacancy migration,[1] but is now considered to be caused by the long-range migration of an interstitial defect.[2,3,4] The migration of monovacancies, which is the mechanism of self-diffusion in iron, requires temperatures well above ambient, in the vicinity of 450 K. The monovacancy mechanism for self diffusion can be represented by the relation[2]:

$$D = fa^2 \nu \exp\left(-\frac{G^F + G^M}{kT}\right) \tag{I-40}$$

where f = correlation factor
a = lattice parameter
ν = attempt frequency for vacancy migration
G^F = Gibbs free energy for vacancy formation $= H^F - TS^F$
G^M = Gibbs free energy for vacancy migration $= H^M - TS^M$

In ferromagnetic iron, the enthalpies (H) and entropies (S) vary with temperature because of spin order, resulting in a curved Arrehenius plot of diffusivity vs. $1/T$, as illustrated in Fig. I-31. In ferromagnetic iron, the preexponential factor D_0, in the equation

$$D = D_0 \exp\left(-\frac{Q}{kT}\right) \tag{I-41}$$

changes with temperature. Q in this equation is equal to $H^F + H^M$. The values for these three quantities are given in Table I-4.[2]

The volume diffusion that can occur in ferromagnetic bcc iron in a given time can be estimated from the Einstein equation

$$s = (2Dt)^{1/2}$$

where s = diffusion distance, cm
D = self-diffusivity, cm^2/s
t = time, s

From Fig. I-31, in 1000 s at 500°C the maximum distance an iron atom can move is about 0.45 nm, or about two lattice parameters. In the core of a dislocation, or within about 1 nm from the center of the dislocation, the activation energy for diffusion is about half of the activation energy for volume diffusion, so diffusion in this domain can be orders of magnitude faster than in the bulk.

At temperatures higher than those at which changes in electrical resistivity can be measured, the dislocations in cold-worked iron begin to decrease in number through

[1] L.J. Cuddy, *Acta Metall.*, 16:23 (1968).
[2] G. Hettich, H. Mehrer, and K. Maier, *Scr. Metall*, 11:795 (1977).
[3] H.-E. Schaefer, K. Maier, M. Weller, D. Herlach, A. Seeger, and J. Diehl, *Scr. Metall.*, 11:803 (1977).
[4] J. Diehl, U. Merbold, and M. Weller, *Scr. Metall.*, 11:811 (1977).

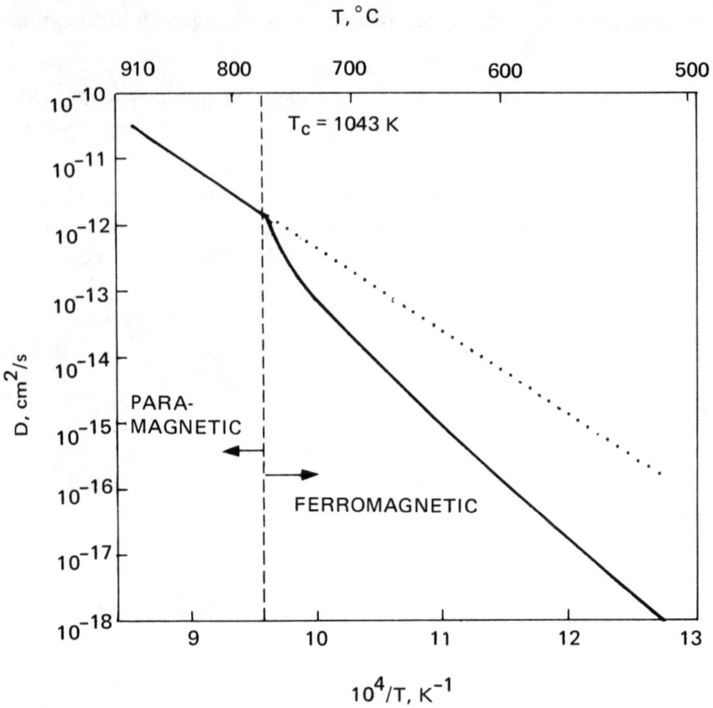

Figure I-31. Arrhenius plot of self diffusion in para- and ferromagnetic alpha iron. (Reprinted with permission from G. Hettich, H. Mehrer, and K. Maier, *Scr. Metall.*, 11:795 (1977).)

mutual annihilation and to arrange themselves by a process of climb into configurations of lower energy. The tangles present in the cold-worked metal are eventually replaced by regular hexagonal networks, planar subboundaries. The temperature at which this process begins is highly dependent upon purity, but it can start in the range 50 to 200°C and can be complete at about 500°C, if recrystallization has not intervened. These hexagonal networks are made up of three sets of dislocations, with Burgers vectors (1/2)⟨111⟩, (1/2)⟨111⟩, and ⟨100⟩, illustrated in Fig. I-32. A pair of $(a/2)\langle 111\rangle$ dislocations react to form an a ⟨100⟩ dislocation. Figure I-33 is a transmission electron micrograph of such a network in recovered iron.

During isothermal annealing, recovery is initially very rapid, and more so as the

Table I-4 Enthalpies of formation, migration, and self-diffusion of monovacancies in iron

	H (formation) (aJ)	*H* (migration) (aJ)	*Q* (aJ)
fcc Fe	0.245 ± 0.24	o.147	0.392
Paramagnetic bcc Fe	0.247 ± 0.24	0.202	0.449
Ferromagnetic bcc Fe	0.256 ± 0.24	0.205 ± 0.040	0.461 (~570 K)

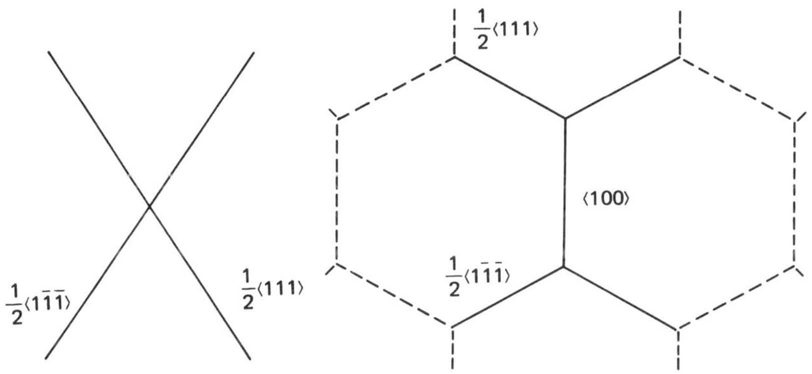

Figure I-32. Formation of hexagonal networks of dislocations (twist boundaries) in bcc iron.

annealing temperature is increased. With increasing time at constant temperature the reaction becomes slower and eventually ceases at a plateau typical of the annealing temperature. The recovery of strain hardening in zone refined iron is shown in Fig. I-34. In this work the measure of recovery employed was the fraction of residual strain hardening

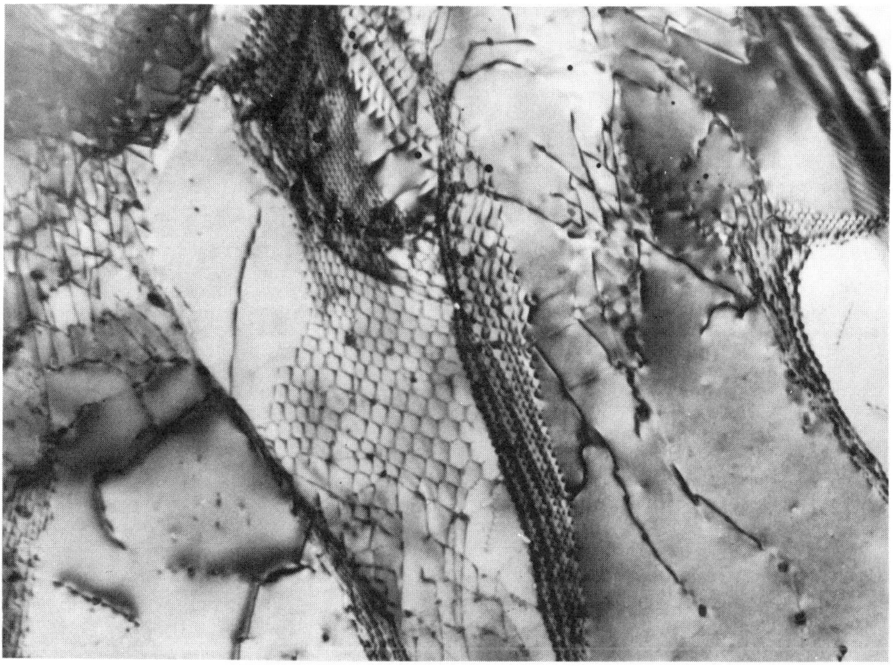

Figure I-33. Transmission electron micrograph of recovered structure in Fe–8%Cu alloy, cold-rolled to 60% reduction in thickness, annealed 3 h at 500°C, then 100 min at 650°C; 54,400×. (W.C. Leslie, J.T. Michalak, and F.W. Aul, in *Iron and its Dilute Solid Solutions,* Interscience, New York, 1963).

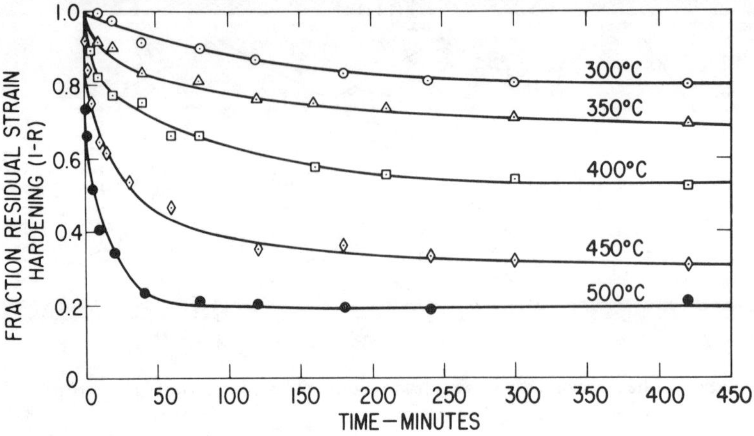

Figure I-34. Recovery of iron after 5% prestrain at 0° C. (J.T. Michalak and H.W. Paxton, *Trans. AIME*, 221:850 (1961), by permission.)

$$1 - R = \frac{\sigma - \sigma_0}{\sigma_m - \sigma_0} \tag{I-42}$$

where R = fraction of recovery
σ = flow stress after recovery anneal
σ_0 = flow stress after complete annealing
σ_m = flow stress before annealing

The greater the initial prestrain—that is, the greater the stored energy—the more rapid is the initial decrease in residual strain hardening. Also, diminishing the grain size increases the rate of recovery because the dislocation density is increased for a given strain.

From Fig. I-34, the time dependence of recovery during isothermal annealing can be described by

$$1 - R = b - a \ln t \tag{I-43}$$

where a and b are constants which depend upon temperature. The rate of recovery can be calculated from the differential form of Eq. I-43

$$\frac{dR}{dt} = \frac{a}{t} \tag{I-44}$$

Arrhenius plots of ln rate for a given fraction of recovery versus reciprocal of absolute temperature indicate that there is no unique activation energy for the recovery of cold-worked iron. The apparent activation energy varied between 100 kJ/mol at 0.1 recovery to 200 kJ/mol at 0.6 recovery. The latter number approaches the value of the activation energy for self-diffusion in ferromagnetic iron (Table I-4). Obviously, there is no single rate-controlling process in the recovery of cold-worked iron.

Residual magnetism is a measure of recovery that may be useful in the production of high-strength, recovery-annealed cold-rolled sheet steel.[1] Because lattice defects which pin domain walls also increase the residual magnetism, this quantity increases with increasing cold work. During isochronal annealing, in the absence of recrystallization, about 60% of the residual magnetization is removed at temperatures below 500°C and about 80% at temperatures below 550°C. The change is similar to that of X-ray line width during recovery. The measurement of residual magnetization is rapid, nondestructive, and reproducible. A unit for measuring residual magnetization can be installed at the outlet of a continuous annealing furnace to measure the degree of recovery in recovery-annealed sheet.

I-11. RECRYSTALLIZATION

The recrystallization of metals has been reviewed most recently by Cotterill and Mould.[2] An earlier report confined to iron[3] is still valid.

If a cold-worked metal is heated to a sufficiently high temperature it will recrystallize. Recrystallization is defined as the reorientation of the crystals in a solid body by the passage of a high-angle boundary. The kinetics of the process resembles that of a phase transformation, in that it often can be described in terms of a nucleation frequency and a rate of growth. When we use the term "nucleation" with respect to recrystallization we do not mean that small, strain-free grains are formed by the classical process of atom-by-atom accretion until a critical size is attained. Rather, new grains are formed by the growth of specific lattice domains already present in the cold-worked metal. When a cold-worked metal is heated, dislocation cells become subgrains by a process of recovery. If cells are not present initially, they can be created by dislocation rearrangement during recovery (Fig. I-33). To serve as a nucleus for recrystallization a subgrain must have the following characteristics:

1. It must be more perfect than its neighbors; that is, it must contain fewer dislocations and thereby a lower strain energy.
2. The boundaries of the subgrain must be mobile by virtue of sufficient lattice misorientation between the subgrain and its neighbors. Sufficient misorientation might be about 15°, referred to a common axis of rotation.
3. It must be large enough so that the additional interfacial energy that must be supplied for growth is less than the volume free energy released when strained cells are replaced by a strain-free subgrain.

When a metal is deformed plastically, the deformation is usually so heterogeneous

[1] B.R. Morrison and D.J. Willis, *Metals 77*, paper 8A-1 (1977).

[2] P. Cotterill and P.R. Mould, *Recrystallization & Grain Growth in Metals*, Surrey Univ. Press, London, 1976.

[3] W.C. Leslie, J.T. Michalak, and F.W. Aul, in *Iron and Its Dilute Solid Solutions*, Interscience, New York, 1963, p. 119.

on a fine scale that the first of these requirements is easily met. The second requirement, sufficient misorientation, is also not difficult to supply. The growth of subgrains occurs at several favored sites, these being preexisting grain boundaries, inclusions or second-phase particles, deformation bands (shear bands) in the interior of grains, and intersections of mechanical twins. The common feature of these sites is that they are all local regions of severe lattice bending.

In heavily deformed polycrystalline metals, recrystallization invariably begins at so-called grain-boundary edges, where three grains meet along a line.[1] Only a small fraction, about 5%, of the grain-boundary edges are sites for nucleation. Nucleation sites develop along the entire length of the active edges—an excellent example of site-saturated nucleation. Therefore, impingement of new grains is almost immediate, and subsequently the grains can grow only two-dimensionally. This characteristic is not limited to cold working and annealing. English and Backofen[2] worked Fe-3%Si in the temperature range from 710 to 910°C and noted that most new grains began at grain edges. One of the characteristics of recrystallization of hot-worked structures is the absence of nucleation within grains, except when inclusions are present. Thus the grain size after recrystallization can be made smaller as the original grain size is reduced.

Although grain edges are much more effective, grain boundary surfaces are also sites for nucleation of grains. It frequently happens that neighboring grains will have two quite different cell (subgrain) sizes. Those from the grain having the larger cells will tend to grow into the grain having the smaller cells. The growing new grain retains the orientation of the cell from which it sprang. This process of grain boundary bulging is illustrated in Fig. I-35. It has been seen to occur during the recrystallization of iron.[3]

A good example of the effectiveness of high-angle boundaries in promoting recrystallization is the fine grain structure and random texture that can be produced in alpha iron by explosive shock loading to pressures above 13 GPa, followed by annealing. This comes about by the introduction of closely spaced high-angle boundaries by twinning and by shear transformations ($\alpha \rightarrow \epsilon \rightarrow \alpha$) during passage of the shock wave. The structure of iron after passage of a shock wave of 22 GPa pressure is shown in Fig. I-36*a*. Because closely spaced high-angle boundaries are already present, they need not be generated by growth of subgrains. The same structure after annealing 5 min at 500°C is presented in Fig. I-36*b*. Considerable boundary migration has occurred, but the resulting very small grains still contain a high density of dislocations. The twin density here is about 3×10^7 cm^{-2}. In Fig. I-36*c* it can be seen that dislocation-free grains have started to grow at the expense of neighboring grains which have higher density of dislocations. The recrystallization of shock-loaded iron which has undergone a pressure-induced phase transformation is very rapid as compared to the same iron that has been heavily cold-worked. This difference is due to two factors:

[1] R.A. Vandermeer and P. Gordon, *Trans. AIME*, 215:577 (1951).

[2] A.T. English and W.A. Backofen, *Trans. AIME*, 230:397 (1964).

[3] W.C. Leslie, J.T. Michalak, and F.W. Aul, in *Iron and Its Dilute Solid Solutions*, Interscience, New York, 1963, p. 119.

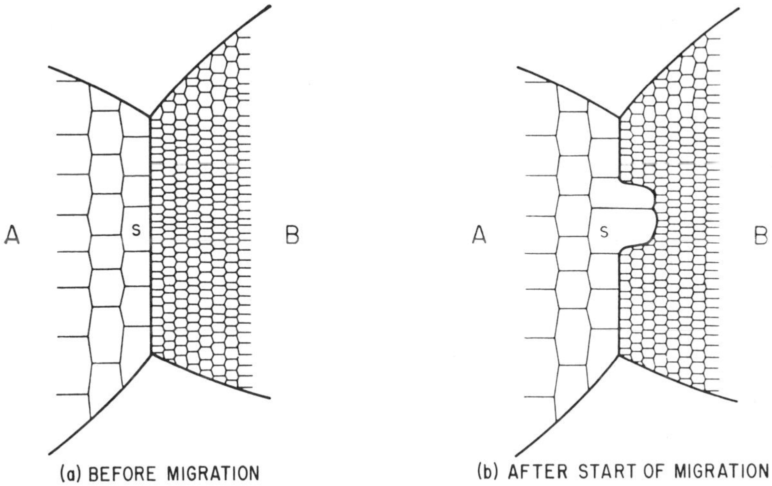

Figure I-35. Recrystallization by migration of a high-angle boundary. (P.A. Beck and P.R. Sperry, *J. Appl. Phys.*, 21:150 (1950), by permission.)

1. The greater energy stored in the pressure-transformed iron; and
2. The vastly greater number of sites at which growth can begin by the migration of an existing high-angle boundary. Because of this nucleation of recrystallization at many sites a large degree of grain refinement can be got by a pressure-induced transformation, followed by recrystallization. A reduction in grain size from ASTM 0 to 7 can be accomplished.

In some alloys, such as Fe–Ni, a similar result can be obtained by quenching to martensite, cold-working the martensite, then annealing at a low temperature.[1]

Nucleation sites of minor consequence lie at the intersections of mechanical twins in iron or at notches in twins (Fig. I-37). These are regions of strain concentration.

The beginning of recrystallization at inclusions in a cold-worked metal, as shown in Fig. I-38, is common knowledge. Around such inclusions are regions of high dislocation density, with the inclusions often lying at the intersection of two dislocation walls. The dislocation density in these walls is much greater than after the same amount of plastic deformation without the inclusions. Cell structures tend to be better developed than when inclusions are absent, and it seems safe to assume that misorientation across cell walls is greater than average in the vicinity of inclusions. These factors all tend to promote subgrain boundary migration in the vicinity of inclusions.

The effects of inclusions or second-phase particles on recrystallization are critically dependent upon particle size and interparticle spacing, as discussed in detail by

[1] R.L. Miller, *Met. Trans.*, 3:905 (1972).

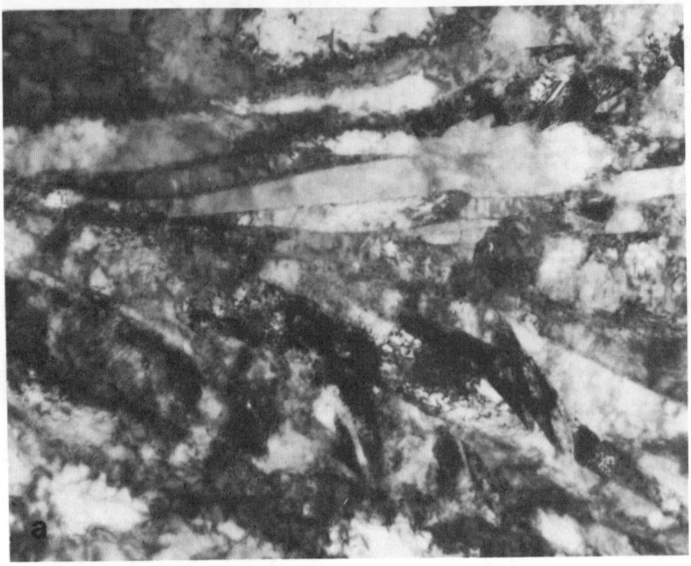

Figure I-36. Development of fine grains in iron by shock loading followed by isothermal recrystallization. (*a*) Structure of iron after passage of shock wave of 22 GPa pressure. (*b*) As in (*a*), then annealed 5 min at 500°C, showing boundary migration and high density of twins. (*c*) As in (*b*), showing new recrystallized grains. Transmission electron micrographs, 40,000X. (W.C. Leslie, J.T. Michalak, and F.W. Aul, in *Iron and its Dilute Solid Solutions,* Interscience, New York, 1963.)

Cotterill and Mould.[1] Large particles (>1 μm) promote nucleation of new recrystallized grains, and if they are widely spaced the rate of recrystallization is accelerated. If particles are fine (<0.1 μm) and closely spaced, the deformation structure consists of a uniform distribution of dislocations, without the regions of high lattice curvature that are required for the development of new recrystallized grains. In such structures recrystallization can be severely inhibited.

Recrystallization occurs in clean single crystals during annealing after cold working, so there must be other nucleation sites in addition to those previously discussed. When a metal is deformed plastically the deformation is often intensely inhomogeneous, especially at large strains, resulting in the formation of transition bands and shear bands, as mentioned before. Because of the large local misorientation, these bands can be the source of new, nearly dislocation-free subgrains during subsequent annealing.

When a metal is heated after cold working, the usual observation is that larger than average subgrains appear that are nearly free of dislocations. The interfaces of such subgrains can advance to consume the neighboring subgrains. The driving force for migration of the boundary is supplied by the difference in energy density across the low-angle subgrain boundary. As the boundary moves, its energy per unit area can

[1] P. Cotterill and P.R. Mould, *Recrystallization and Grain Growth in Metals,* Surrey Univ. Press, London, 1976.

Figure I-36. *(Continued)*

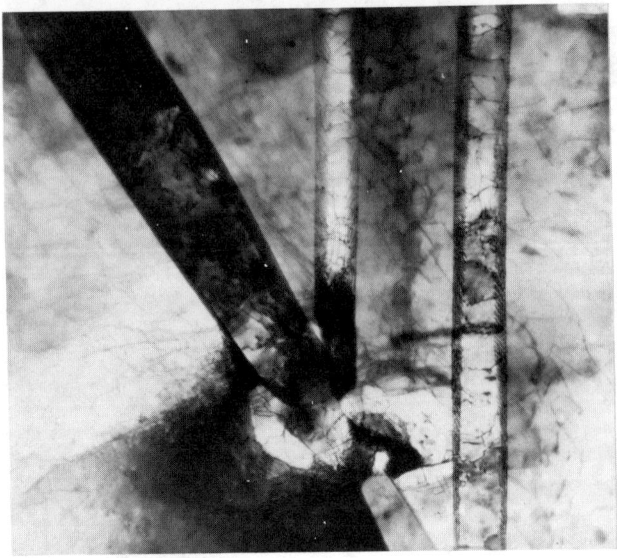

Figure I-37. Nucleation of recrystallized grains at an intersection of mechanical twins in iron. Transmission electron micrograph, 40,000×. (W.C. Leslie, J.T. Michalak, and F.W. Aul, in *Iron and its Dilute Solid Solutions,* Interscience, New York, 1963.)

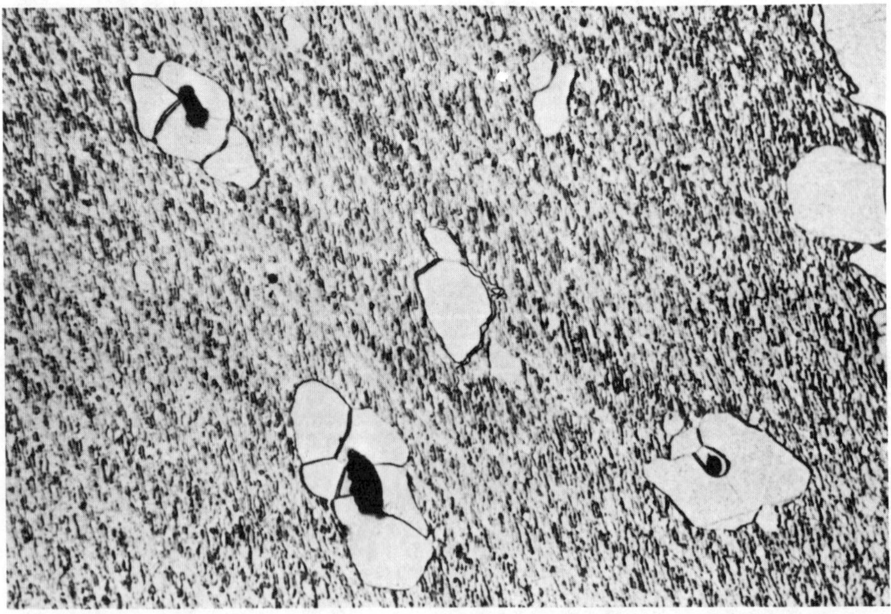

Figure I-38. Recrystallized grains formed at oxide inclusions in iron. Nital etch, 550×. (W.C. Leslie, J.T. Michalak, and F.W. Aul, in *Iron and its Dilute Solid Solutions,* Interscience, New York, 1963).

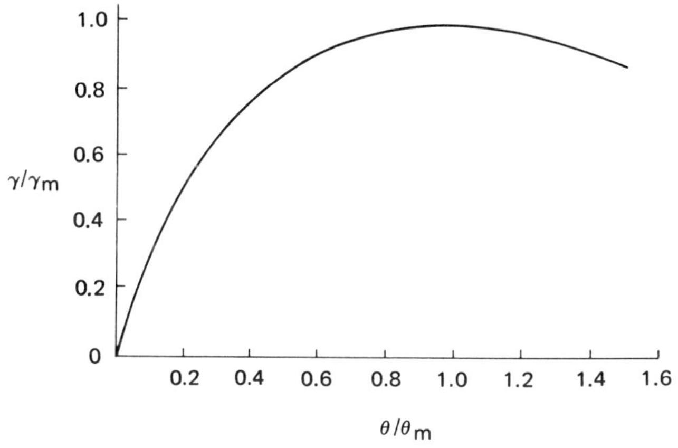

Figure I-39. Relation between misorientation and boundary energy in low-angle boundaries.

be decreased by an increasing misorientation, according to the Read-Shockley equation[1]

$$\gamma = \gamma_0 \theta (A - \ln \theta) \tag{I-45}$$

where γ = boundary energy per unit area
γ_0 = constant depending on the geometrical properties of the dislocations making up the boundary
θ = misorientation
A = integration constant

The equation can be written in terms of the dimensionless quantities γ/γ_M and θ/θ_M, where γ_M is the maximum of the γ vs. θ curve and θ_M is the misorientation at which the curve reaches the maximum, such that

$$\frac{\gamma}{\gamma_M} = \frac{\theta}{\theta_M}\left(1 - \ln \frac{\theta}{\theta_M}\right) \tag{I-46}$$

Figure I-39 is a plot of Eq. I-46. At small values of θ, the boundary energy increases at a decreasing rate as θ increases. Therefore, a boundary of misorientation θ_3 formed from subboundaries θ_1 and θ_2 will have an energy less than the sum of the energies of θ_1 and θ_2. The process of subboundary migration is illustrated in Fig. I-40. The boundaries marked with the arrows will move toward their centers of curvature, making subgrain 1 wider. This process can occur by a succession of migrations across subgrains, until a high-angle boundary is developed and we can describe the nucleus as a recrystallized grain.

A third mechanism for development of a recrystallized grain is subgrain coales-

[1] W.T. Read and W. Shockley, *Phys. Rev.*, 78:275 (1950).

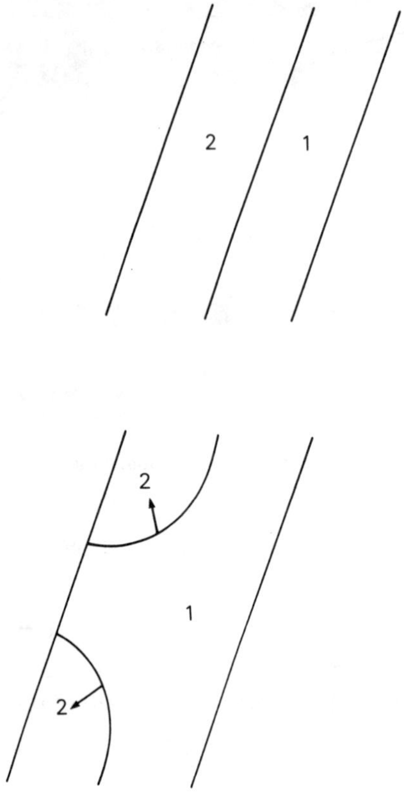

Figure I-40. Formation of a nucleus for recrystallization by subboundary migration.

cence, as proposed by Hu[1] and supported by Li.[2] Subgrains can grow by elimination of the boundary between two adjacent subgrains. The common boundary disappears by a movement of dislocations out of it and into the other boundaries of the two subgrains. The orientation of the coalesced subgrain will differ from that of the two parent subgrains—that is, a rotation of the lattice will result when the common boundary is eliminated. The process is presented schematically in Figs. I-41 and I-42. It involves dislocation climb along the disappearing boundary. If the annealing temperature is high enough, the subgrain boundaries will not retain long-range stresses and the dislocation distribution in the boundary will remain uniform. This requires that the dislocations climb cooperatively so that the spacings between them remain equal and so that the boundary retains its low-energy configuration.

Li's work[2] indicates that coalescence does not violate the second law of thermodynamics, which requires that for an isothermal process the free energy of the system must decrease with time. The relevant free energy is the total free energy of all the boundaries of the rotating subgrain. For an entire subgrain, the boundary with the greatest area containing the largest number of dislocations for the same angle of misfit will furnish the largest portion of the driving force. The rotation of a subgrain

[1] H. Hu, *Trans. AIME,* 224:75 (1962).
[2] J.C.M. Li, *J. Appl. Phys.,* 33:2958 (1962).

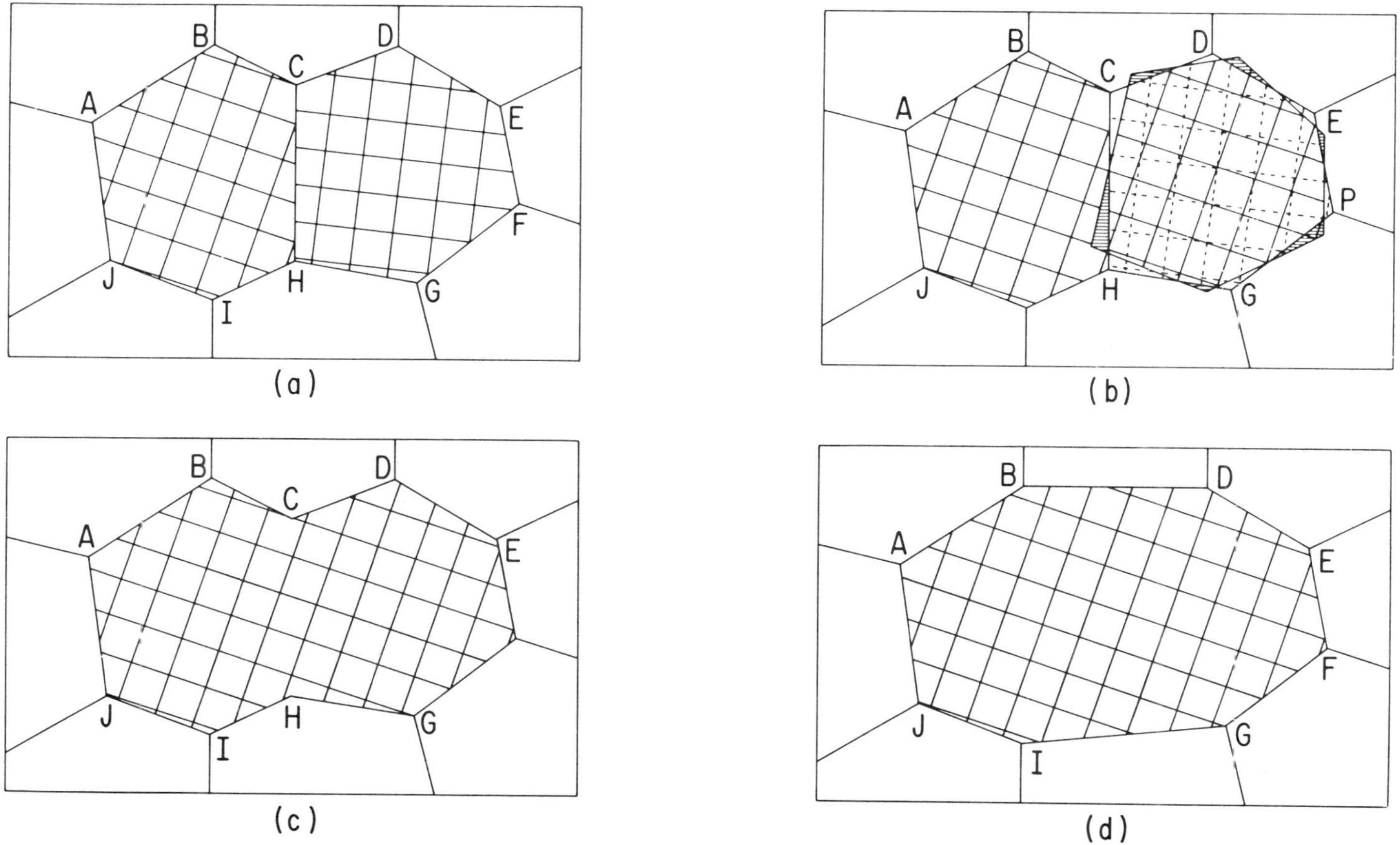

Figure I-41. Coalescence of two subgrains by rotation of one of them. (*a*) The original subgrain structure before coalescence. (*b*) One subgrain undergoing rotation. (*c*) Subgrain structure just after coalescence. (*d*) The final subgrain structure after some subboundary migration. (J.C.M. Li, *J. Appl. Phys.* 33:2958 (1962), by permission.)

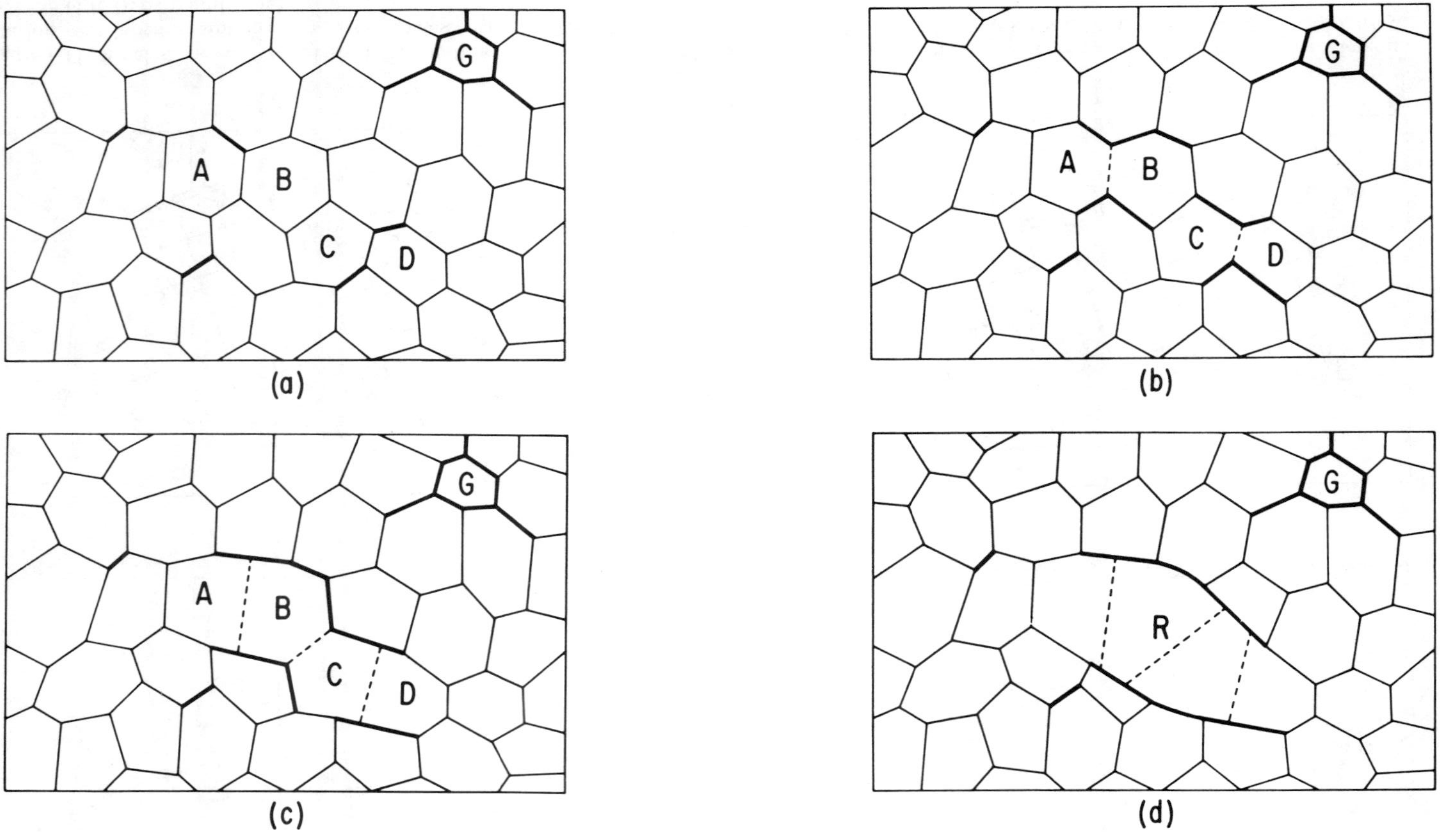

Figure I-42. Schematic representation of the formation of a recrystallized grain by the coalescence of subgrains. (H. Hu, in *Electron Microscopy and the Strength of Crystals.* (G. Thomas and J. Washburn, eds., Interscience, New York, 1962, by permission.)

favors the elimination of such a boundary, leading to coalescence of the two subgrains initially separated by this boundary.

The coalescense model for nucleation of recrystallization is as follows. The cold-worked structure after some recovery consists of well-defined subgrains. Some of these increase in size by elimination of boundaries, thus lowering the free energy of the system. As a result, the angles of misfit across other boundaries defining the subgrains may become greater. Two sets of coalesced subgrains may then also undergo additional coalescense, producing a recrystallized grain defined by high-angle boundaries.

Three mechanisms for the nucleation of recrystallization in a metal have been described:

1. Movement of a preexisting grain boundary
2. Subgrain boundary migration
3. Subgrain coalescence

The first two of these (which are similar in that they both involve boundary migration) require that the orientation of the newly formed recrystallized grain must have been a component of the texture of the cold-worked metal. If nucleation occurs by coalescence, the new grain, since it formed by rotation of existing subgrains, can have an orientation that differs from any component of the texture of the cold-worked metal. For polycrystalline iron, the preponderance of evidence indicates that movement of preexisting grain boundaries is the most common mode for creation of new, strain-free grains.

Recrystallization of metals by isothermal annealing follows a sigmoidal pattern with respect to time, typified by Fig. I-43. For the same prior deformation, the time

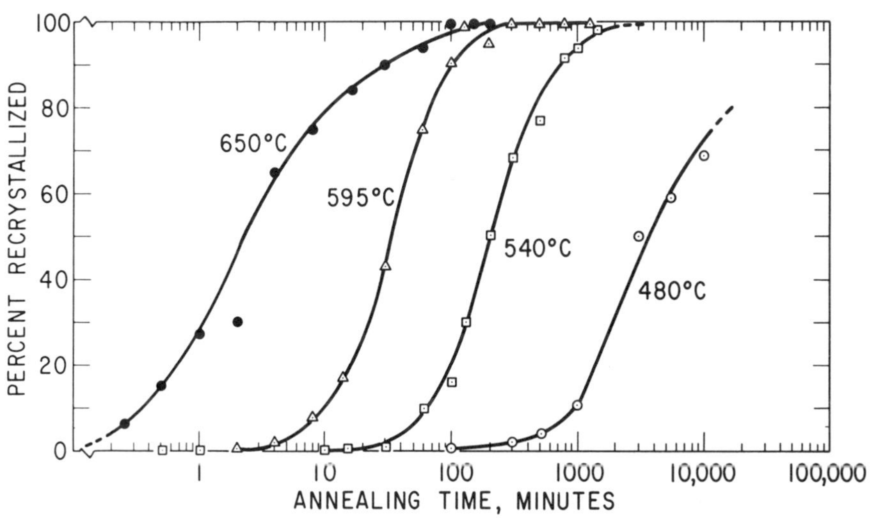

Figure I-43. Isothermal recrystallization of zone-melted Fe–0.60%Mn alloy. (W.C. Leslie, F.J. Plecity, and J.T. Michalak, *Trans. AIME,* 221:691 (1961).)

required for complete recrystallization is reduced by increasing the annealing temperature. There is an initial period (incubation time) during which no recrystallization occurs; then the process begins slowly, accelerates to a maximum rate, and finally comes slowly to completion. This pattern lends itself to a formal description in terms of a nucleation frequency and a linear rate of growth, the classical presentation of recrystallization developed by Avrami,[1] Johnson and Mehl,[2] and Burke and Turnbull.[3] An assumption is made that the linear rate of growth is constant with time, and, as will be seen later, this is true only when recovery and recrystallization are separated in time so that the driving force for recrystallization is constant.

Then

$$D = G(t - t_0) \tag{I-47}$$

where D = diameter of recrystallized grain growing isothermally into a strained matrix
G = linear rate of growth
t = time
t_0 = nucleation or incubation period

If the total volume of the metal is unity and X is the volume recrystallized in time t, the problem is to find X as a function of time.

At time t the volume of a given recrystallized grain is

$$v = fG_xG_yG_z(t - t_0)^3 \tag{I-48}$$

where G_x, G_y, and G_z are orthogonal linear growth rates and f is a shape factor. The number of grains originating in a given time is

$$dn = \dot{N}(1 - X)\,dt \tag{I-49}$$

where $\dot{N}$ = nucleation frequency per unit volume
dt = time interval

As grains grow, they impinge upon one another. To take this into account, another number is defined:

$$dn' = dn + \dot{N}X\,dt = \dot{N}\,dt \tag{I-50}$$

where $\dot{N}X\,dt$ = number of "phantom" grains that would have nucleated in the volume X if it had not recrystallized

Another concept is needed, that of extended volume.

$$X_{ex} = \int_0^t v\,dn' = fG_xG_yG_z \int_0^t (t - t_0)^3\dot{N}\,dt \tag{I-51}$$

[1] M. Avrami, *J. Chem. Phys.*, 7:1103 (1939).
[2] W.A. Johnson and R.F. Mehl, *Trans. AIME*, 135:416 (1939).
[3] J.E. Burke and D. Turnbull, *Prog. Met. Phys.*, 3:220 (1952).

This extended volume is calculated without regard for impingement and by counting the phantom grains.

The equation relating X and X_{ex} is

$$1 - X = \frac{dX}{dX_{ex}} \tag{I-52}$$

That is, the ratio of the rate of growth of an actual grain to the rate of growth of its extended volume is equal to the fraction of unrecrystallized metal. Substituting I-52 into I-51 gives the basic equation for the rate of a nucleation and growth process.

$$\int_0^X \frac{dX}{(1-X)} = \ln\left(\frac{1}{1-X}\right) = fG_x G_y G_z \int_0^t (t-t_0)^3 \dot{N}\, dt \tag{I-53}$$

$$X = 1 - \exp\left(-\int_0^t v\, dn'\right) \tag{I-54}$$

In a cold-worked metal matrix there are a number of nucleation sites $\bar{N}$ each having a nucleation frequency ν. During recrystallization the sites are exhausted by becoming nuclei, so the number of potential sites N after time t is

$$N = \bar{N} \exp(-\nu t) \tag{I-55}$$

Thus, $$\dot{N} = \bar{N} \exp(-\nu t) \tag{I-56}$$

Substituting in Eq. I-54 and assuming that growth is isotropic ($G_x = G_y = G_z$), we obtain

$$X = 1 - \left[\exp \bar{N}\nu f G^3 \int_0^t (t-t_0)^3 \exp(-\nu t)\, dt\right] \tag{I-57}$$

Integrating and simplifying, if the nucleation rate does not vary with time,

$$X = 1 - \exp\left(\frac{fG^3 \dot{N} t^4}{4}\right) \tag{I-58}$$

For recrystallization, Avrami proposed the general equation

$$X = 1 - \exp(Bt^k) \tag{I-59}$$

where B and k are constants. When recrystallization is three-dimensional, k lies between 3 and 4. When recrystallization is two-dimensional, as in a thin sheet, k is between 2 and 3, and when recrystallization is one-dimensional as in a wire, k is between 1 and 2. Therefore, if the proper conditions are fulfilled, Avrami's equation can be used to determine the dimensionality of growth of recrystallized grains. The usual procedure is to plot log ln $[1/(1 - X)]$ vs. log time, where X is the fraction recrystallized, to give a straight line of slope k. Vandermeer and Gordon[1] studying recrystallization of zone-refined aluminum containing 0.008% Cu, found that the

[1] R.A. Vandermeer and P. Gordon, *Trans. AIME*, 215:577 (1951).

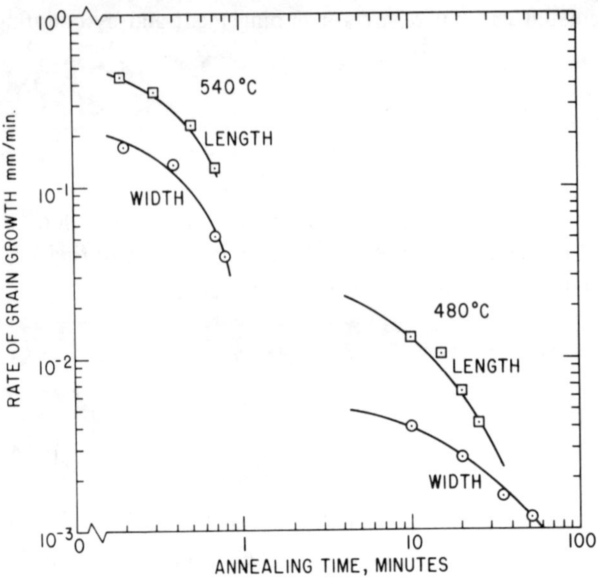

Figure I-44. Effect of time and temperature on rate of grain growth in zone-melted iron cold-rolled to 60%. (W.C. Leslie, J.T. Michalak, and F.W. Aul, in *Iron and its Dilute Solutions,* Interscience, New York, 1963).

rate of growth of recrystallized grains was constant; the recrystallized grains began to grow at essentially zero time and no additional nucleation occurred. Under these conditions, the constant k in Avrami's equation was equal to 2, indicating that growth of grains was two-dimensional. This was verified by observation.

Most commercial annealing is done on fine-grained, heavily deformed metal. Under these conditions the rate of growth of recrystallized grains usually decreases rapidly with time, because of the depletion of the stored energy of cold work by concurrent recovery. Also, the rate of growth is anisotropic: that is, $G_x \neq G_y \neq G_z$. Under these common conditions, straight lines are not obtained from plots of log ln $[1/(1 - X)]$ vs. log time, and no conclusions can be drawn regarding the dimensionality of grain growth. The very sharp decrease in rate of grain growth with time during recrystallization of iron is shown in Fig. I-44, which also illustrates the anisotropy of growth.

The kinetics of grain growth noted in the recrystallization of heavily cold-worked iron are similar to the kinetics of recovery (Fig. I-34); the rate of recovery was high initially, then decreased with time, according to Eq. I-43. The decrease in rate of growth with time indicates that the driving force for growth is depleted by concurrent recovery. The situation is consistent with a model of a growth-controlled process in which nuclei were present at zero time.

Let us assume that rate of growth is proportional to the residual strain energy in the matrix, or

$$G = K_1 F \tag{I-60}$$

where G = instantaneous linear rate of growth
F = stored energy per unit volume

The stored energy is reduced by concurrent recovery; therefore

$$F = F_0 - R \tag{I-61}$$

where F_0 = stored energy per unit volume at completion of cold working
R = stored energy removed by recovery

In the period t_1 to t_2, where $t_1 > 0$, the removal of stored energy by recovery follows the relationship

$$R = b \ln t \tag{I-62}$$

Then $$F = F_0 - b \ln t \tag{I-63}$$

and $$G = K_1 F_0 - K_2 b \ln t \tag{I-64}$$

At periods greater than t_2, recovery ceases, the stored energy of the unrecrystallized matrix remains constant, and the rate of linear growth is constant. Equation I-63 calls for a linear relation between rate of growth and ln t. This is shown in Fig. I-45. At longer periods the rate seems to become constant; but before that point is reached, extensive impingement of growing grains prevents further determinations of growth rates.

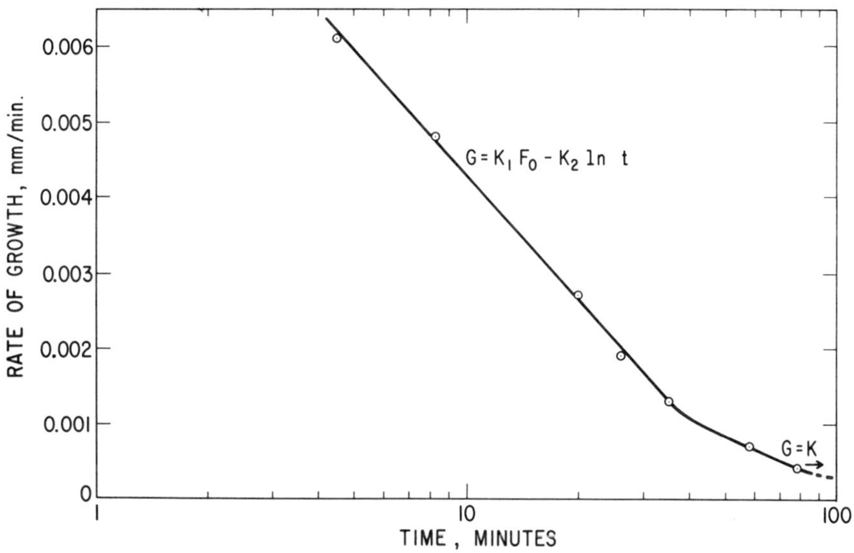

Figure I-45. Time dependence of rate of grain growth in zone-melted iron, cold-rolled to 60%, then annealed at 480°C. (W.C. Leslie, F.J. Plecity, and J.T. Michalak, *Trans. AIME*, 221:691 (1961).)

The observations that growth of recrystallized grains in heavily cold-worked iron or aluminum decreases rapidly with increasing annealing time are not in conflict with earlier observations that rate of growth in lightly deformed iron or aluminum was constant. In a cold-worked metal, recovery and recrystallization are both driven by the stored energy of cold work, and whether they are competing processes depends upon their relative rates at a given annealing temperature. These rates will depend upon a variety of factors, including the amount of type of deformation, grain size, cleanliness and purity of the metal, the orientation of grains with relation to the plastic strain, and the annealing temperature. Thus, after small strains, recovery can be substantially complete before recrystallization begins, leaving a constant driving force for recrystallization and producing a constant rate of growth of new grains into the strained and recovered matrix.

No unique activation energy can be assigned to the recrystallization of heavily cold-worked iron (Fig. I-46) or to the growth of individual grains into the unrecrystallized matrix (Fig. I-47), as indicated by the non-linearity of these Arrhenius plots.

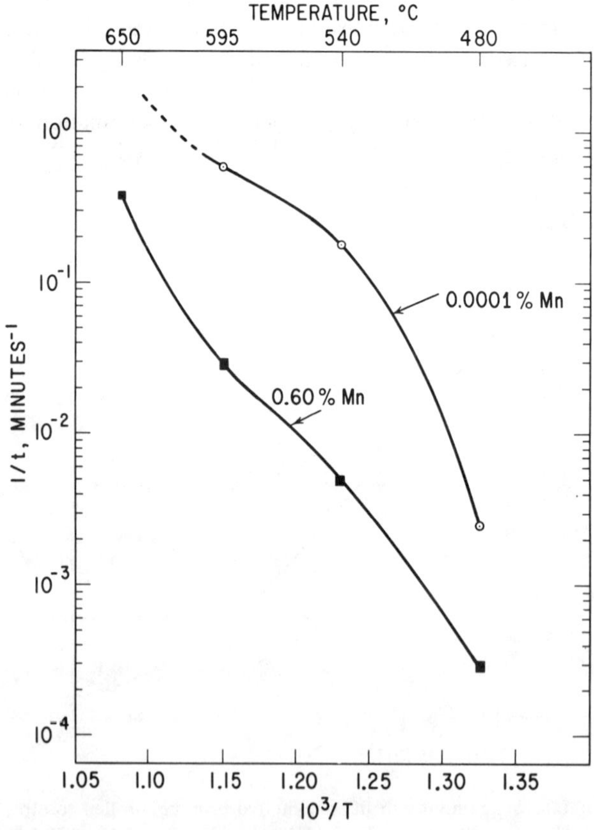

Figure I-46. Temperature dependence of rate of recrystallization (50%) in iron and in an Fe–Mn alloy. (W.C. Leslie, F.J. Plecity, and J.T. Michalak, *Trans. AIME*, 221:691 (1961).)

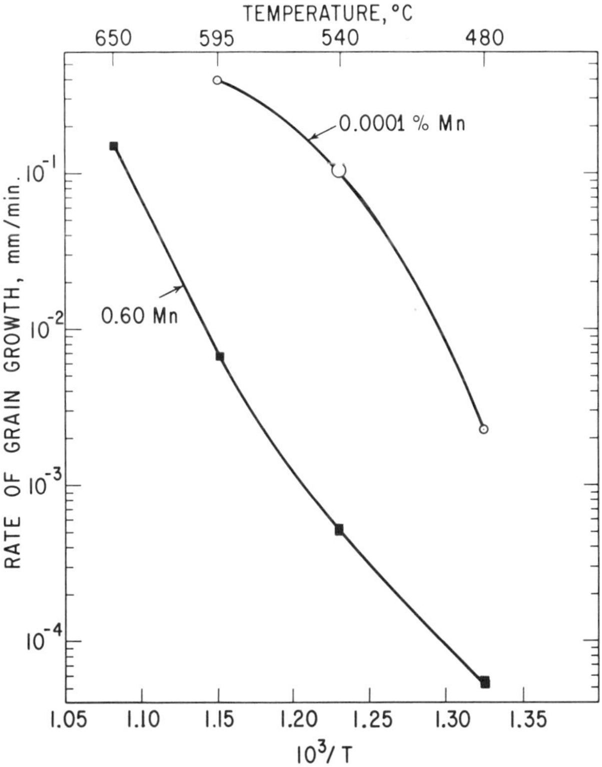

Figure I-47. Temperature dependence of rate of grain growth at 5% recrystallization in iron and in Fe–Mn alloy. (W.C. Leslie, F.J. Plecity, and J.T. Michalak, *Trans. AIME,* 221:691 (1961).

Depending upon the temperature range selected, the activation energy for recrystallization can vary between 125 and 370 kJ/mole.

I-12. NORMAL GRAIN GROWTH

Normal grain growth in an annealed metal is driven by the energy contained in grain boundaries. It is equivalent to photographic enlargement, in that the distributions of grain size and grain shape do not change with time. Normal grain growth in α iron has not been extensively studied for two obvious reasons: it is impossible to employ high temperature because of the $\alpha \leftrightarrow \gamma$ phase transformations, and the ultra-high purity essential for such studies was not available until recently. To obtain consistent data it is necessary to begin with a structure of uniformly fine grains, randomly oriented.

Normal grain growth can be described approximately by the relation, as developed by Beck,[1]

[1] P.A. Beck, *J. Appl. Phys.*, 19:507 (1948).

$$D = Kt^n \tag{I-65}$$

where D = average instantaneous grain diameter
t = time at constant annealing temperature
K, n = parameters dependent upon material and temperature, independent of grain size

The value of n usually increases with increasing annealing temperature, approaching a maximum of 0.5 as the melting temperature T_M is approached.

The simple Eq. I-65 can be used when the starting grain size is small as compared to the grain size after growth; a more precise form is

$$D^{1/n} - D_0^{1/n} = ct \tag{I-66}$$

where D_0 is the average grain size at the start of annealing.

Studies of grain boundary migration in zone-refined aluminum crystals[1] have shown that the boundary velocity is a power function of the driving force:

$$V = M(\Delta G)^m \tag{I-67}$$

where V = boundary velocity
M = boundary mobility (velocity at unit driving force)
ΔG = driving force/unit volume

Hu and Rath[2] have shown that the time exponent n and the driving force exponent m are related by the expression

$$m = \frac{1}{n} - 1 \tag{I-68}$$

Therefore, a parabolic growth law $n = 0.5$ corresponds to a linear dependence of boundary velocity on driving force ($m = 1$). The temperature dependence of the mobility term M is given by

$$M = M_0 \exp\left(-\frac{Q}{RT}\right) \tag{I-69}$$

where M_0 is a rate constant and Q is the activation energy for boundary migration.

In zone-refined iron, Eq. I-65 was followed, with n increasing with temperature, from 0.26 at 550°C to 0.5 at 850°C ($0.6T_M$). Therefore, m decreased from about 3 at 550°C to 1 at 850°C. When n varies with temperature, no unique activation energy can be obtained because the rate of grain growth then depends not only upon temperature but also upon grain size. When n reaches 0.5 and is thereafter independent of temperature, the activation energy for boundary migration is equal to 250 kJ/mol, just slightly less than the activation energy for volume self-diffusion of iron (Table I-4).

[1] B.B. Rath and H. Hu, *Trans. AIME*, 245:1577 (1969).
[2] H. Hu and B.B. Rath, *Met. Trans.*, 1:3181 (1970).

I-13. ANNEALING TEXTURES

The textures of cold-worked iron are altered by primary recrystallization which is driven by the stored energy of cold work. During subsequent heating the material can be recrystallized again, if the boundaries of the primary recrystallized grains have been pinned by second-phase particles. This is called secondary recrystallization and is of great importance in the production of grain-oriented silicon steels. After secondary recrystallization, and removal of the pinning particles, the texture of a thin sheet can be changed again under the influence of gas–metal interfacial energies. In a bcc metal exposed to high vacuum, the surface of lowest free energy will be the plane of closest packing, {110}. However, if contaminants are present in the atmosphere or in the metal, the close-packed plane may no longer be the plane of lowest specific surface energy.

To return to the topic of primary recrystallization: There were two theories of the development of texture during the annealing of cold-worked metals, the oriented nucleation theory and the oriented growth theory, but it is now generally conceded that textures are developed during primary recrystallization because of the orientation dependence of grain boundary mobility, i.e., oriented growth. If the deformation texture is complex, as in cold-rolled iron, with large spreads around particular orientations, it is difficult to predict the resulting annealing texture because we do not have a detailed knowledge of grain boundary mobility.

The difference in grain boundary mobility need only be slight for the oriented growth mechanism to operate.[1] The critical factor is the increase of volume of recrystallized metal as a function of orientation; therefore, the controlling factor is the third power of boundary mobility.

As Beck points out, there is another factor that exaggerates small differences in boundary mobility, and that is the impingement of recrystallized grains. Fast-growing grains pinch off those grains whose boundaries have lower mobility, and grains with boundaries of maximum mobility grow to a disproportionate size.

As in deformation textures, textures in bcc iron after primary recrystallization are complex, with intensity peaks in pole figures usually being no more than three times the random intensity. The texture after 90% cold reduction followed by annealing consists of the cold-rolled texture rotated 15° about the sheet normal to each side of the rolling direction. It has been described by Barrett[2] as (111) $[1\bar{1}2]$, (100) 15° from [011] and (112) 15° from $[1\bar{1}0]$. The annealing texture is quite sensitive to the degree of prior cold work. In the range of 60 to 75% cold reduction by rolling, there is a (111) $[1\bar{1}0]$ component of the annealing texture, which, as will be seen later, has important consequences for the deep drawing of low-carbon sheet steels. A typical pole figure of iron annealed after heavy cold rolling is shown in Fig. I-48.

Because they are dependent upon factors that are not present in high-purity iron, the textures developed during secondary and tertiary recrystallization will be discussed in sections dealing with steels.

[1] S. Kohara, M.N. Parathasarathi, and P.A. Beck, *Trans. AIME,* 212:875 (1958).
[2] C.S. Barrett and J.B. Massalski, *Structure of Metals,* 3d ed., McGraw-Hill, New York, 1966.

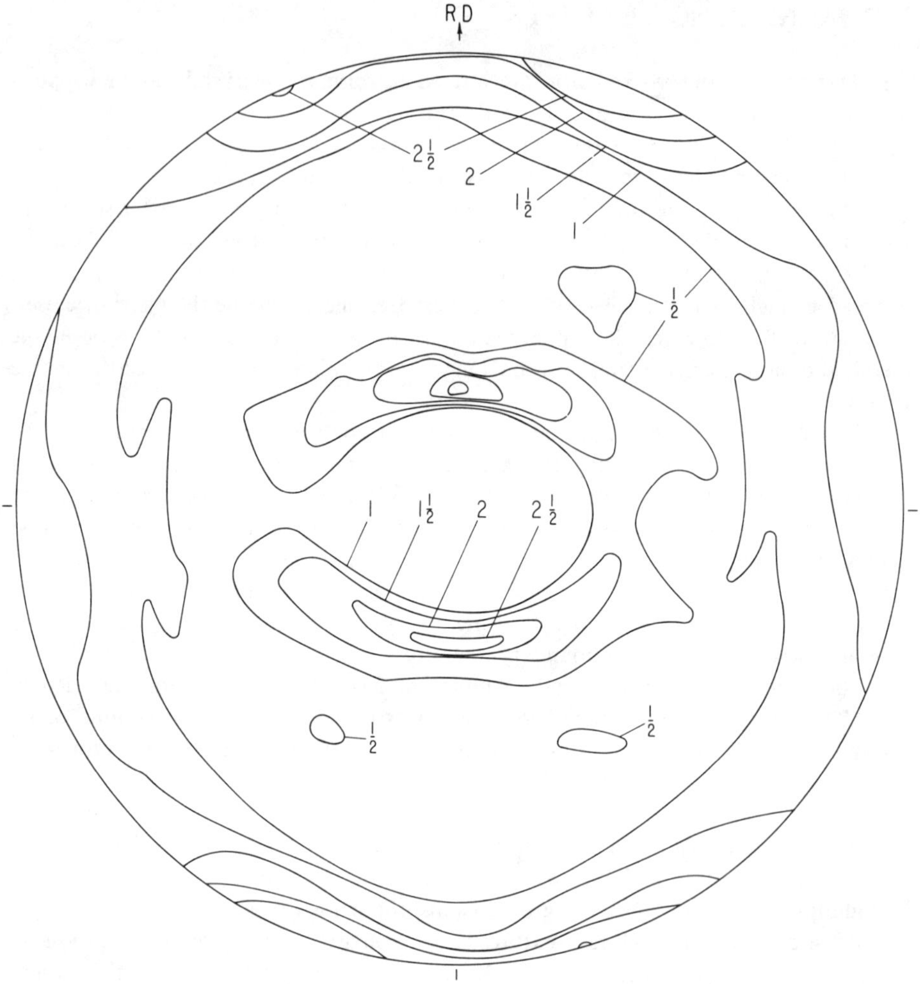

Figure I-48. (110) pole figure of recrystallized iron annealed 3 h at 500°C, then 5 h at 700°C, after 90% cold reduction by rolling. (W.C. Leslie, *Trans. AIME,* 221:752 (1961).)

I-14. PHASE TRANSFORMATIONS IN HIGH-PURITY IRON

Although the details are still far from being clear, it now seems possible that most of the types of phase transformations that occur in steels also can occur in high-purity iron.[1–4] This text does not purport to give an adequate treatment of phase transforma-

[1] M.J. Bibby and J.G. Parr, *J. Iron Steel Inst.,* 202:100 (1964).

[2] O.P. Morozov, D.A. Mirzayev, and M.M. Shteynberg, *Phys. Met. Metallogr. (USSR),* 34:114 (1972).

[3] E.A. Wilson, *Scr. Metall.,* 12:961 (1978).

[4] E.A. Wilson, S.M. Vickers, C. Quixall, and A. Bradshaw, *Proc. Conf. Phase Transformations,* Institution of Metallurgists (London), 1979, paper F-67.

tions in steels—its length would be at least doubled if it did—but some discussion is required. In order of increasing rates of reaction, the mechanisms of the transformation of austenite at subcritical temperatures are (1) nucleation and growth, (2) massive, (3) bainitic, and (4) martensitic. As the cooling rate from above 912°C is increased, the transformation temperature is reduced and transformations can occur at a series of constant temperatures, as indicated by thermal arrests (Fig. I-49). In some instances, two arrests occur during the cooling of one specimen, an indication that two mechanisms of transformation are operating. The data in Fig. I-49 were obtained from iron containing 0.01%C; other work[1] has shown that it will be more difficult to obtain such clear separation of transformation mechanisms in iron of higher purity containing only 0.001%C.

When a transformation occurs by nucleation and growth, a nucleus of the new phase forms by atom-by-atom accretion until it reaches a size that is stable—that is, until its size is sufficient for the decrease in volume free energy to be greater than the sum of the energy required to form a new surface and the volume strain energy involved in size mismatch between the nucleus and the matrix. In a polymorphic transformation of a high-purity metal, such as the $\gamma \rightleftarrows \alpha$ transformation in iron, no long-

[1] M. Izumiyama, M. Tsuchiya, and Y. Imai, *J. Jpn. Inst. Met.*, 34:286 (1970).

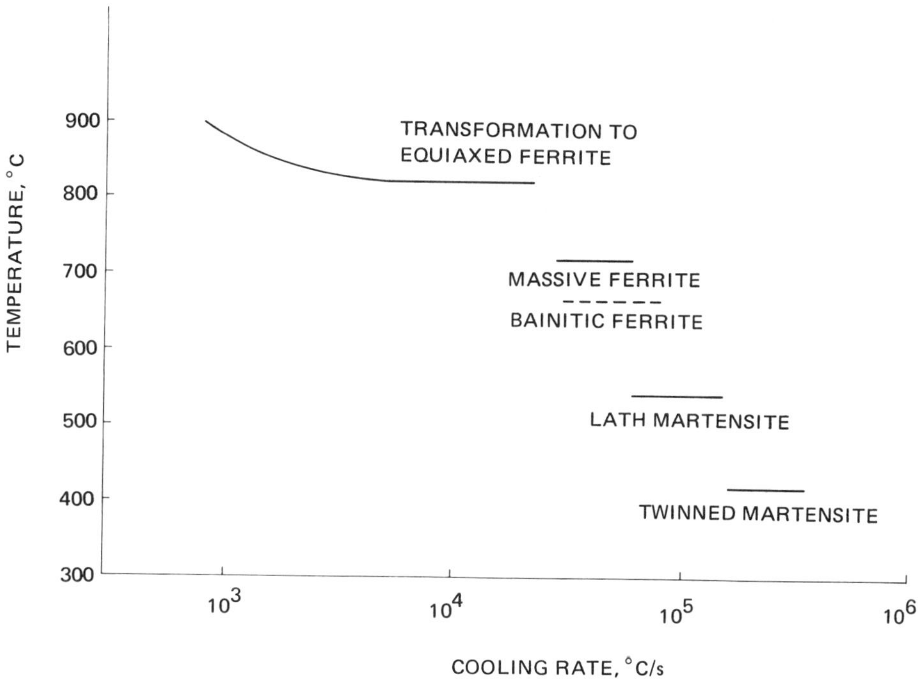

Figure I-49. Variation of transformation temperature with cooling rate for iron containing 0.01%C. (E.A. Wilson, S.M. Vickers, C. Quixall, and A. Bradshaw, *Proc. Conf. on Phase Transformations*, Institution of Metallurgists (London), 1979, by permission.)

range diffusion is required; single iron atoms need only move a fraction of an interatomic distance across the interface to shift from one lattice to another. This mechanism can operate during slow cooling to temperatures below 912°C, or isothermally at temperatures just below the critical. The product would be grains of alpha iron with smooth boundaries and a low dislocation density. These new grains can grow across prior γ boundaries.[1]

When the cooling rate is increased to something in excess of about 10^4 °C/s, the transformation product changes from ferrite grains with smooth boundaries and a low dislocation density to irregularly shaped ferrite grains with a dislocation density of about 5×10^9 /cm^{-2}. The transformation also takes place at a lower temperature, ~740°C (Fig. I-49). This is presumed to indicate a change in the mechanism of transformation from nucleation and growth to massive. A massive transformation also occurs by short-range diffusion across the α-γ interface, which may be moving at about 1 mm/s. In these respects, in a polymorphic transformation in a pure metal, the massive transformation is indistinguishable from transformation by nucleation and growth, but the product is different. Wilson[2] has attempted to resolve this difficulty by suggesting that the rate-controlling step in both processes is nucleation and the nuclei differ. For the transformation to equi-axed ferrite, the nucleus is presumed to have two noncoherent boundaries; the nucleus for the massive transformation has one coherent boundary.[3] The massive transformation does not produce upheaval of a polished surface.

The mode of transformation of austenite to bainite is still enmeshed in controversy.[4,5] When upper bainite forms in a steel containing carbon, carbides are precipitated *between* the lath-like particles of ferrite; in lower bainite the carbides form *within* the laths—that is, they appear to precipitate from ferrite supersaturated with carbon. The growth of the ferrite component of upper bainite (and of Widmanstätten ferrite plates) may occur by the movement of ledges on the partially coherent α-γ interface. This process can produce relief on a polished surface. The alternative proposal is that the ferrite component of both upper and lower bainite forms by shear. According to Wilson,[6] the plateau of transformation temperature in iron at about 640°C (Fig. I-49) is produced by a bainitic transformation, perhaps by ledge movement on growing alpha iron plates.

The name "martensite" does not define a structure; martensite is a generic term applied to any product of a martensitic transformation, which is a phase transformation in solids that occurs by a cooperative movement of atoms.[7] Such movements involve neither compositional change nor diffusion; the transformation produces relief

[1] R.H. Goodenow and R.F. Hehemann, *Trans. AIME,* 223:1777 (1965).

[2] E.A. Wilson, *Scr. Metall.*, 12:961 (1978).

[3] W.C. Johnson, C.L. White, P.E. Marth, P.K. Ruf, S.M. Tuominen, K.D. Wade, K.C. Russell, and H.I. Aaronson, *Met. Trans. A,* 6A:911 (1975).

[4] R.F. Hehemann, K.R. Kinsman, and H.I. Aaronson, *Met. Trans.*, 3:107 (1972).

[5] H.K.D.H. Bhadesia and D.V. Edmonds, *Met. Trans. A,* 10A:895 (1979).

[6] E.A. Wilson, *Scr. Metall.*, 4:309 (1970).

[7] N. Zishiyama, *Martensite Transformation* (English translation), Academic Press, New York, 1978.

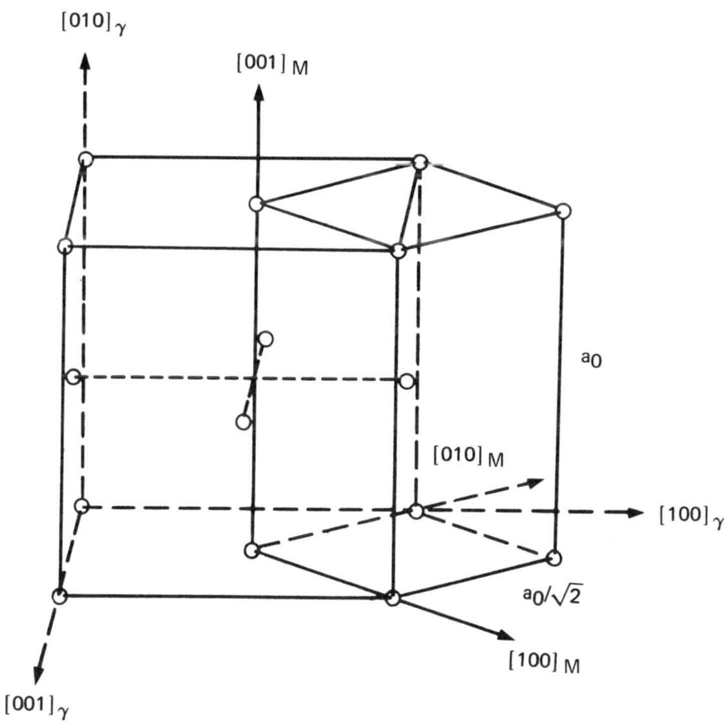

Figure I-50. Body-centered tetragonal unit cell (M) having axial ratio $\sqrt{2}$ outlined in cubic cell. a_0 = length of cube edge of γ.

on a polished surface, and the product phase usually contains a high density of lattice defects. There is a specific orientation relationship between the parent phase and the product and a definite habit plane.

The martensitic transformation of fcc to bcc iron can be accomplished with the minimum expenditure of strain energy by a homogeneous distortion of the austenite, called the Bain strain[1] (see Fig. I-50). This involves a compression of about 17% along $[100]_\gamma$ and an expansion of about 12% along both $[010]_\gamma$ and $[001]_\gamma$. In a martensitic transformation, the interface between the parent and the product must be coherent, or the cooperative motion of atoms from one phase to another could not occur. The interface, therefore, must be an undistorted and unrotated plane, called an invariant plane, but the Bain strain is unable to produce such a plane. A shear on the invariant plane, the invariant plane strain, must accompany the Bain strain. This microscopically inhomogeneous strain in the $\gamma \rightarrow \alpha$ transformation can occur either by slip or by fine twinning. Slip occurs when the Ms (martensite start) temperature is high and twinning when the Ms is low. Martensite in iron has a specific orientation relationship to the parent austenite. The close-packed planes and directions in both lattices are parallel. This is the Kurdjumov–Sachs relation, one variant of which is

[1] E.C. Bain, *Trans. AIME*, 70:25 (1924).

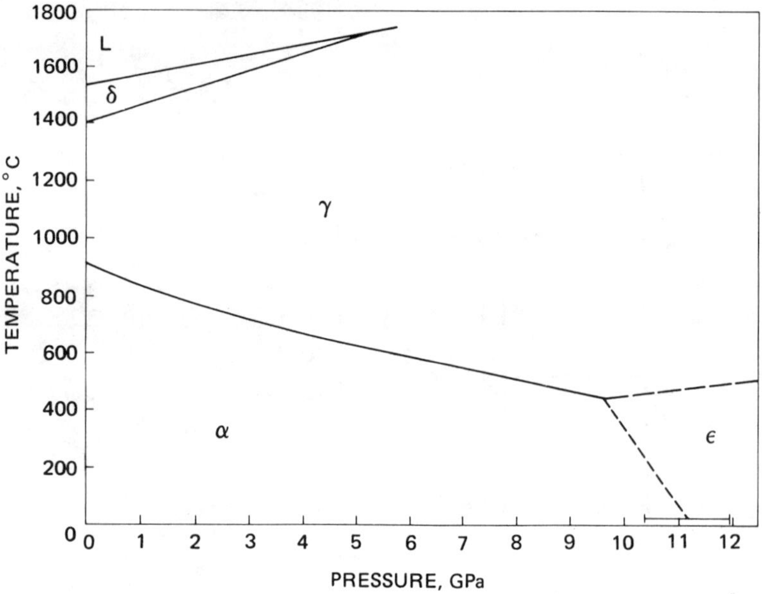

Figure I-51. Equilibrium phase diagram for iron. (Reprinted with permission from J.F. Cannon, *J. Phys. Chem. Ref. Data.* 3:871 (1974), copyright 1974, American Chemical Society.)

$$(111)_\gamma \parallel (011)_\alpha \qquad [\bar{1}01]_\gamma \parallel [\bar{1}\bar{1}1]_\alpha$$

As a consequence, martensite crystals do not cross austenite boundaries.

Work by Morozov et al.[1] indicates that at cooling rates between about 6×10^4 and 1.5×10^5 °C/s, transformation of high-purity austenite to lath martensite occurs at 540°C. In the formation of lath martensite, the lattice invariant shear is by slip. At 420°C and cooling rates between 1.5 and 3.4×10^5 °C/s, the plateau (Fig. I-49) corresponds to the formation of twinned (or plate) martensite.

It may be some time before all the details of the several modes of transformation of high-purity iron are established. The experimental difficulties are formidable and the effects of impurity content may be difficult to determine.

Iron can also undergo a pressure-induced phase transformation, as mentioned earlier. The pressure required, whether applied statically or dynamically by explosive shock loading, is 12.8 ± 0.1 GPa at ambient temperature, at which point the iron transforms martensitically to the denser nonferromagnetic ϵ form, with a hexagonal close-packed structure.[2] When the pressure is released, the iron transforms back to α by a second shear transformation, at a pressure of 9.8 ± 0.4 GPa. The equilibrium phase transition pressure, the pressure at which ϵ and α have the same Gibbs free energy, may be about 11.3 GPa (Fig. I-51).

[1] O.P. Morozov, D.A. Mirzayev, and M.M. Shteynberg, *Phys. Met. Metallogr. (USSR)*, 34:114 (1972).

[2] J.F. Cannon, *J. Phys. Chem. Ref. Data*, 3:781 (1974).

The pressure-induced $\alpha \rightarrow \epsilon$ transformation probably occurs by way of an invariant plane strain on $\{112\}_\alpha$ ($\{10\bar{1}0\}_\epsilon$), plus a small uniform dilation.[1] Usually a small, inhomogeneous shear on {0001} must be included, apparently in the form of slip. The transformation $\epsilon \rightarrow \alpha$ appears to be a reversal of the prior $\alpha \rightarrow \epsilon$ transformation.

[1] H.G. Bowden and P.M. Kelly, *Acta Metall.*, 15:1489 (1967).

CHAPTER

TWO

INTERSTITIAL ATOMS IN ALPHA IRON

II-1. INTERSTITIAL SITES AND ATOMS

Iron is unique among metallic elements in its combination of abundance—it constitutes about 5.8% of the continental crust—relative ease of reduction from its ores, high modulus of elasticity, and low strength when pure. It would be a highly valuable metal even without the great strengthening effects that can arise from the difference in the solubility of carbon in austenite and in ferrite, but with this feature it becomes our most versatile and useful engineering material. This difference in solubility comes from the difference in size of the interstitial spaces in the fcc and bcc lattices. In fcc austenite the largest void at an octahedral site is 0.052 nm in radius (Fig. II-1*a*); the next largest is 0.028 nm in radius. Carbon atoms have a radius of 0.08 nm and nitrogen atoms one of 0.07 nm. These atoms reside in octahedral sites, with an expansion of the austenite lattice. The maximum solubility of carbon in austenite, in equilibrium with Fe_3C, is 2.11 wt % at 1148°C (Fig. II-4), decreasing to 0.77% at 727°C.

The interstitial voids in bcc iron are smaller than those in austenite; the tetrahedral site is 0.036 nm in radius and the octahedral site only 0.019 nm. Despite this considerable difference in void size, carbon and nitrogen atoms apparently prefer to reside in octahedral sites where they have only two nearest-neighbor iron atoms rather than four, as in the tetrahedral site. One result is that each of these two nearest-neighbor iron atoms is displaced 0.053 nm in one of the ⟨100⟩ directions, causing a tetragonal distortion of the lattice.[1] Another result of the small size of the interstitial voids in ferrite is a very low solubility of carbon in equilibrium with Fe_3C, a maximum of 0.0218% at 727°C, decreasing to essentially zero at room temperature (Fig. II-4). Therefore, when a solid solution of carbon in fcc iron is cooled rapidly to a temperature below 727°C, the result is a highly supersaturated solution of carbon in bcc or body-centered tetragonal (bct) iron.

[1] S.C. Moss, *Acta Metall.*, 15:1815 (1967).

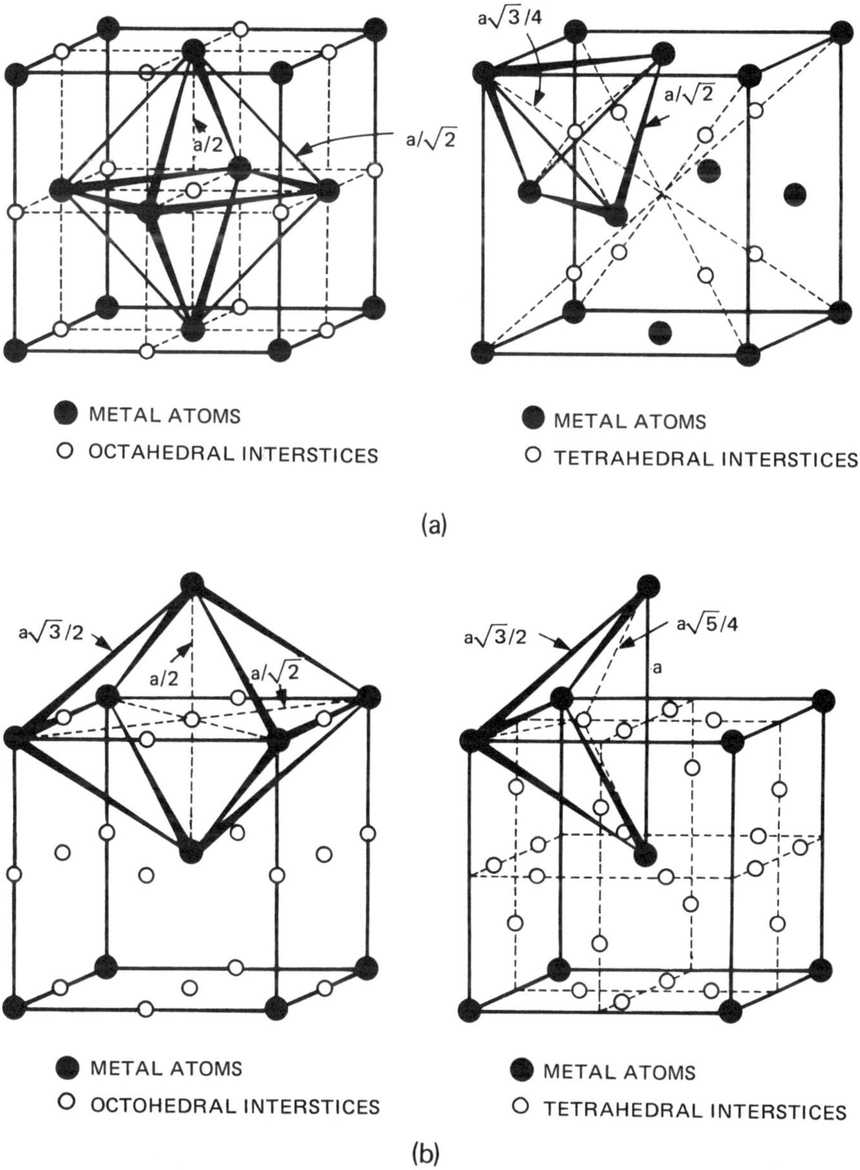

Figure II-1. Interstitial voids in iron. (*a*) Interstitial voids in the fcc structure, octahedral (1) and tetrahedral (2). (*b*) Interstitial voids in the bcc structure; octahedral (1) and tetrahedral (2). (From C.S. Barrett and T.B. Massalski, *Structure of Metals,* 3d ed., copyright 1966, used with the permission of McGraw-Hill Book Co., New York.)

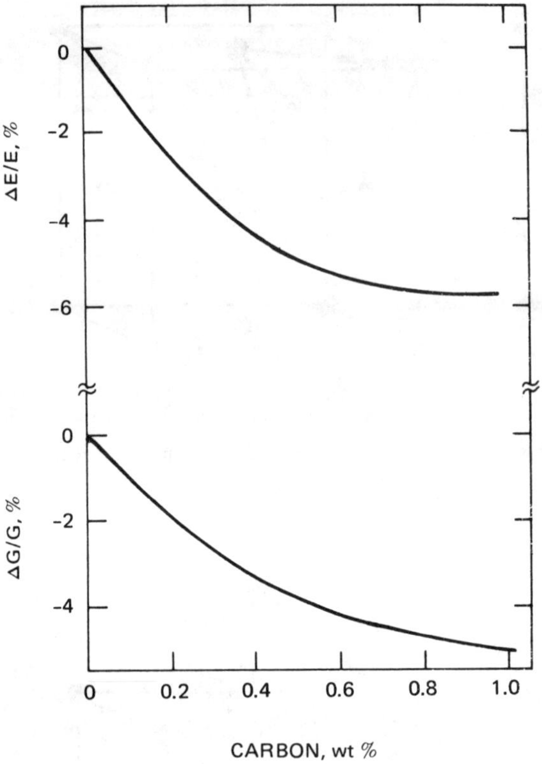

Figure II-2. Effect of carbon content on elastic constants of martensite. (G.R. Speich and W.C. Leslie, *Met. Trans.*, 4:1873 (1973).)

When a carbon atom with a radius of 0.08 nm is added to alpha iron, the result is an expansion of the lattice. The magnitude of this expansion at room temperature is given by Fasiska and Wagenblast[1] as

$$a_{(Fe+C)}\,(\text{nm}) = (0.28664 \pm 0.0001) + (0.84 \pm 0.08) \times 10^{-3} \text{ at \% C} \qquad \text{(II-1)}$$

Nitrogen, the other common interstitial solute, produces a slightly smaller expansion

$$a_{(Fe+N)}\,(\text{nm}) = (0.28664 \pm 0.0001) + (0.79 \pm 0.07) \times 10^{-3} \text{ at \% N} \qquad \text{(II-2)}$$

according to Wriedt and Zwell.[2] This lattice expansion by carbon causes a sharp decrease in Young's modulus and in the shear modulus, as shown in Fig. II-2.

The very low solubility of carbon and nitrogen in alpha iron and their high diffusivity at low temperatures (Fig. II-6) cause these interstitial solutes to segregate to point, linear, and planar lattice defects, giving rise to the various phenomena that we call strain aging. These factors also lead to precipitation of carbide phases at low temperatures, causing precipitation strengthening, called quench aging. Both topics will be discussed in some detail.

[1] E.J. Fasiska and H. Wagenblast, *Trans. AIME*, 239:1818 (1967).
[2] H.A. Wriedt and L. Zwell, *Trans. AIME*, 224:1242 (1962).

When a supersaturated solid solution of carbon in alpha iron (martensite) is formed by rapid cooling from austenite, the presence of carbon, especially when it is in excess of 0.20 wt %, causes the lattice to become tetragonal. The relation between carbon content and the *c/a* ratio in tetragonal martensite is given by[1,2]

$$c/a = 1.000 + 0.045 \text{ wt \% C} \tag{II-3}$$

The determination of the solubility and diffusivity of carbon and nitrogen in alpha iron has been greatly aided by the development of several physical techniques, the most important of which has been the measurement of internal friction. This topic will be discussed before we consider the magnitude of the several effects of interstitial solutes in alpha iron.

II-2. INTERNAL FRICTION

In some circumstances, the interstitial solutes C and N can migrate in alpha iron under the influence of an applied stress that is much lower than the yield stress. These short-range motions can cause strain to lag behind stress, giving an anelastic effect. Internal friction is commonly measured by a torsional pendulum. A small weight is suspended by a thin wire of the test material, in a furnace with accurate temperature control. The weight and wire are twisted to a strain well within the elastic range, then released. The consequent series of cycles of untwisting and twisting, of continuously decreasing amplitudes, is recorded. The angle by which strain lags behind stress is called δ, and $\tan \delta$ is the usual measure of internal friction. The ratio of successive amplitudes is called the logarithmic decrement, and when $\tan \delta$ is small,

$$\tan \delta = \frac{\log \text{dec}}{\pi} \equiv Q^{-1} \tag{II-4}$$

Tan δ is usually established by determining the time t required for the original amplitude to decay to $1/n$ of its original value

$$\tan \delta = \frac{\ln (1/n)}{\pi \nu t} \tag{II-5}$$

where ν = vibration frequency

The short-range, stress-induced migration of interstitial atoms producing internal friction is called the Snoek effect, after its discoverer.[3] The octahedral interstices where carbon or nitrogen atoms prefer to reside are located at the centers of the cube edges and at the centers of the cube faces, which are equivalent spaces, as shown in Fig. II-1. If a stress is applied to a crystal in a ⟨100⟩ direction as indicated in Fig. II-3, the octahedral interstices along the x and y axes will shrink in size whereas those along

[1] C.S. Roberts, B.L. Averbach, and M. Cohen, *Trans. Am. Soc. Met.*, 45:576 (1953).
[2] P.G. Winchell and M. Cohen, *Trans. Am. Soc. Met.*, 55:347 (1962).
[3] J.L. Snoek, *Physica*, 8:711 (1941).

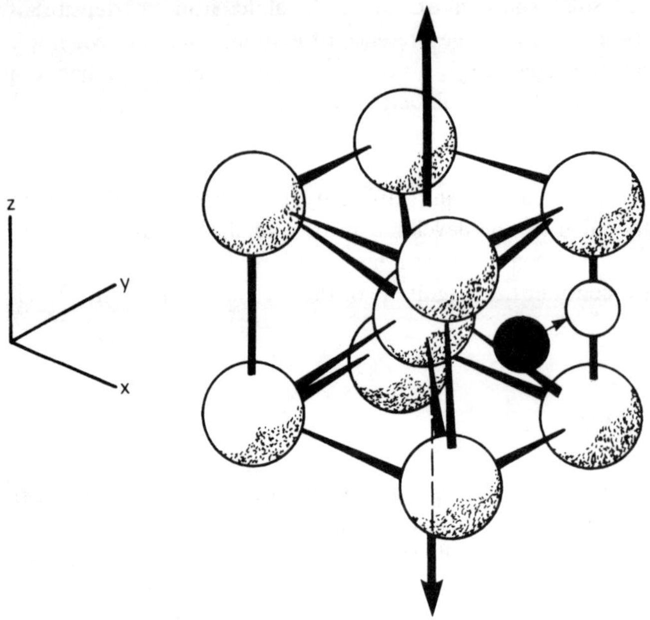

Figure II-3. Snoek ordering of a carbon atom in bcc iron under the influence of an applied stress.

the z axis will expand. The latter will then become favored sites for interstitial atoms and if time and temperature allow, the atoms will move to these sites. The greater the number of atoms that jump, the larger will be the value of Q^{-1}. The maximum internal friction is attained when the period of application of stress equals the jump time. Fortunately, the maximum value of Q^{-1} at ambient temperature is close to a 1-s pendulum period. At temperatures below ambient the atoms cannot move rapidly enough to accompany the pendulum oscillations, and at higher temperatures the atoms move too rapidly. The result is a curve of Q^{-1} vs. temperature with a maximum (Snoek peak) at a temperature typical of each solute.

It is obvious from Figs. II-1 and II-3 that the maximum value of Q^{-1} will depend upon the texture of the specimen, since stress in the ⟨111⟩ direction will be equally inclined to all three tetragonal axes and no particular interstitial site would be favored.

The principal value of internal friction in the physical metallurgy of steels has been in the determination of the diffusivity of carbon and nitrogen in alpha iron and of the concentration of these elements retained in solid solution. The diffusuvity is related to the jump distance of the solute and to the time for one cycle,

$$D \cong \frac{(\text{jump distance})^2}{\text{time for one cycle}} \tag{II-6}$$

and the Snoek peak height is proportional to the concentration of interstitial atoms in solid solution,

$$\text{concentration of interstitials} = \alpha Q_{\max}^{-1} \tag{II-7}$$

The magnitude of α is usually not far from 1, but as mentioned earlier, it varies with specimen texture. It also varies with grain size; as grains become finer, more solute is segregated to grain boundaries and α increases. In the range of grain sizes from ASTM 0 to 10, the variation of α is from about 0.9 to 2.5.

The interstitial solutes that contribute to internal friction in the manner described are those that occupy normal sites in the undistorted lattice. Atoms within the strain field of a dislocation, those neighboring a substitutional solute atom, or those at grain boundaries will not react in the same manner to an applied stress. For this reason, internal friction measurements have not been outstandingly successful in determining concentrations of carbon or nitrogen atoms in solid solution in ferrite in commercial steels.

Equation II-6 can be rewritten as

$$t = \frac{l^2}{D} \tag{II-8}$$

where t is the jump time of the interstitial atom, l is the jump distance, and D is the diffusivity. From the usual relation

$$D = D_0 \exp\left(-\frac{\Delta H}{RT}\right) \tag{II-9}$$

where ΔH is the enthalpy for activation of diffusion, then

$$t = \frac{l^2}{D_0} \exp\left(\frac{\Delta H}{RT}\right) \tag{II-10}$$

II-3. SOLUBILITY OF INTERSTITIAL SOLUTES

Two procedures have been employed to measure solubilities in ferrite at low temperature, calculations based on the free energies of the reactants[1] and internal friction,[2] and the results have not always been in agreement. Now, however, the major discrepancies between the results of the two methods seem to have been resolved. Chipman[1] lists the following equations for the solubility of carbon in alpha iron in equilibrium with graphite, cementite, or ϵ carbide ($Fe_{2.4}C$, hcp):

$$\text{graphite:} \qquad \log\,[C]_{ppm} = 7.81 - \frac{5550}{T} \tag{II-11}$$

$$\text{cementite:} \qquad \log\,[C]_{ppm} = 6.38 - \frac{4040}{T} \tag{II-12}$$

[1] J. Chipman, *Met. Trans.*, 3:55 (1972), *Metals Handbook*, 8th ed., vol. 8, Am. Soc. Met., Metals Park, Ohio, 1973, p. 275.

[2] J.C. Swartz, *Trans. AIME*, 245:1083 (1969).

Table II-1 Solubility of C in α Fe, in equilibrium with cementite (wt ppm)

Temperature, °C	200	300	400	500	600	700	727
Internal friction			6.3	23.2	68	168	218
Calculated (from free energies)	0.007	0.21	2.3	13	57	160	218

$$\epsilon \text{ carbide:} \qquad \log\,[\mathrm{C}]_{\mathrm{ppm}} = 4.06 - \frac{1335}{T} \qquad \text{(approximate)} \tag{II-13}$$

The solubility of carbon in equilibrium with cementite, from Eq. II-12, is compared in Table II-1 with the solubility determined by Swartz by internal friction.

A plot of the ferrite phase field with the solubility limits for graphite, cementite, and epsilon carbide is shown in Fig. II-4*c*, a portion of the entire C–Fe phase diagram of Fig. 4*a–d*.

Chipman also points out that the χ or Hägg carbide ($Fe_{2.2}C$) becomes more stable than cementite at temperatures below about 350°C; that is, it is less soluble than cementite in ferrite below that temperature.

Fast and Verrijp[1] determined the solubilities of $Fe_{16}N_2$, Fe_4N, and N_2 gas in alpha iron, and give the following equations:

$$Fe_{16}N_2,\ \alpha'',\ \text{wt \% N} = 3.3 \times 10^2 \exp\left(-\frac{41.4\ \text{kJ/mol}}{RT}\right) \tag{II-14}$$

$$Fe_4N,\ \gamma',\ \text{wt \% N} = 12.3 \exp\left(-\frac{34.7}{RT}\right) \tag{II-15}$$

$$N_2,\ \text{wt \% N} = 9.8 \times 10^{-2} \exp\left(-\frac{30.1}{RT}\right) \tag{II-16}$$

These translate to the data listed in Table II-2 and shown in Fig. II-5.

The solubility of Fe_4N at the eutectoid temperature (585°C) is 0.095 wt % N. The greater solubility of nitrogen than carbon in ferrite should be noted. At 400°C the solubility of carbon in equilibrium with cementite is between 3 and 6 ppm by weight. At the same temperature, the solubility of nitrogen in equilibrium with Fe_4N is 250 ppm by weight. This greater solubility has very important practical consequences, in that nitrogen in solution in ferrite is the principal cause of the strain aging of low-carbon steels.

II-4. DIFFUSIVITY OF INTERSTITIAL SOLUTES

As indicated in Fig. II-6, adapted from an unpublished diagram by L.S. Darken, the diffusivities of C and N in alpha iron at a given temperature are vastly greater than

[1] J.D. Fast and M.B. Verrijp, *J. Iron Steel Inst.*, 180:337 (1955).

those of substitutional elements and very much less than that of hydrogen (H). The comparatively high rate of diffusion of interstitial atoms in iron makes possible some very important technological processes such as carburizing and nitriding, and often controls the kinetics of phase transformations in steels, but it also creates problems such as strain aging, hydrogen embrittlement, and sensitization of stainless steels.

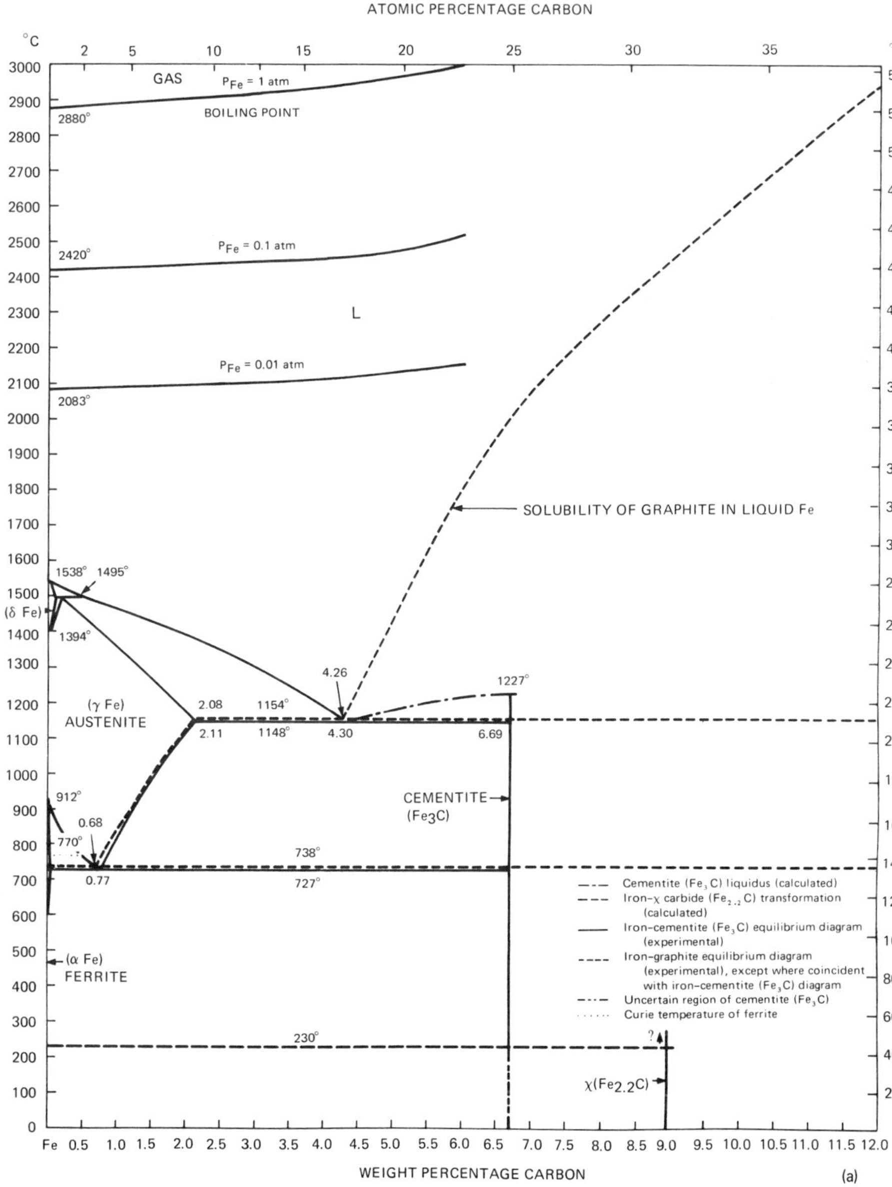

Figure II-4. (*a*) The carbon–iron phase diagram. (*b*) The eutectoid region. (*c*) Ferrite field, showing equilibrium between Fe–C, Fe–Fe_3C, and Fe–$Fe_{2.4}C$. (*d*) Delta ferrite field. (J. Chipman, *Metals Handbook,* 8th ed., vol. 8, Am. Soc. Met., Metals Park, Ohio, 1973, by permission.)

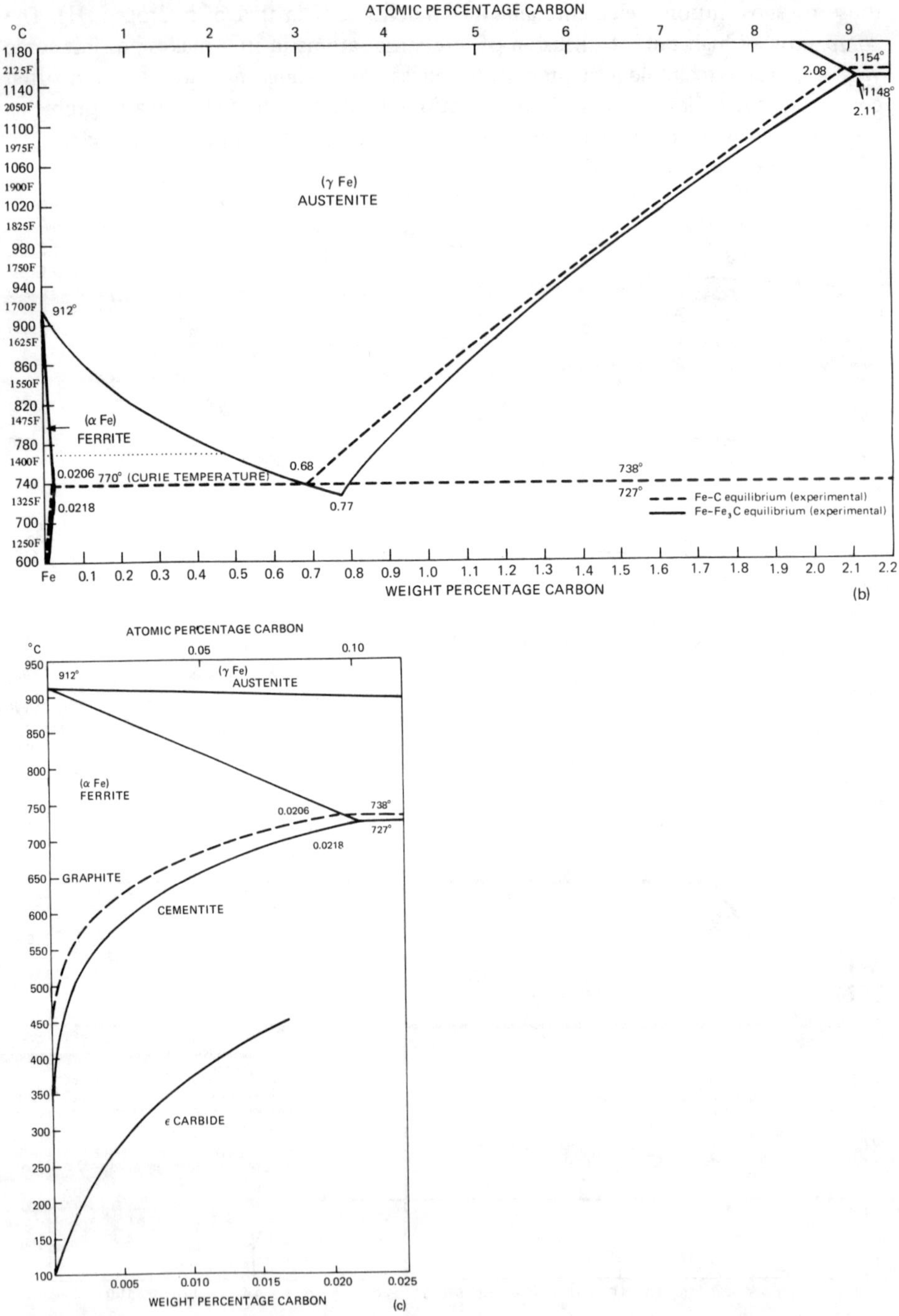

Figure II-4. *(Continued)*

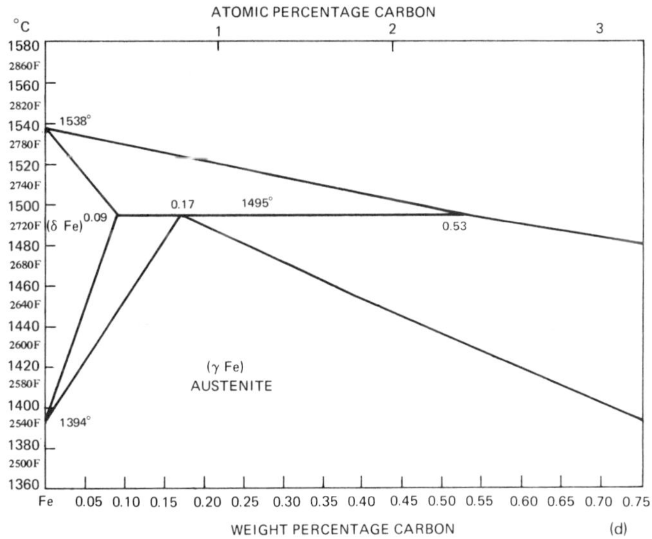

Figure II-4. *(Continued)*

As also indicated in Fig. II-6, the diffusivity of substitutional solutes is affected much more by the transformation from ferromagnetism to paramagnetism than is the diffusivity of interstitial solutes. There are differences in the diffusivities of various substitutional solutes—P, for example, diffuses comparatively rapidly—but for the most part these differences are minor, so substitutional elements are represented by a single curve.

The three columns on the right in Fig. II-6 allow estimation of the jump time of solute atoms, the time for 50% completion of migration of solutes to the boundaries

Table II-2 Solubilities of nitrogen in alpha iron (wt ppm)

Temp., °C	In equil. with $Fe_{16}N_2$	In equil. with Fe_4N	In equil. with N_2, at 1 atm
20	0.14		
100	5.2		
200	88		
300	500	(84)	
400	(2000)	250	
500		550	9.0
585		950	14
700			24
800			33
900			45

of grains 10^{-3} cm in radius, and the time for essentially complete diffusion ($Dt/L^2 = 1$, $L = \frac{1}{2}$ thickness) in slabs 2 mm or 2 m thick.

When the logarithm of the diffusivity of carbon or nitrogen in alpha iron is plotted vs. $1/T$, there is a slight positive deviation from linearity; that is, the diffusivity is enhanced as the Curie temperature is approached.[1] Such plots are shown in Figs. II-7*a* and 7*b*. The diffusivity data from which these plots were made were obtained by a variety of techniques: mass flow, internal friction, elastic after-effect, permeability after-effect, and magnetic after-effect. We are still uncertain whether or not the jump frequency of carbon or nitrogen atoms is increased by changes in the magnetic exchange energy.

It should be noted from Figs II-7*a* and 7*b* that there is no large difference between

[1] J.R.G. da Silva and R.B. McLellan, *Mater. Sci. Eng.*, 26:83 (1976).

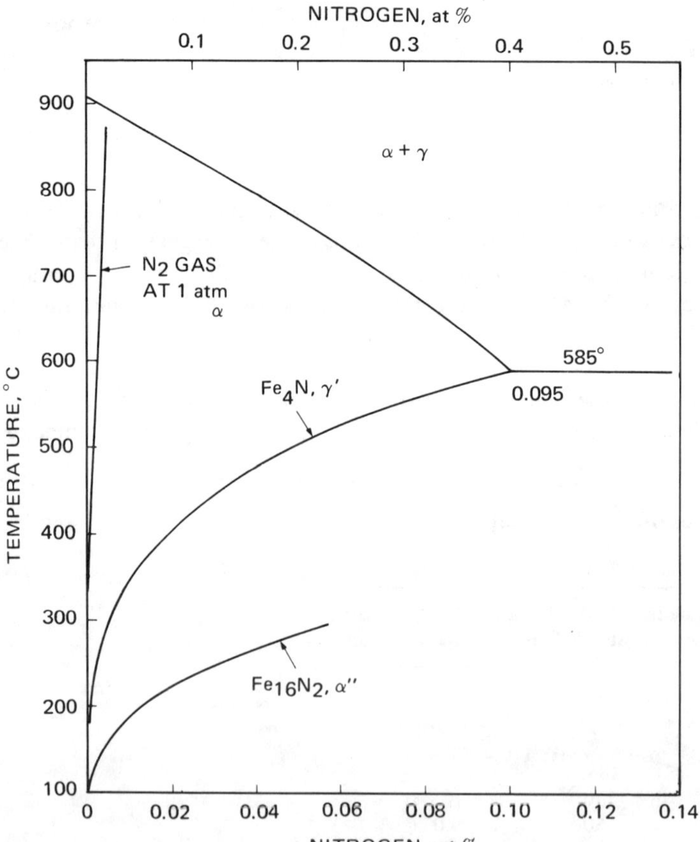

Figure II-5. Ferrite field, showing equilibrium between Fe–N_2, Fe–Fe_4N and Fe–$Fe_{16}N_2$. (J.D. Fast and M.B. Verrijp, *J. Iron Steel Inst.*, 180:337 (1955), by permission.)

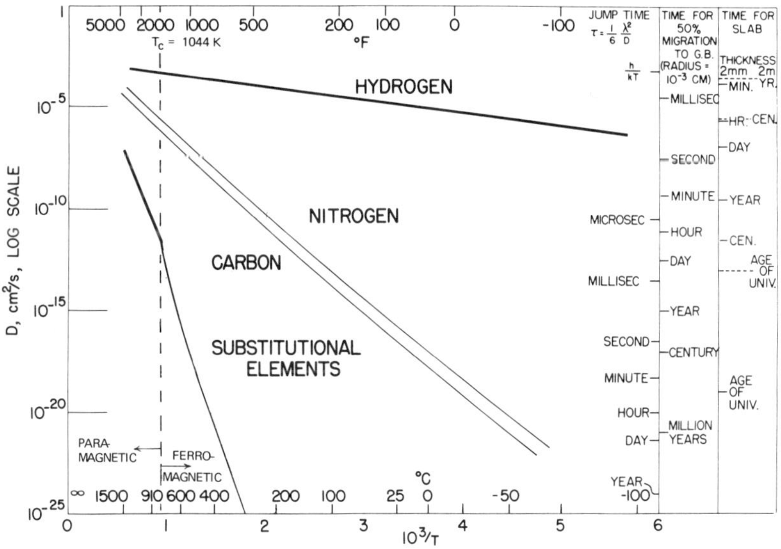

Figure II-6. Comparison of diffusivities of substitutional and interstitial solutes in ferrite: h = Planck's constant, k = Boltzmann's constant, $\hbar \cong 10^{-13}$ s at ambient temperature (h/kT), λ = distance between adjacent interstitial atom positions. (Adapted from an unpublished plot by L.S. Darken.)

the diffusivities of carbon and nitrogen in alpha iron. At room temperature, the diffusivity of nitrogen is about 1×10^{-16} cm²/s and that of carbon about 3×10^{-17} cm²/s. This is a further indication that the critical factor in the strain aging of commercial steels at ambient temperature is the greater solubility of nitrogen rather than an enhanced diffusivity.

The diffusivity of C in austenite is shown in Fig. II-8. At 900°C, the diffusivity of C in austenite containing 0.1%C is about 6×10^{-8} cm²/s. In ferrite at the same temperature the diffusivity is about 5×10^{-6} cm²/s. This greater diffusivity in ferrite affects, among other things, the formation and growth of interphase precipitates in steels. The diffusivity of C in austenite increases with C concentration; the change in log D with change in C content is 0.3 (Δ wt % C).[1]

II-5. STATIC STRAIN AGING

Strain aging is defined as the changes in properties of a metal that occur by the interactions of point defects—principally interstitial solute atoms—and dislocations, during or after plastic deformation. When the property changes occur after plastic deformation, the process is called *static strain aging*; when they are concurrent with plastic strain, the process is called *dynamic strain aging.* These changes in properties can be

[1] C. Wells, W. Batz, and R.F. Mehl, *Trans. AIME,* 188:553 (1950).

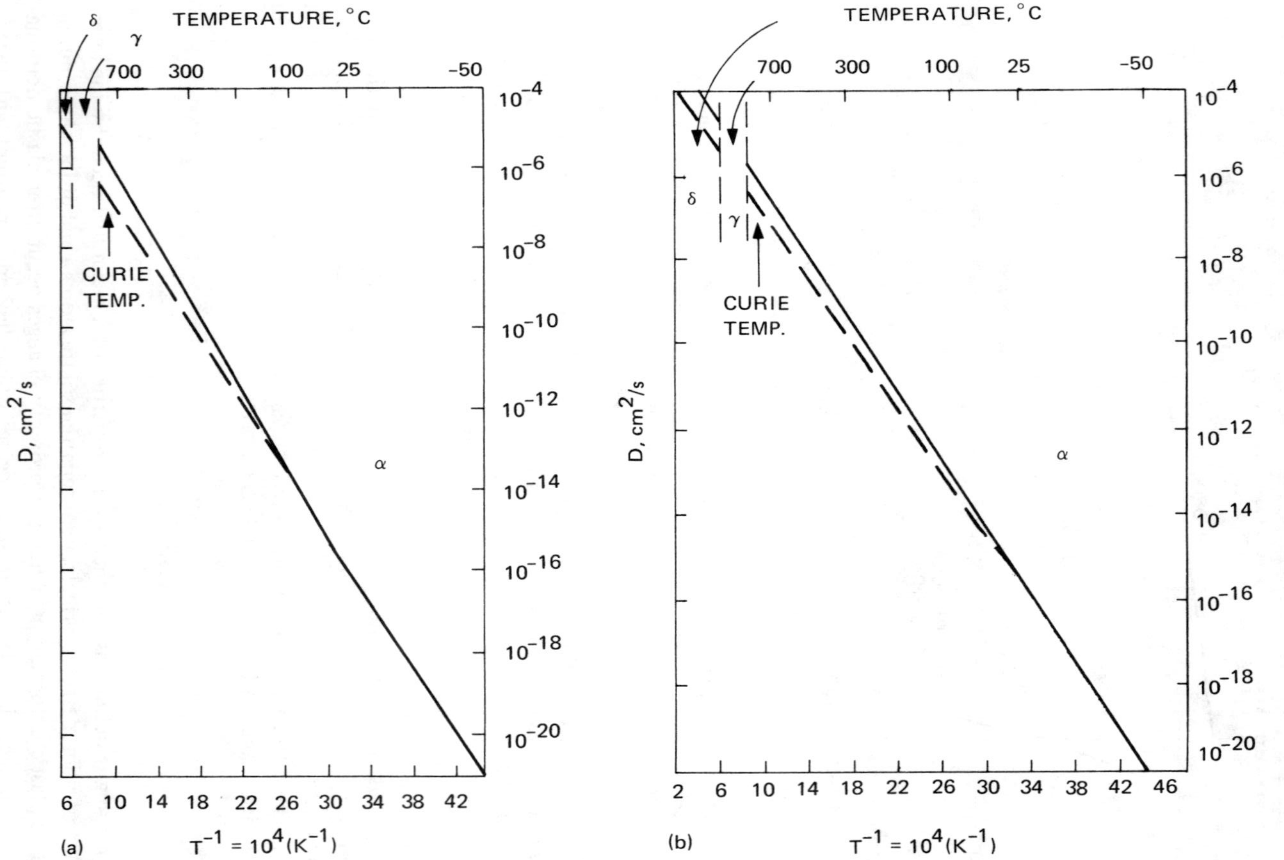

Figure II-7. Temperature dependence of diffusivities in alpha iron. (*a*) Carbon. (*b*) Nitrogen. (J.R.G. da Silva and R.B. McLellan, *Mater. Sci. Eng.*, 26:83 (1976), by permission.)

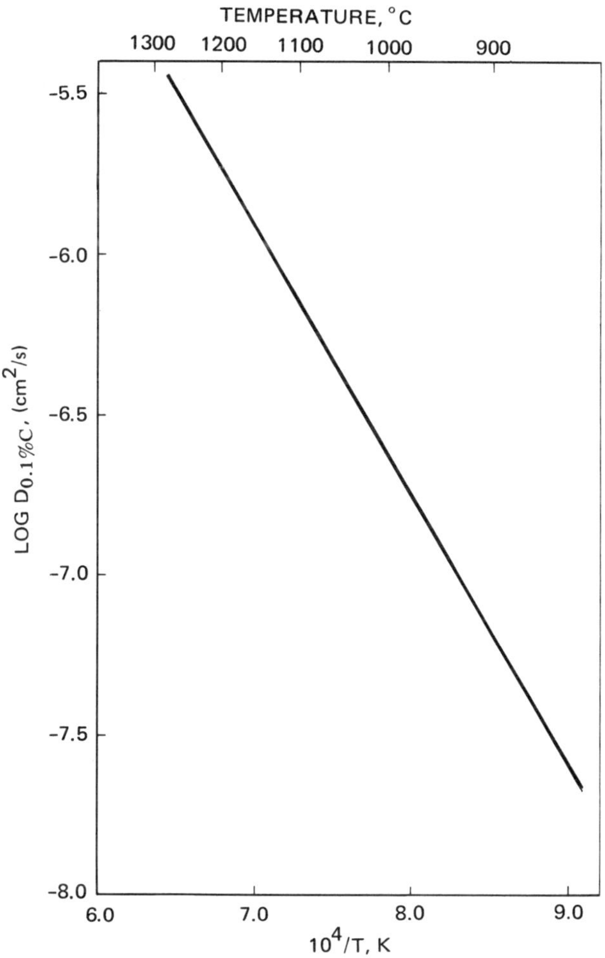

Figure II-8. Diffusivity of carbon in γ iron containing 0.1 wt % carbon. (R.P. Smith, *Trans. AIME,* 230:476 (1964), by permission.)

detrimental and even calamitous, but when used with knowledge and discretion, strain aging can be a valuable and economical means of strengthening steels. The effects of the strain aging process have been known as long as steel has been made, but it seems that every generation must learn to recognize them through hard experience. To wit, in 1874 John Fritz[1] wrote:

> All violent disturbances of the molecular arrangement of the particles of steel, as in straightening, punching, notching, etc., tends to make it brittle. The advantage of drilling holes in rails over punching is very marked.

[1] J. Fritz, *Trans. AIME,* 3:91 (1874-75).

In the same journal W.A. Sweet[1] wrote:

> It is an established fact that punching holes for fish-plate bolts weakens the rail 75%.

The language is archaic but the facts are correct.

In 1932 Samuel Epstein[2] published on excellent paper on embrittlement of hot-dipped galvanized steel, made by either the Bessemer or the open-hearth process. His study was begun after an incident in 1927. During the unloading of a shipment of galvanized electric power transmission tower parts, an angle section fell to the ground from a railroad car and broke through the punched rivet holes. Epstein found that the embrittlement was caused by cold work followed by rapid aging during hot-dip galvanizing. His remedies were simple; drill the holes instead of punching them and use open-hearth instead of Bessemer steels.

In 1965 a newly erected 175-m television mast at Mt. Gambier, South Australia, collapsed when its guy wires were being tightened. The cause was the brittle failure of a 2-mm-thick steel plate. Fracture began at a sheared edge of the plate, which had been cold sheared, then hot-dip galvanized.

Any serious study of strain aging should begin with a reading of two comprehensive reviews by Baird.[3] The most common measures of property changes during strain aging are the effects produced on load-elongation curves in a uniaxial tension test (Fig. II-9). Conventionally, a specimen is strained an arbitrary amount in excess of the yield point elongation (Lüders strain), unloaded, aged at a constant temperature, then reloaded to fracture. The effects, as indicated on the figure, are:

1. An increase in yield stress during aging, ΔY
2. A return of the yield point elongation (Lüders strain) and an increase in this strain with increased aging time
3. An increase in the ultimate tensile stress, ΔU
4. A decrease in total elongation, $\Delta\epsilon$

The theory of strain aging proposed by Cottrell and Bilby[4] in 1949 has withstood the test of time and is generally accepted as correct in concept. They examined the elastic interaction between a solute atom and a positive edge dislocation, beginning by considering a *solvent* atom as an elastic sphere of radius r with its center at point O in Fig. II-10 and a point P on its surface. Assume a stress field σ_{ij} in the body which creates forces on the surface of the spherical atom. The stress normal to the sphere at point P is given by

$$\sigma_{rr} = l_{rx}^2\sigma_{xx} + l_{ry}^2\sigma_{yy} + l_{rz}^2\sigma_{zz} + 2l_{rx}l_{ry}\sigma_{xy} + 2l_{ry}l_{rz}\sigma_{yz} + 2l_{rz}l_{rx}\sigma_{zx} \qquad \text{(II-17)}$$

[1] W.A. Sweet, *Trans. AIME,* 3:92 (1874–75).
[2] S. Epstein, *Proc. Am. Soc. Test. Mater.*, 32:293 (1932).
[3] J.D. Baird, Iron and Steel, 36:186, 326, 368, 400, 450 (1963), *Met. Rev.*, 16:1 (1971).
[4] A.H. Cottrell and B.A. Bilby, *Proc. Phys. Soc.*, A62:49 (1949).

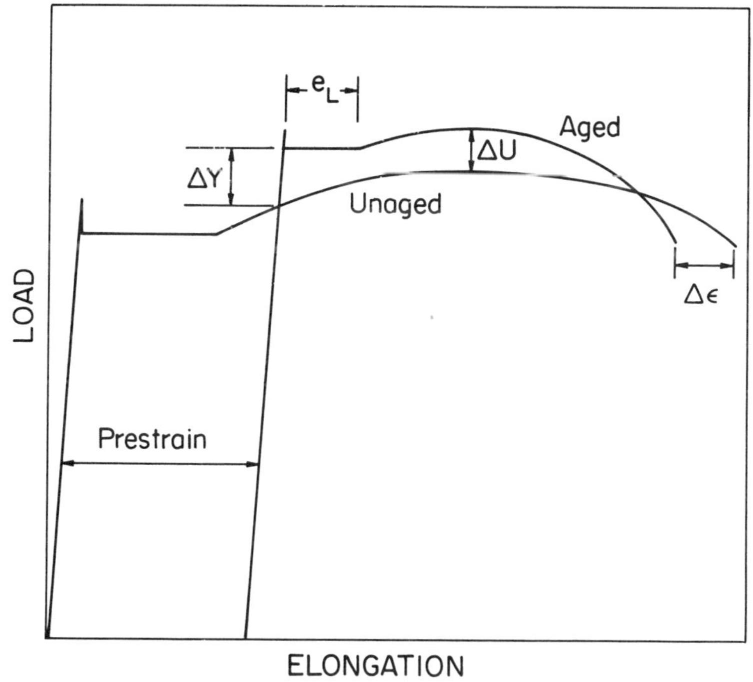

Figure II-9. Effects of static aging on the load-elongation curve of iron containing interstitial solutes: ΔY = change in yield stress due to strain aging, e_L = Lüders strain after strain aging, ΔU = increase in ultimate tensile strength due to strain aging, $\Delta\epsilon$ = decrease in elongation due to strain aging.

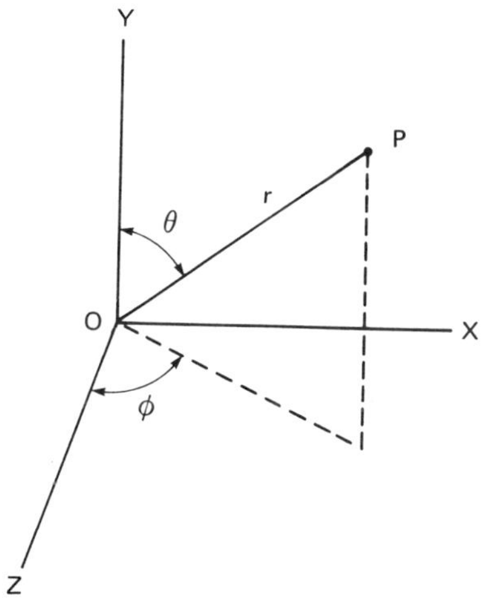

Figure II-10. Representation of sphere of radius r, center at O, and a point on the surface, P

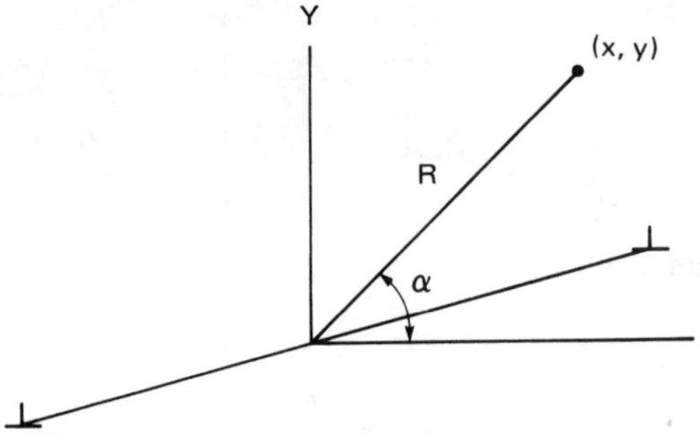

Figure II-11. Solute atom at (x, y) in strain field of a positive edge dislocation.

where $l_{rx} = \sin\theta \sin\phi$
$l_{ry} = \cos\theta$
$l_{rz} = \sin\theta \cos\phi$

The normal force on an element of the sphere is

$$\sigma_{rr} r^2 \sin\theta \, d\theta \, d\phi$$

Replace the *solvent* atom by a *solute* atom of radius r'. This changes the radius of the sphere by a small amount, ϵr, where

$$\epsilon = \frac{r' - r}{r} \tag{II-18}$$

The work done by the normal stress, σ_{rr}, when this replacement occurs is

$$\begin{aligned} W &= -\int \epsilon r(\sigma_{rr} r^2 \sin\theta)\, d\theta \, d\phi \\ &= \tfrac{4}{3}\pi\epsilon r^3(\sigma_{xx} + \sigma_{yy} + \sigma_{zz}) \\ &= U_I \end{aligned} \tag{II-19}$$

This is the interaction energy between the strain field and the solute atom. It is set by the change in volume of the sphere, $4\pi\epsilon r^3$, and by the hydrostatic pressure of the stress field, $-\frac{1}{3}(\sigma_{xx} + \sigma_{yy} + \sigma_{zz})$.

They then considered that the solute atom was located within the strain field of a positive edge dislocation, at polar coordinates R, α (Fig. II-11). The stresses associated with such a dislocation are

$$\sigma_x = -\frac{Gb}{2\pi(1-\nu)}\, y\, \frac{3x^2 + y^2}{x^2 + y^2} \tag{II-20}$$

$$\sigma_y = \frac{Gb}{2\pi(1-\nu)}\, y\, \frac{3x^2 + y^2}{(x^2 + y^2)^2} \tag{II-21}$$

$$\sigma_z = \nu(\sigma_x + \sigma_y) \tag{II-22}$$

where G = shear modulus
ν = Poisson's ratio
b = Burgers vector magnitude

Substituting into Eq. II-19,

$$U_{\mathrm{I}} = \frac{4\,(1+\nu)}{3\,(1-\nu)} \frac{Gb\epsilon r^3 \sin\alpha}{R} \tag{II-23}$$

$$= \frac{A \sin\alpha}{R} \qquad \text{for} \qquad R \geqslant r \tag{II-24}$$

It should be carefully noted that this treatment assumes an elastically isotropic medium. In such a medium there can be no elastic interaction between a solute atom producing a spherically symmetric distortion of the lattice and a pure screw dislocation, because the strain field of the dislocation has no hydrostatic component. Iron, of course, is not elastically isotropic, and Chou[1] has shown that when anisotropy of elasticity is considered, screw dislocations in bcc metals have dilatational fields. Even in an isotropic medium, a solute atom producing a tetragonal distortion of the lattice (C or N in α Fe) can interact with a screw dislocation since it is able to relieve shear stress. Also, if screw dislocations in bcc iron are slightly dissociated, they could interact with the hydrostatic component of stress from a solute atom. As will be seen later, it now appears that under the proper conditions, *all* solute atoms in bcc iron can interact with *all* types of dislocations.

Considering Eq. II-24, the term A can be assumed to have a value of about 3×10^{-21} Pa. Then at $R = 0.2$ nm, $U_{\mathrm{I}} \cong 0.15$ aJ. The usual value now quoted for interaction energy between a carbon atom and a dislocation is 0.08 to 0.12 aJ, determined by comparing the binding energy with that in Fe_3C.

Strain aging is unusual because although it is a process by which a plastically deformed metal reduces the energy contained in its strained lattice, the mechanical properties move farther away from their values in the annealed condition. It is a time-dependent process, for the interstitial solutes must diffuse to the vicinity of a dislocation. A problem arises in determining the distribution of carbon atoms around the dislocation. As previously stated, the interaction energy between a carbon atom and a dislocation in iron is about 0.1 aJ. At room temperature, the thermal energy of the solute atom, kT (k is Boltzmann's constant), is 0.004 aJ, so the "atmosphere" of carbon atoms around the dislocation becomes saturated. If one assumes a Maxwell–Boltzmann distribution of solute atoms,

$$C = C_0 \exp\left(\frac{Q}{kT}\right) \tag{II-25}$$

where C_0 is the concentration of solute in the lattice, an impossibly high concentration is obtained as R in Eq. II-24 approaches zero. To avoid this difficulty, Cottrell and

[1] Y.T. Chou, *Acta Metall.*, 13:251 (1965).

Bilby suggested that the core region of the dislocation, where $R < 1$ nm, reached saturation when it contained one carbon atom per atom plane of the dislocation. For a positive edge dislocation, this would produce a line of carbon atoms just below the extra half-plane of iron atoms. Surrounding this "condensed" atmosphere would be a dilute atmosphere having a Maxwell–Boltzmann distribution.

One can use Eq. II-25 to calculate the maximum temperature at which such a condensed atmosphere can exist. Assume that the quantity C is unity, C_0 is 10^{-4}, and Q is equal to 0.08 aJ. Then

$$\frac{C}{C_0} = \exp\left(\frac{Q}{kT}\right) \tag{II-26}$$

and

$$T = \frac{Q}{k \ln (C/C_0)} \tag{II-27}$$

$$= \frac{0.08 \times 10^{-18}\ J}{1.38 \times 10^{-21}\ \mathrm{J/K}\,[\ln (1/1 \times 10^4)]}$$

$$= 629\ \mathrm{K}\ (356°\mathrm{C})$$

Obviously, the number obtained depends upon the value chosen for the interaction energy.

Equation II-24 can be used to estimate the diameter of an atmosphere around a dislocation. The carbon atoms will be affected by the stress field around the dislocation as long as Q is greater than kT. Then,

$$R = \frac{A}{kT} \tag{II-28}$$

Assuming, as before, that $A = 3 \times 10^{-21}$ Pa, then at room temperature R is about 7.5 nm.

Later, Louat[1] attempted to improve the description of the distribution of solute atoms around a dislocation by using Fermi–Dirac statistics rather than Maxwell–Boltzmann, to give

$$C = \frac{C_0 \exp (Q/kT)}{1 + C_0 \exp (Q/kT)} \tag{II-29}$$

Li and Chou[2] used Fermi–Dirac statistics to eliminate the need for an empirical energy of interaction of interstitial atoms with the dislocation core.

Cottrell and Bilby considered the problem of diffusion of an interstitial atom in the stress field of a dislocation during the early stages of atmosphere formation and obtained a solution for the total number of atoms supplied to the dislocation in the period $t = 0$ to $t = t$, per unit length of dislocation:

[1] N.P. Louat, *Proc. Phys. Soc. London,* B69:459 (1956).
[2] J.C.M. Li and Y.T. Chou, *Trans. AIME,* 245:1606 (1969).

$$n_t = 3n_0\left(\frac{ADt}{kT}\right)^{2/3} \tag{II-30}$$

where n_0 = number of solute atoms in a unit volume of matrix
D = diffusion coefficient of interstitial solute
A = as given in Eqs. II-23 and 24

This is usually referred to as the "$t^{2/3}$ law." It does not apply to the completion of aging because it does not consider the effects of saturation and back-diffusion from the saturated atmosphere.

Harper[1] modified Eq. II-30 in an effort to widen its range of applicability. He made the reasonable assumption that the rate of segregation of interstitial solutes, dW/dt, was proportional to the fraction remaining in solution, or

$$\frac{dW}{dt} = (1 - W)f(t) \tag{II-31}$$

where W is the fraction of segregated solute.

Equation II-30 can be rewritten as

$$W = \frac{n_t}{n_0} = 3\left(\frac{ADt}{kT}\right)^{2/3} \tag{II-32}$$

Solving Eq. II-31 with the boundary condition that $t \doteq 0$, defined by Eq. II-30, Harper obtained

$$W = 1 - \exp\left[-3L\left(\frac{ADt}{kT}\right)^{2/3}\right] \tag{II-33}$$

where L is the total length of dislocation line per unit volume. Harper found good agreement with experiments for fractions of solute depletion up to 0.90. However, he was dealing with the kinetics of atmosphere formation, precipitation on dislocations, and precipitation in the matrix. For segregation to dislocations alone, Eq. II-33 probably will not hold to fractions of completion in excess of 0.5. Equation II-33 can be reduced to

$$\ln(1 - W) = -\left(\frac{t}{\tau}\right)^{2/3} \tag{II-34}$$

where τ is a time constant. A plot of $\ln(1 - W)$ vs. $t^{2/3}$ should be linear if the relation is valid, with a slope proportional to the diffusivity of the interstitial solute.

As discussed by Rosinger,[2] strain aging can be divided into two parts, caused by two distinct processes. The initial part is caused by the strain-induced, short-range ordering of interstitial solutes in the stress field of a dislocation. The activation energy

[1] S. Harper, *Phys. Rev.*, 83:709 (1951).
[2] H.E. Rosinger, *Met. Sci.*, 9:1 (1975).

for this process is 59 ± 9.6 to 62.4 ± 2.1 kJ/mol. It is complete in about the time required for a single jump of an interstitial atom in the stress field of a dislocation. It follows a $t^{2/3}$ law, as does the second and slower stage, which is the formation of Cottrell atmospheres by long-range diffusion of interstitial solutes outside the strained region. The activation energy for the second stage is 87.1 ± 10 kJ/mol, in agreement with the activation energy for bulk diffusion of interstitials.

Most attempts to measure kinetics of strain aging have been done with highly supersaturated solid solutions of carbon or nitrogen in alpha iron. In such conditions, precipitates form rapidly on dislocations. One is then measuring the rate of growth of second-phase particles rather than the kinetics of strain aging. The experiments of Wilson and Russell[1] are noteworthy exceptions to this generalization.

Strain aging in iron and steel at temperatures below about 100°C is almost entirely due to nitrogen; at these temperatures the solubility of carbon is too low to produce any appreciable aging effects. Therefore, if nitrogen is removed from solution by precipitation as AlN, a steel can be made "nonstrain aging" even when the carbon content is normal, provided that the aging temperature is not much above ambient.

In static strain aging after prestraining in tension, the first change to occur is the reappearance of discontinuous yielding, as indicated in Fig. II-12. With increased time the Lüders strain increases, up to about 0.02, and the yield stress increases to a maximum. After a considerable increase in the yield stress, the ultimate tensile strength and the work-hardening coefficient increase and the elongation to fracture decreases.

The rate at which these changes occur at a given aging temperature is a function of the interstitial solute content, as demonstrated by Wilson and Russell.[1] The concentration of interstitial solute atoms in *solid solution* must be reduced to about 0.0001% (1 ppm) or less to eliminate the effects of strain aging. Strain-aging effects can approach their maxima at concentrations of only about 0.002%.

The change in yield stress is the most consistent indication of strain aging at all solute contents and at all aging times. The ultimate tensile strength and elongation only change at later stages of aging when high solute contents are present; the Lüders strain increases rather irregularly.

It should be clearly understood that the process of strain aging does not necessarily involve precipitation from solid solution; that is, quench aging is not an essential part of strain aging. The concept that precipitation is a necessary part of the process grew out of the observation that strain aging involved not only a return of the yield point and an increase in the yield stress but also

1. an increase in the ultimate tensile stress
2. sharp decreases in ductility
3. an increased rate of work hardening

The Cottrell–Bilby model provides a mechanism for the return of the yield point and an increase in the yield stress, but it does not provide for these other three factors, which must be explained without recourse to precipitation. The generally accepted conclusion is that once firmly pinned by solute atoms during strain aging, dislocations

[1] D.V. Wilson and B. Russell, *Acta Metall.*, 8:468 (1960).

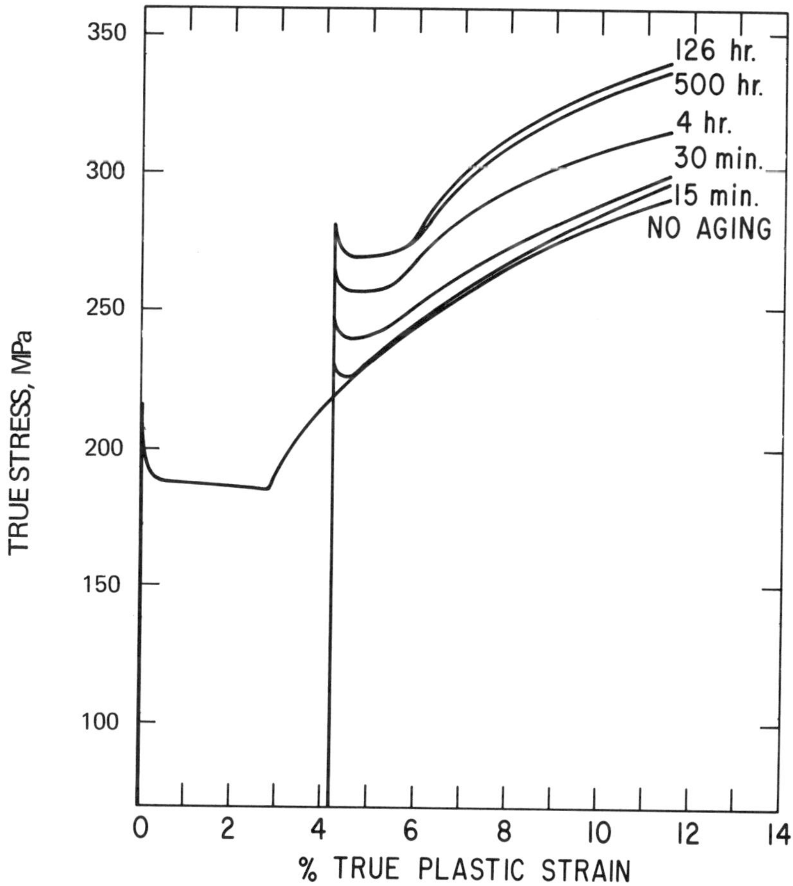

Figure II-12. Strain aging of 0.03%C rimmed steel at 60°C. (A.S. Keh and W.C. Leslie, in *Materials Science Research,* vol. 1, Plenum Publishing Corp., New York, 1963, by permission.)

are not unpinned at the upper yield stress during retesting. Rather, new dislocations are generated at sites of stress concentration, such as grain boundary edges or inclusion interfaces. For a given total strain, therefore, the dislocation density is greater if an aging step has been interposed than if the strain is applied continuously, and consequently the rate of work hardening is increased.

It has just been stated that precipitation from solid solution is not a necessary part of the strain aging process. However, the property changes that accompany strain aging can be enhanced if a highly supersaturated solid solution is strained and aged—that is, if quench aging is superimposed upon strain aging. This is illustrated in Fig. II-13. The Fe–0.022%N alloy was strained in the supersaturated, as-quenched condition and also after quenching followed by aging at 100°C.[1] In the as-quenched alloy, two stages of aging were observed. In the first stage, no precipitates were observed and the hardening

[1] A.S. Keh and H.A. Wriedt, *Trans. AIME,* 224:560 (1962).

must be attributed to the segregation of nitrogen to dislocations. After aging 10 min at 100°C the first precipitates appeared on dislocations and an increment of strength due to quench aging was added to that due to strain aging. As the precipitate particles grew, the strengthening from quench aging decreased, but the increment from strain aging did not decrease. This illustrates another characteristic of strain aging: it is not susceptible to overaging.

II-6. DYNAMIC STRAIN AGING

Dynamic strain aging is aging that occurs during the deformation process. This implies that a solute atom must be able to make at least one diffusive jump while the alloy is being deformed and that therefore the process is sensitive to temperature and to strain rate. Dynamic strain aging has four distinctive characteristics:

1. It is accompanied by a high rate of work-hardening.
2. Flow stress during dynamic strain aging has a negative strain-rate dependence; that is, if strain rate is increased the flow stress decreases.
3. The stress–strain curve is serrated.
4. The hardness, or stress at a given plastic strain and constant strain rate, goes through a maximum as a function of temperature.

A comparison between the effects produced by static and dynamic strain aging of low-carbon steel is given in Fig. II-14. Even though the strain employed at 200°C was less than the prestrain at room temperature, the dynamic strain-aging process was much more effective in strengthening the steel. Figure II-15 illustrates the serrated

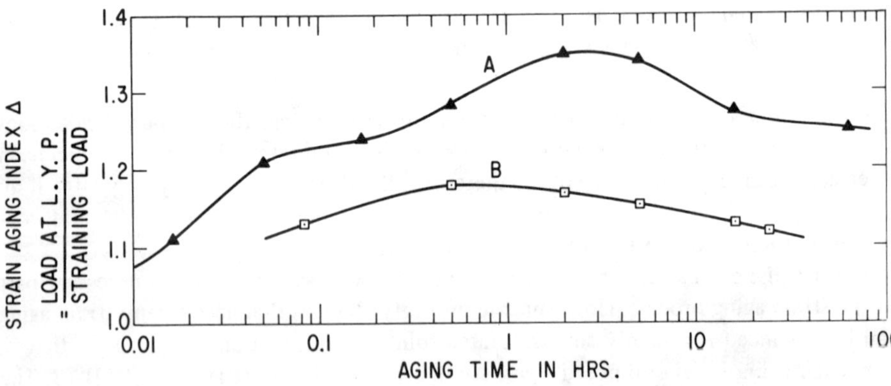

Figure II-13. Strain aging of an Fe–0.022%N alloy. (*a*) Quenched from 500°C, strained 3%, aged at 100°C. (*b*) Quenched from 500°C, aged 48 h at 100°C, strained 3%, re-aged at 100°C. (A.S. Keh and H.A. Wriedt, *Trans. AIME*, 224:560 (1962), by permission.)

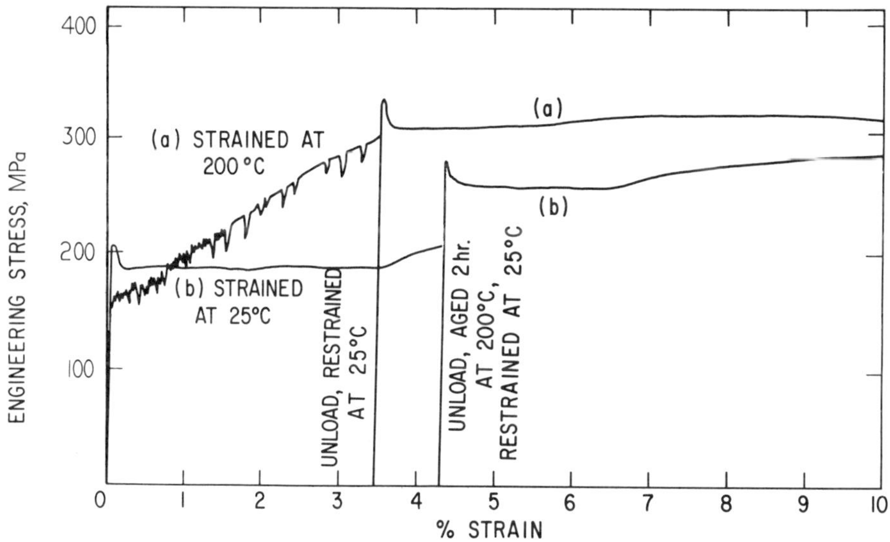

Figure II-14. (*a*) Dynamic strain aging vs. (*b*) static strain aging at 200°C of 0.03%C rimmed steel. (A.S. Keh and W.C. Leslie, in *Materials Science Research,* vol. 1, Plenum Publishing Corp., New York, 1963, by permission.)

flow, the high rate of work-hardening, and the temperature dependence of the flow stress for a given strain, typical of dynamic strain aging.

The mechanism of the increased work-hardening is clear enough (Fig. II-16). When the temperature of the specimen is raised into the dynamic strain-aging region, the dislocations generated are quickly pinned so that others must be generated to allow deformation to continue. As a result, the dislocation density at a given strain is higher than when the straining is done outside the dynamic strain-aging range.

The temperature–strain rate regime in which serrations appear in a low-carbon steel is shown in Fig. II-17. From the slope of the line for the initiation of serrations, the activation energy for the process is calculated to be 75–84 kJ/mol, which corresponds to the activation energy for volume diffusion of carbon in alpha iron. From the slope of the line for termination of serrations, the calculated activation energy is 128 to 153 kJ/mol. It has been suggested[1] that this may be the sum of the activation energy for solute diffusion plus the binding energy between carbon and a dislocation. In other work on dynamic strain aging of bcc and fcc alloys, the activation energy for the initiation of serrations has been found to be much less than the activation energy for volume diffusion and the activation energy for termination of serrations has been found to be equal to the activation energy for volume diffusion, so the situation is confused.

[1] A.S. Keh, Y. Nakada, and W.C. Leslie, *Dislocation Dynamics,* McGraw-Hill Book Co., New York, 1968, p. 381.

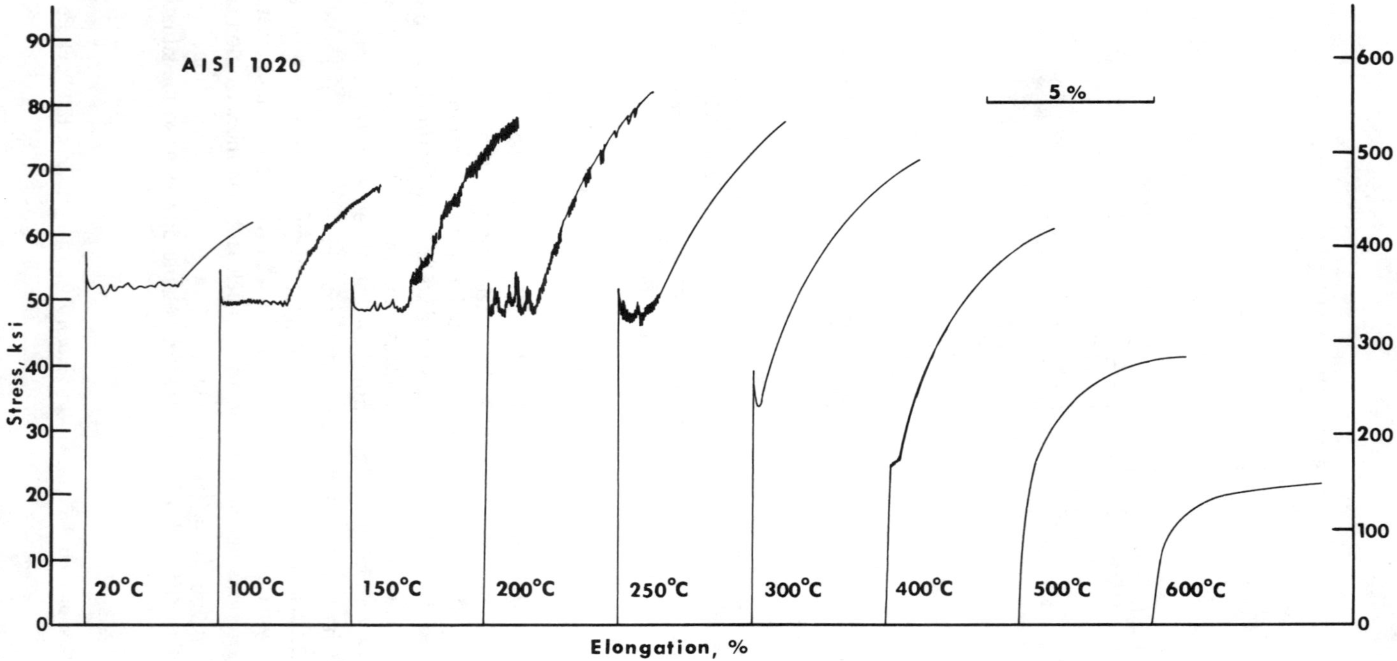

Figure II-15. Stress–elongation curves of a carbon steel strained in tension, $\dot{\epsilon} = 1.75 \times 10^{-4}\ s^{-1}$. (C.-C. Li and W.C. Leslie, *Met. Trans. A*, 9A:1765 (1978).

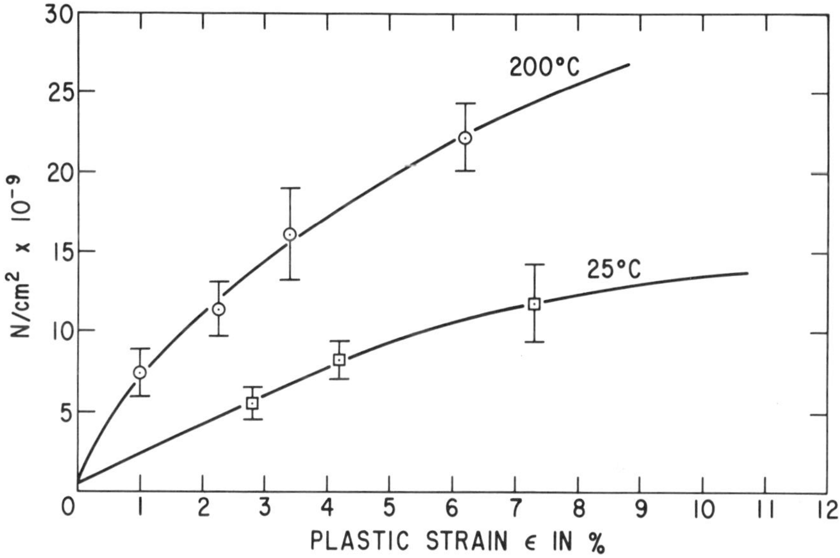

Figure II-16. Average dislocation density (N) vs. plastic strain (ϵ) for a 0.03%C rimmed steel, deformed at two temperatures at the same strain rate, $3.3 \times 10^{-4}\ s^{-1}$. (W.C. Leslie and A.S. Keh, in *Mechanical Working of Steel 2,* AIME, New York, 1965, p. 337, by permission.)

Another characteristic of dynamic strain aging is the abrupt termination of serrations with a very slight increase in temperature, which does not seem to be compatible with a diffusive mechanism.

It has been suggested that a critical strain is required before serrations can begin,[1] to generate vacancies to assist diffusion of substitutional solute atoms. However, in some instances this critical strain is of uncertain magnitude or nonexistent, so the general validity of this concept is in doubt.

The line for initiation of serrations (Fig. II-17) can be moved to the right (toward lower temperatures) as the concentration of interstitial solute atoms is increased (Fig. II-18). The location of the line for the disappearance of serrations has little or no concentration dependence. As for static strain aging, dynamic strain aging requires a very low concentration of interstitial solutes, less than 10 ppm by weight.

Serrations on flow curves can take several forms. One form, Type A, results from the repeated propagation of Lüders bands which move smoothly through the gauge section. In Type B, more common in interstitial alloys,[2] a Lüders band is initiated, but moves only a short distance before it is pinned. A new band is started in the region adjacent to the halted band, and so on.

One of the most prominent aspects of strain aging is a reduction in ductility and

[1] A.H. Cottrell, in *Relation of Properties to Microstructure,* Am. Soc. Met., Metals Park, Ohio, 1966, p. 131.

[2] J.S. Blakemore and E.O. Hall, *J. Iron Steel Inst.,* 204:817 (1966).

in fracture toughness. It was this loss of toughness, as indicated by the rise in notch impact transition temperature, that more than any other factor spelled the doom of the high-nitrogen acid Bessemer steels. The magnitude of the loss of toughness is about proportional to the concentration of interstitial solutes. Although this embrittlement must always be kept in mind, it should not prevent consideration of strain aging as a strengthening mechanism, particularly in sheet steels because the transition temperature of thin sheets is low.

Strain aging increases coercivity, but to nothing like the extent caused by quench aging.

II-7. SOLID SOLUTION HARDENING AND SOFTENING BY INTERSTITIAL SOLUTES

Many studies in recent years have shown that if an element in solid solution in a bcc metal hardens that metal in the temperature range in which the yield strength is

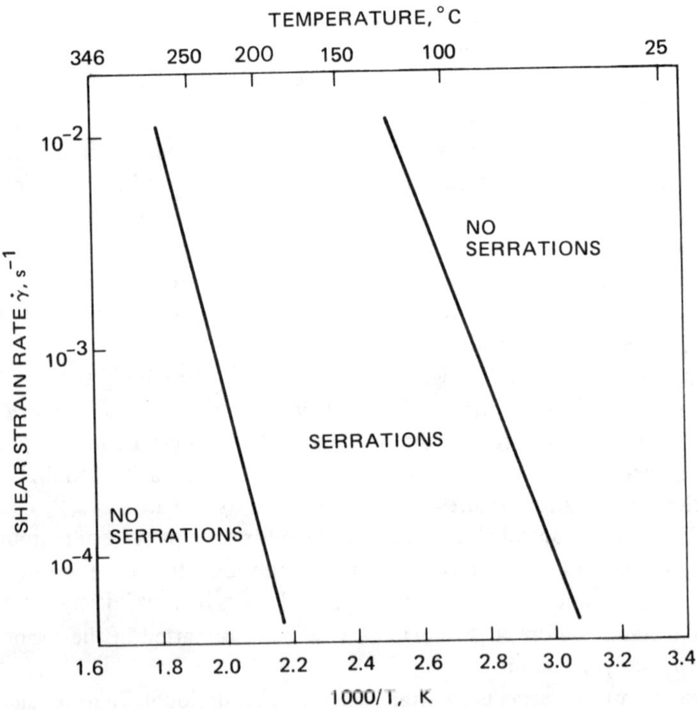

Figure II-17. The shear strain rate ($\dot{\gamma}$)–($1/T$) regime required for the presence of serrations on the flow curves of a 0.03%C steel, quenched from 250°C. (A.S. Keh, Y. Nakada, and W.C. Leslie, in *Dislocation Dynamics,* copyright 1968, used with the permission of the McGraw-Hill Book Co., New York.)

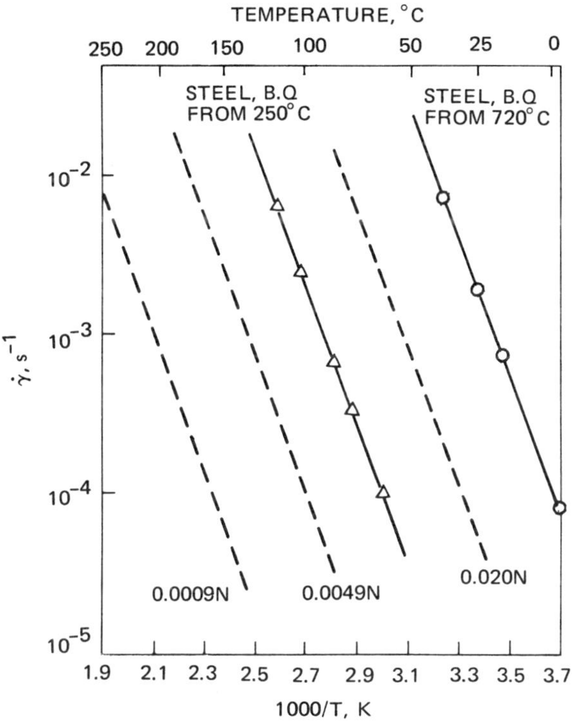

Figure II-18. Temperature and strain-rate dependence for the onset of serrations in a 0.035%C steel and in three Fe–N alloy crystals. (A.S. Keh, Y. Nakada, and W.C. Leslie, in *Dislocation Dynamics*, copyright 1968, used with the permission of the McGraw-Hill Book Co., New York.)

athermal, it will inevitably soften that metal in the temperature region in which the yield strength is sharply dependent upon temperature. Let us first consider strengthening. At ambient temperature, the strengthening of bcc iron by C or N atoms in solid solution arises from dynamic processes (Snoek ordering and Cottrell atmosphere formation) as well as from interactions between stationary solute atoms and dislocations. The latter process is usually what is meant by solid solution strengthening. The relative magnitudes of these dynamic and static effects are difficult to sort out of the available data; but if the strength is measured at 0°C, at a fairly rapid strain rate of $5 \times 10^{-2}\ s^{-1}$, when dynamic strengthening by carbon is minimal, one finds that the change in shear stress with concentration of carbon, $d\tau/dC$, is about equal to $G/2$, where G is the shear modulus.[1]

At temperatures in the neighborhood of 200 K, the flow stress of alpha iron containing carbon in solid solution can be less than that of high-purity iron[2,3] as shown in Fig. II-19. At temperatures between about 100 and 250 K, the presence of interstitial solute atoms aids the nucleation of double kinks on screw dislocations. The presence of 300 atom ppm carbon in high-purity iron reduces the activation energy for double

[1] W.C. Leslie and R.J. Sober, *Trans. Am. Soc. Met.*, 60:459 (1967).
[2] D.J. Quesnel, A. Sato, and M. Meshii, *Mater. Sci. Eng.*, 18:199 (1975).
[3] J.P. Cottu, J.P. Peyrade, P. Chomel, and P. Groh, *Acta Metall.*, 26:1179 (1978).

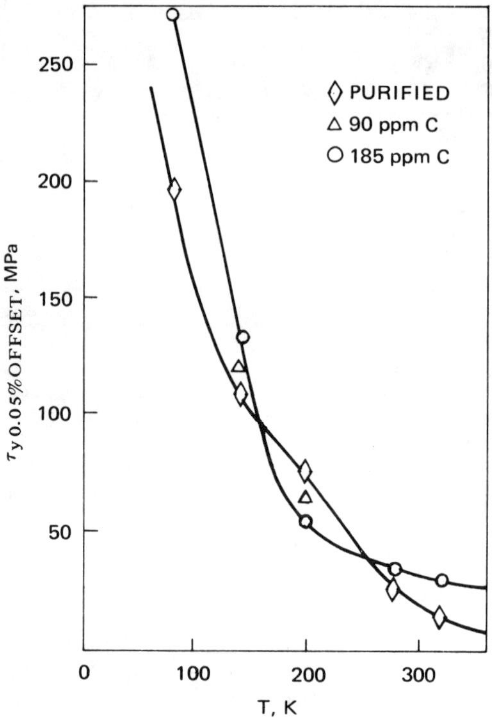

Figure II-19. The temperature dependence of resolved shear yield stress, measured at 0.05% plastic strain. (D.J. Quesnel, A. Sato, and M. Meshii, *Mater. Sci. Eng.*, 18:199 (1975), by permission.)

kink nucleation from $0.64\text{–}0.70 \times 10^{-19}$ J to 0.50×10^{-19} J, or by about one-third.[1] At temperatures above about 250 K, the flow stress of Fe–C or Fe–N alloys is greater than that of high-purity iron because of the dynamic effects previously mentioned.

II-8. EFFECTS OF INTERSTITIAL SOLUTES ON RECOVERY AND RECRYSTALLIZATION

The presence of carbon and nitrogen atoms in iron, whether they are in solid solution or precipitated as carbides, increases the work-hardening for any given strain. Therefore the energy stored in the metal, which is to say the driving force for subsequent recovery and recrystallization, is increased. However, this increased driving force is offset by the strong interactions between vacancies and interstitial atoms and between dislocations and such atoms. During subsequent annealing, motion of vacancies and dislocations is strongly inhibited, and recovery is delayed by these interactions, at least at low annealing temperatures.[2,3] Because little of the stored energy of cold work is

[1] L.P. Kubin and F. Louchet, *Acta Metall.*, 27:337 (1979); L.P. Kubin, F. Louchet, J.P. Peyrade, P. Groh, and J.P. Cottu, *Acta Metall.*, 27:343 (1979).

[2] G. Venturello, C. Antonione, and F. Bonaccorso, *Trans. AIME,* 227:1433 (1963).

[3] C. Antonione, G. Della Gatta, and G. Venturello, *Trans. AIME,* 230:700 (1964).

consumed by recovery, softening of the cold-worked iron is almost directly proportional to the amount of recrystallization, as shown in Fig. II-20. It is for this reason that the recrystallization of low-carbon steels can be measured quite well by determining hardness.

Because of the interaction of interstitial atoms with dislocations and cell walls, the rate of nucleation of new grains during recrystallization is reduced, but the rate of migration of grain boundaries is not greatly affected.

II-9. PRECIPITATION FROM SOLID SOLUTION (QUENCH AGING)

The subject of quench aging will be discussed in some detail, for it is both commercially important and a good example of precipitation of an interstitial solute from a supersaturated metallic solid solution. Although quench aging is not often deliberately used as a strengthening method, it is the principal reason why continuous annealed sheet is not as soft and ductile as box (batch) annealed sheet. Continuous galvanized sheet, which enters a bath of zinc maintained at a temperature of over 480°C, is struck with a blast of cold air as it emerges from the liquid zinc. As a result, carbon is

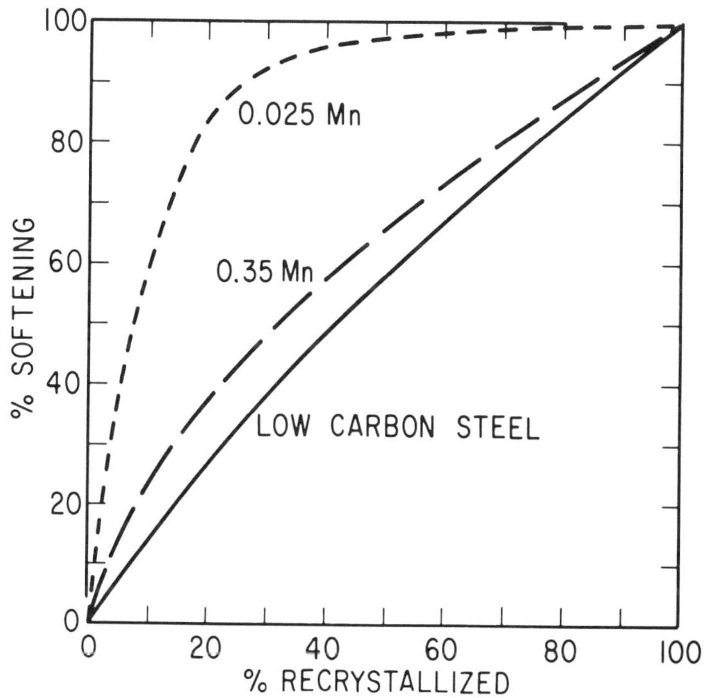

Figure II-20. The softening of three iron alloys recrystallized at 595°C. (W.C. Leslie, F.J. Plecity, and J.T. Michalak, *Trans. AIME,* 221:691 (1961).)

retained in supersaturated solid solution and precipitates during subsequent holding at ambient temperature. If the maximum ductility is required in the sheet, a supplementary annealing treatment at about 315°C may be required.

In principle, the quench aging of iron can occur by the precipitation of either carbon or nitrogen from solid solution, but in practice nitrogen is almost never present in sufficient quantity to make any sizeable contribution to quench aging, which is almost always due to carbon.

In common with other systems, the initial precipitation in Fe–C and Fe–N alloys follows the path of minimum activation energy rather than that of the maximum release of free energy. At low temperatures the precipitates which form most rapidly are γ'' iron nitride, $Fe_{16}N_2$, which has a body-centred tetragonal structure,[1,2] and a metastable carbide, which most investigations identify as the hexagonal ϵ carbide, $Fe_{2.4}C$. Both the carbide and the nitride phase form on {100} planes of the ferrite matrix. The nitride forms with $(100)_{\gamma''} \| (100)_{\alpha}$. If strain energy were unimportant in determining the configuration and habit of low-temperature precipitates, then such particles might form as spheres to minimize their interfacial area. In some instances, as in dilute alloys of copper in iron, where the two kinds of atoms have nearly the same size, the initial precipitates are spherical. In most precipitation hardening systems, however, strain energy is important and can be minimized by formation of plate- or rod-shaped particles. For all cubic metals except molybdenum, the modulus of elasticity is a minimum in ⟨100⟩ directions. Therefore plate-shaped precipitates on {100} planes are generally observed when aging is done at low temperatures.[3]

The solubility relations governing precipitation of carbides from supersaturated solid solution in alpha iron are shown in Fig. II-4*c*. At aging temperatures below the metastable solvus, the transition precipitate ϵ may form first, rather than Fe_3C, because of its better matching with the ferrite lattice, but the decrease in free energy that accompanies the precipitation is less than for Fe_3C.

The metastable carbide in iron forms as arrays of parallel needles growing in $\langle 100 \rangle_{\alpha}$ directions.[4] As growth continues, these needles can coalesce into lenticular disks on $\{100\}_{\alpha}$, as shown in Fig. II-21. In some instances, this carbide can grow on all three $\{100\}_{\alpha}$ planes from a common point.

Cementite also grows as needles, but on {110} planes of the ferrite with the *b* axis of the orthorhombic unit cell, $[010]_{cem}$, aligned along the length of the needle, parallel to $[111]_{\alpha}$.[5] Growth always occurs in more than one $\langle 111 \rangle_{\alpha}$ direction, resulting in a dendrite-like structure, as illustrated in Figs. II-22 and II-23. The individual needles are single crystals, 3.5 to 5.0 nm in diameter, or about five to seven unit cells of cementite. With continued growth, the needles coalesce into massive particles.

These modes of growth are adopted because of the good fit of Fe-to-Fe interatomic distances across the carbide–ferrite interfaces, as indicated in Fig. II-24. The

[1] K.H. Jack, *Proc. Roy. Soc. (London)*, A208:216 (1951).

[2] G.R. Booker, J. Norbury, and A.L. Sutton, *J. Iron Steel Inst.* 187:205 (1957).

[3] E. Hornbogen, in *Precipitation from Iron-Base Alloys,* Gordon and Breach, New York 1965, p. 1.

[4] E.W. Langer, *Met. Sci. J.*, 2:59 (1968).

[5] W.C. Leslie, R.M. Fisher, and N. Sen, *Acta Metall.*, 7:632 (1959).

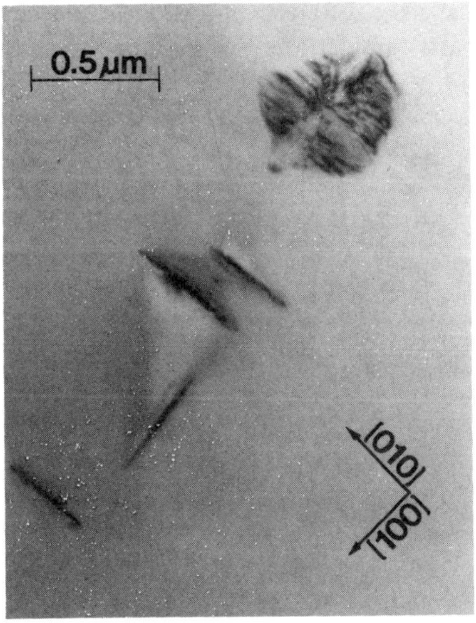

Figure II-21. Metastable ϵ carbide precipitates on $(100)_\alpha$ planes in an Fe–0.013%C, alloy quenched from 700°C and then aged 6 h at 200°C. (H. Wagenblast and R.C. Glenn, *Met. Trans.*, 1:2299 (1970), by permission.)

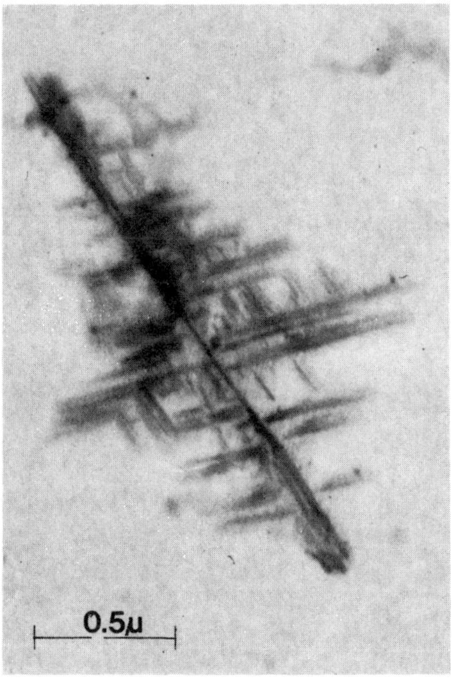

Figure II-22. Dendritic cementite in Fe–0.014%C alloy, brine quenched from 740°C and then aged 10 min at 260°C. Extraction replica. (Reprinted with permission from W.C. Leslie, R.M. Fisher, and N. Sen, *Acta Metall.*, 7:632 (1959), copyright Pergamon Press, Ltd.)

difference in habit plane, $\{100\}_\alpha$ vs. $\{110\}_\alpha$, and the difference in the angle between the needles, 90° vs. 70.5°, serve to distinguish ϵ carbide from cementite in the early stages of growth.

The orientation relations between ferrite and the two carbides have been summarized by Pitsch and Schrader[1,2]:

ϵ carbide $\quad (101)_\alpha \parallel (101)_\epsilon$

$$(001)_\alpha \parallel (001)_\epsilon$$

$$[100]_\alpha \parallel [110]_\epsilon$$

[1] W. Pitsch and A. Schrader, *Arch. Eisenhuettenwes,* 29:485 (1958).
[2] W. Pitsch and A. Schrader, *Arch. Eisenhuettenwes,* 29:715 (1958).

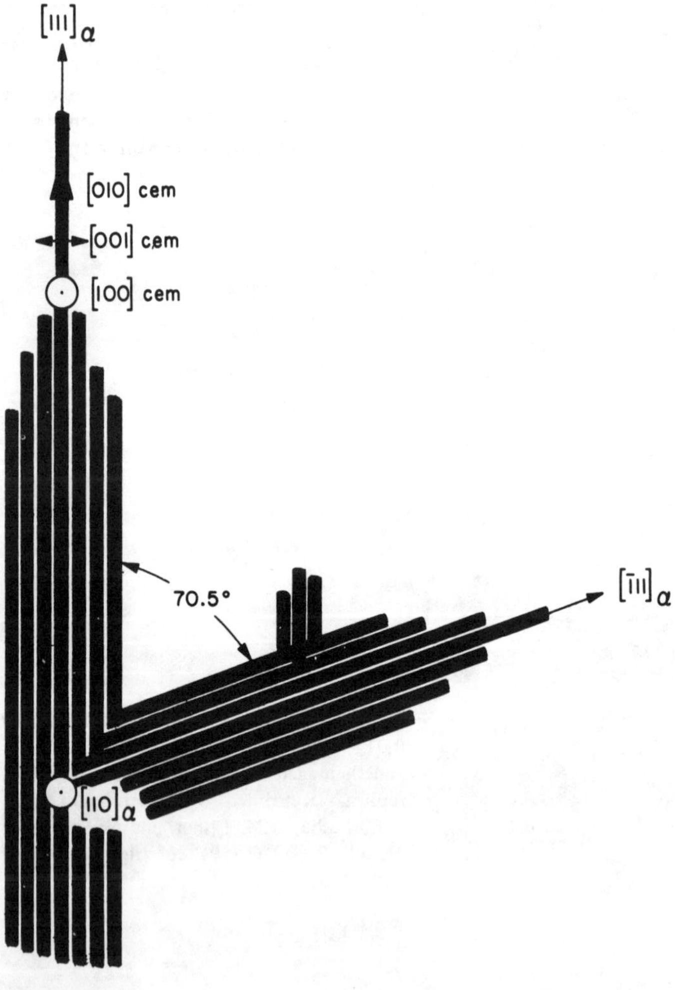

Figure II-23. Orientation of growth of dendritic cementite in ferrite. (Reprinted with permission. W.C. Leslie, R.M. Fisher, and N. Sen, *Acta Metall.,* 7:632 (1959), copyright Pergamon Press, Ltd.)

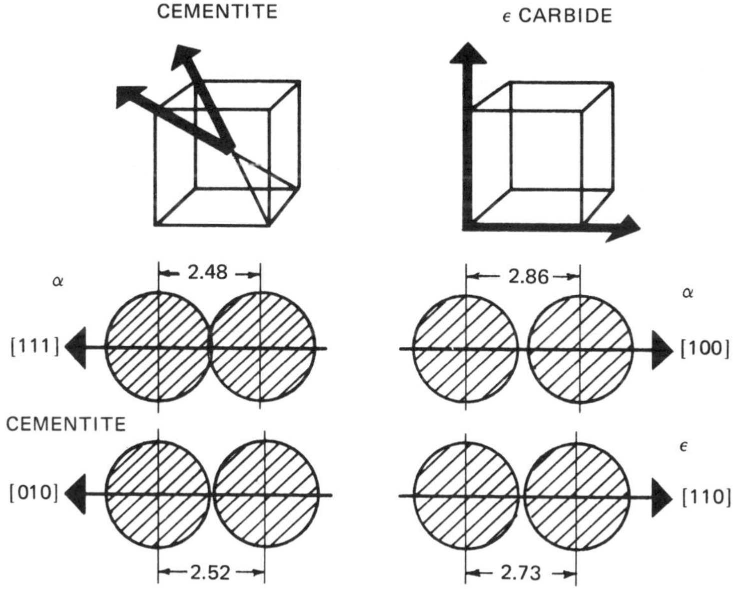

Figure II-24. Orientation relationship between ferrite/cementite and ferrite/ϵ carbide. (W. Pitsch and A. Schrader, *Archiv für das Eisenhüttenwesen,* Dusseldorf, 29:717 (1958), by permission.)

$$\text{cementite } (\theta) \qquad (211)_\alpha \parallel (001)_{\text{cem}}$$
$$[00\bar{1}]_\alpha \parallel [100]_{\text{cem}}$$
$$[\bar{1}11]_\alpha \parallel [010]_{\text{cem}}$$

In the quench aging of Fe–C alloys at low temperatures, dislocations are always favored nucleation sites; but they are not the only ones. The effectiveness of carbide precipitates in hardening depends upon closely spaced particles nucleated within the alpha iron matrix. Although direct evidence of vacancies as nucleation sites is lacking, it is known that mono- or divacancies are excellent traps for carbon atoms,[1–3] and true homogeneous nucleation of carbides seems improbable; therefore, vacancies remain the most likely nucleation sites provided they are not present in such numbers as to trap all carbon atoms. At low aging temperatures, from ambient to about 150°C, vacancies are much less mobile than are carbon or nitrogen atoms and are effective traps. At higher aging temperatures, 200°C and above, vacancies are highly mobile and the driving force for precipitation (the supersaturation of carbon) is reduced, so dislocations become the only operative nucleation sites. In contrast to these ideas, Abiko and Kimura[4] propose that impurity atoms in ferrite serve as nucleation sites for precipitation of ϵ carbide.

[1] H. Wagenblast and A.C. Damask, *J. Phys. Chem. Solids,* 23:221 (1962).
[2] T. Takayama and H. Takahashi, *J. Phys. Soc. Jpn.,* 35:939 (1973).
[3] M. Mondino and A. Seeger, *Scr. Metall.,* 11:817 (1977).
[4] K. Abiko and H. Kimura, *Jpn. Inst. Met.,* 17:384 (1976).

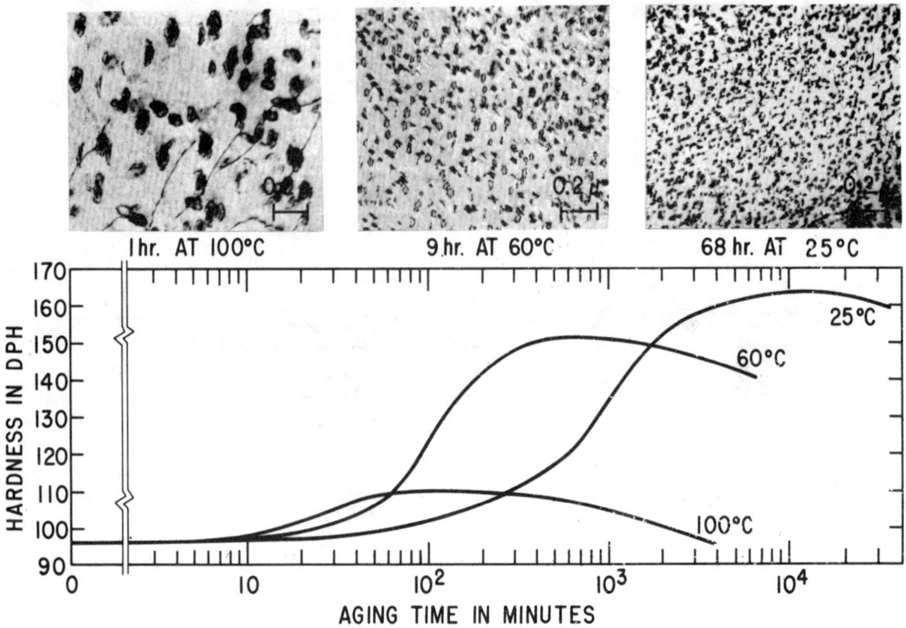

Figure II-25. Quench-aging kinetics of Fe–0.02%N alloy with microstructures corresponding to hardness maxima. (A.S. Keh and W.C. Leslie, in *Materials Science Research,* vol. 1, Plenum Publishing Corp., New York, 1963, by permission.)

The nucleation site does not appear to influence the structure of the precipitated carbide. For a given aging treatment, the habit of the carbide was either $\{110\}_\alpha$ or $\{100\}_\alpha$, regardless of whether the particles originated in the matrix or at dislocations.

Keh[1] made a very complete study of the quench-aging process in iron containing 0.02%N, quenched from 500°C, then aged at 25, 60, or 100°C for various periods. As expected, the maximum hardness attained and the time to reach this maximum increased with decreasing aging temperature (Fig. II-25). In specimens aged at 100°C the average particle size and the interparticle spacing were large as compared with those in specimens aged at 25° and 60°C. The precipitate formed at all three aging temperatures was $Fe_{16}N_2$. This work has been supplemented by Abiko and Imai,[2] who examined precipitation during isothermal aging in the temperature range from 100 to 300°C.

The kinetics of aging and the hardness obtained in the Fe–N system and in other systems can be controlled by pre-aging treatments that establish a fine distribution of stable nuclei. The Fe–0.02%N alloy was used in such experiments, the results of which are given in Fig. II-26. Specimens quenched from 500°C were pre-aged at 25°C, then heated to 100°C for various periods. When held 2 h at 25°C before aging at

[1] A.S. Keh and W.C. Leslie, *Mater. Sci. Res.*, 1:208 (1963).

[2] K. Abiko and Y. Imai, *Trans. Jpn. Inst. Met.*, 18:113(1977).

100°C, the hardness suffered an initial drop; then the remainder of the aging curve corresponded to that of specimens given no pre-aging treatment. Particles formed during the early stage of pre-aging at 25°C are not stable at 100°C; they redissolve, accounting for the initial drop in hardness. After 6 h at 25°C the particles formed were subsequently stable at 100°C, for a much higher maximum hardness was developed after aging 1 h at 100°C. The extra hardening due to pre-aging passed through a maximum as the pre-aging time was lengthened. Relatively long pre-aging at 25°C led to a high hardness which decreased immediately during subsequent aging at 100°C. This immediate decrease is probably due to a decrease in the number of the smallest particles during aging at 100°C. The hardening is determined by particle size and spacing, which are controlled to a large extent by the number of stable nuclei created by the pre-aging treatment.

Quench rate has several effects on the quench aging of Fe–C or Fe–N alloys, and neglect of this variable can cause disparities between the results of experiments by different investigators. The most thorough study was done by Vynhal and Radcliffe,[1] on an Fe-0.019%C alloy, cooled from 710°C at one of three rates, fast (3000 to 24,000°C/s), intermediate (ice brine quench), and slow (delayed transfer from furnace to brine).

[1] R.F. Vynhal and S.V. Radcliffe, *Acta Metall.*, 20:435 (1972).

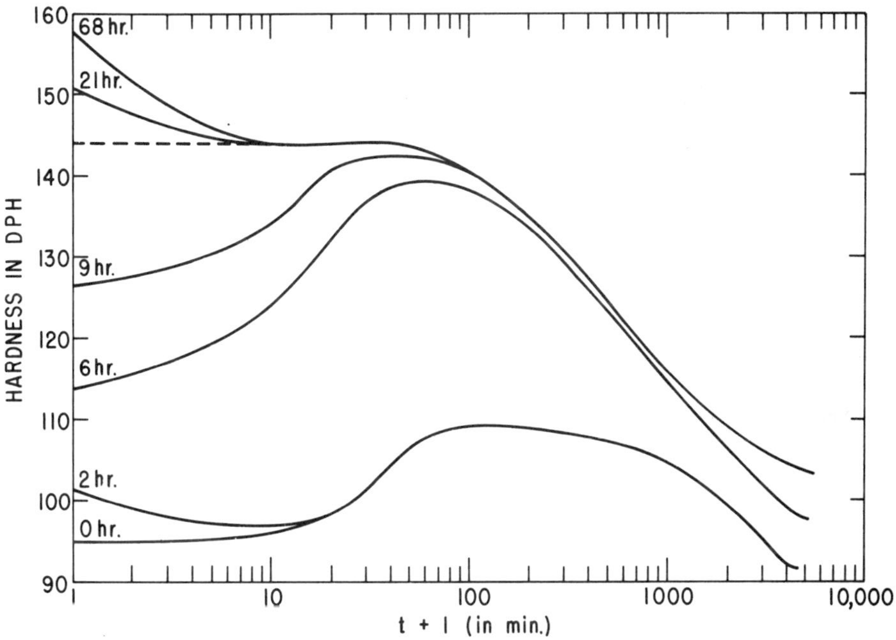

Figure II-26. Effects of pre-aging at 25°C on the aging kinetics of an Fe–0.02%N alloy at 100°C. (A.S. Keh and W.C. Leslie, in *Materials Science Research*, vol. 1, Plenum Publishing Corp., New York, 1963, by permission.)

During quenching, vacancies trap C atoms. An increased rate of quenching retains a higher concentration of vacancies as carbon-vacancy pairs. If a low aging temperature is used after quenching, the carbon-vacancy pairs provide a high density of nucleation sites, leading to a rapid rate of precipitation and a fine dispersion of precipitate particles. At higher aging temperatures, a high quenching rate produces an opposite effect. The principal role of the carbon-vacancy couple at higher aging temperatures is to aid precipitate particle growth by relieving coherency strains developed during growth. This produces a low density of large carbide particles, in contrast to the smaller and more numerous precipitates in a slowly quenched alloy.

When an Fe–0.014%C alloy is quenched from 730°C, then aged at 200°C, nucleation occurs only on dislocations and only Fe_3C is precipitated.[1] Under these conditions the rate of growth of particles can be represented by an equation of the form

$$R = Kt^{-n} \qquad \text{(II-35)}$$

where t is time at constant aging temperature. In this instance, n had a value of 1.16 and the relation was valid at least to the time at which about 80% of the carbon was precipitated. In several instances, it has been reported that the fraction of carbon precipitated during isothermal aging can be represented by the equation

$$W = Kt^{n} \qquad \text{(II-36)}$$

so it appears that under the conditions chosen, the rate of growth of carbide particles was inversely proportional to the fraction precipitated.

II-10. MAGNETIC AGING

The change of magnetic properties of iron or iron–silicon alloys with time is simply another manifestation of quench aging, but the terms used, the experimental techniques employed, and the processes involved are generally not familiar to metallurgists, so a separate treatment is desirable. Let us begin by defining terms, taken from ANSI/ASTM A340-77.

Aging, magnetic—the change in the magnetic properties of a material resulting from metallurgical change, which implies a deterioration of the magnetic properties of magnetic materials for electronic and electrical applications.

Bohr Magneton—a constant, 9.274×10^{-24} J/s, equal to the magnetic moment of an electron due to its spin.

Coercive force, H_c—the dc magnetizing force at which the magnetic induction is zero units (A/m).

Coercivity, H_{cs}—the maximum value of coercive force.

Core loss, total, P_c—the active power (watts) expended in a magnetic circuit in which there is a cyclically alternating induction.

Domains, ferromagnetic—magnetized regions within ferromagnetic materials. Each

[1] W.C. Leslie, *Acta Metall.*, 9:1004 (1961).

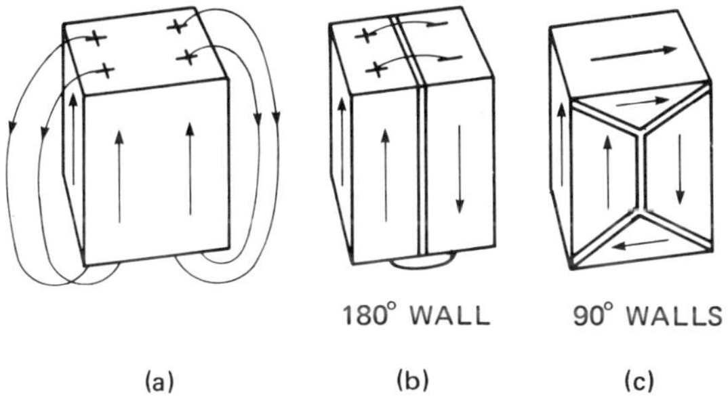

Figure II-27. Elimination of external lines of force (*a*) by creation of domain walls at (*b*) 180° and (*c*) 90°.

domain is magnetized to intrinsic saturation at all times and this saturation induction is unidirectional within the domain.

Domain wall—a boundary region between two adjacent domains within which the orientation of the magnetic moment of one domain changes into a different orientation of the magnetic moment in the other domain.

Ferromagnetic material—a material that exhibits the phenomena of hysteresis and saturation and whose permeability is dependent on the magnetizing force.

Induction, intrinsic, B_i—the vector difference between the magnetic induction in a magnetic material and the magnetic induction that would exist in a vacuum under the influence of the same magnetizing force.

Induction, saturation, B_s—the maximum intrinsic induction possible in a material. The external field has aligned all possible domains.

Iron is by far the most important magnetic material. It is ferromagnetic below the Curie temperature of 770°C and contains permanently magnetized domains whose width is in the range 0.1 to 1 mm. Below the Curie temperature these domains can be aligned by a weak magnetic field. Domains exist because they tend to minimize the total energy of the system, as shown schematically in Fig. II-27. The external lines of force (II-27*a*) can be reduced (II-27*b*) or eliminated (II-27*c*) by the subdivision of the material by domain walls. The domains in Fig. II-27*c* are termed domains of closure.

The domain walls must have a finite thickness because of the necessity for a transition in the antiparallel alignment of atomic magnets in adjacent domains (Fig. II-28). The transition zone (domain wall) is often about 120 nm thick.

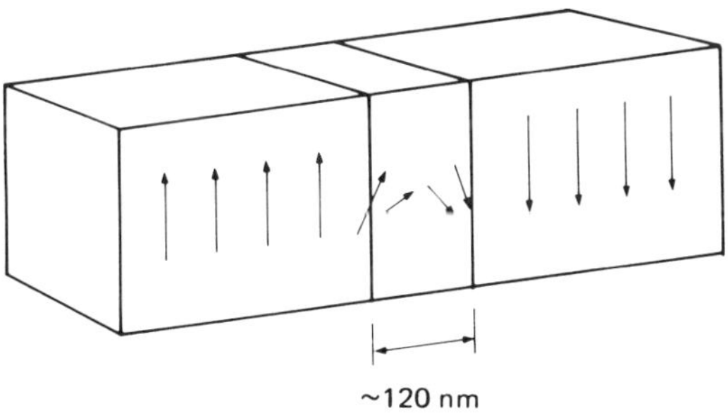

Figure II-28. Change of direction of magnetization across a 180° domain wall.

Domain walls will move under the influence of an applied field, increasing the size of domains with magnetization parallel to the applied field and reducing the size of those with magnetization in the antiparallel direction. Soft magnetic materials (iron, low-carbon steels, and silicon iron) are those whose magnetization can be reversed with little expenditure of energy. Their coercivity is low, less than 150 A/m (2 oersteds). In transformers and rotating electrical machinery, energy of magnetization reversal is lost as heat and thereby reduces the efficiency of the process. One need only consider the enormous amount of electric power that is generated and transmitted to realize the importance of minimizing energy loss in soft magnetic materials.

Domain walls can be pushed past obstacles such as precipitate particles and inclusions only with a loss of energy. To determine how this energy loss occurs it is necessary to make direct observations of the interaction of domain walls with precipitate particles. This can be done by transmission electron microscopy, provided the magnetic field in the microscope at the specimen position is low. By under-focussing or over-focussing, the domain walls in the specimen become visible,[1] and if the magnetic field at the specimen can be varied, the domain wall can be moved at will and the dynamics of the domain wall–particle interaction can be seen.

If iron supersaturated with carbon is isothermally aged to precipitate carbides, the coercivity of the specimen increases with aging time, passes through a maximum, and then declines, in a manner analogous to the change in hardness during aging. However, the maximum coercivity is reached after much longer aging than is required to reach maximum hardness.[2,3] At the peak of coercivity the carbide particles have a maximum dimension of at least 600 nm. Up to that peak, domain walls are not noticeably affected by cementite particles (Fig. II-29), but the coercivity increases steadily with increasing particle size. At maximum coercive force, domains of closure are seen to form around the carbide particles (Fig. II-30).

Although cementite is itself ferromagnetic, the complex shape of the dendritic particles and lack of correspondence between the directions of easy magnetization in the ferrite and in the cementite make the cementite behave like a nonmagnetic inclusion. Free poles form at the particle–matrix interfaces. The coercive force increases to a maximum because the magnetostatic energy of the demagnetizing fields around the particles increases more rapidly than the opposing effect due to a decreasing number of particles. At the peak of aging, when the average particle size is about 1200 nm, domains of closure form around the particles, reducing the magnetostatic energy. Further aging causes particles to become more equi-axed, to decrease in number, and to develop domains of closure, all being processes leading to a net reduction in coercive force.

Magnetic aging is not a simple process because it varies with

1. The volume fraction of the precipitate
2. The size of precipitate particles

[1] J.T. Michalak and R.C. Glenn, *J. Appl. Phys.*, 32:1261 (1961).
[2] W.C. Leslie and D.W. Stevens, *Trans. Am. Soc. Met.*, 57:261 (1964).
[3] P. Schwaab and W. Schwartz, *Arch. Eisenhuetenwes*, 36:145 (1965).

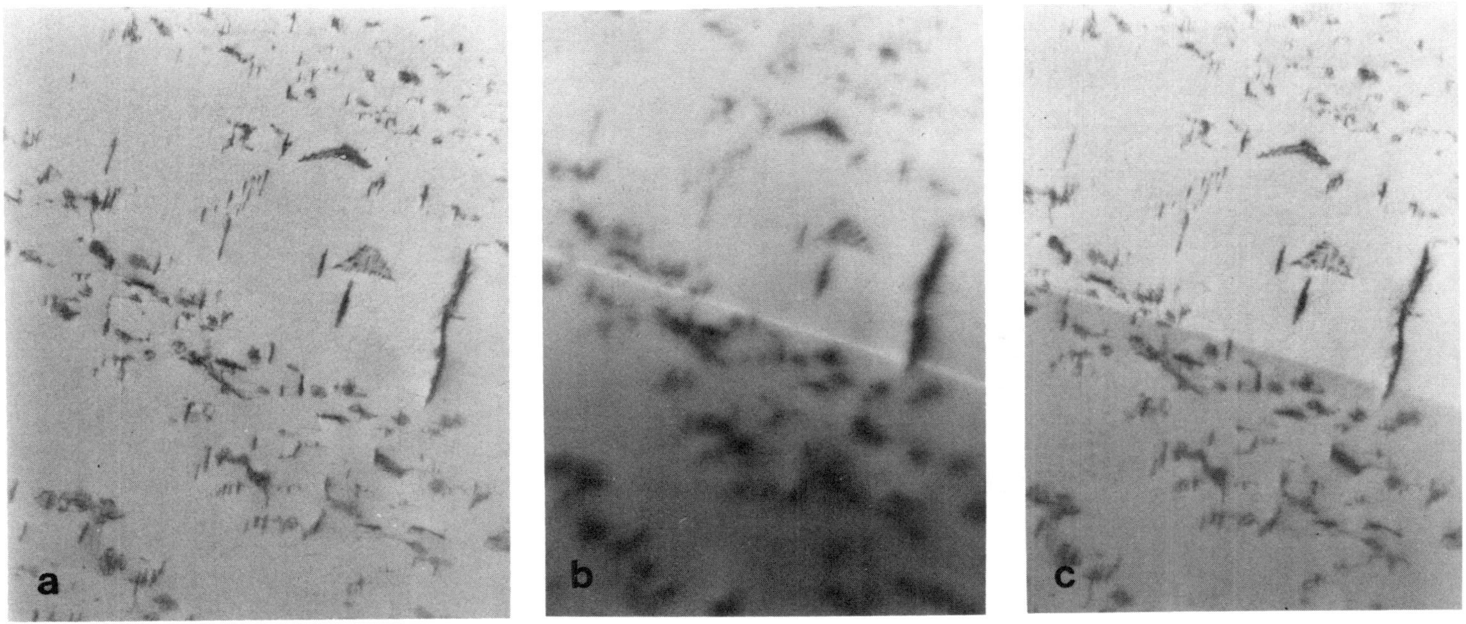

Figure II-29. Lack of apparent interaction between domain walls and small cementite particles in Fe–0.45%Mn–0.017%C alloy aged 50 min at 200°C. (*a*) In focus. (*b*) Overfocus. (*c*) Aperture off-center. (W.C. Leslie and D.W. Stevens, *Trans. Am. Soc. Met.*, 57:261 (1964).)

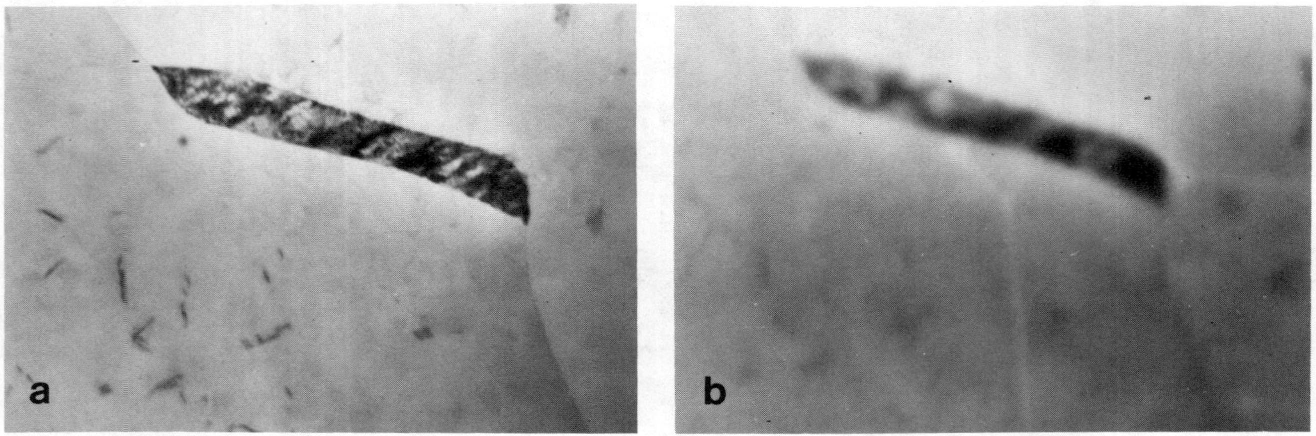

Figure II-30. Domains of closure around a large cementite particle in Fe–0.45%Mn–0.017%C alloy aged 4,500 min at 200°C (overaged). (*a*) In focus. (*b*) Underfocus. (W.C. Leslie and D.W. Stevens, *Trans. Am. Soc. Met.*, 57:261 (1964).)

3. Precipitate morphology
4. Precipitate habit
5. Magnetic properties of the precipitate

In principle, prevention of magnetic aging in iron or low-carbon steels is simple. It can be done either by removing carbon or by heat-treating the material so that all carbides are present as large (>2000 nm) particles, widely separated, and little or no carbon is left in solid solution.

Magnetic aging will be mentioned again in the section on silicon steels.

CHAPTER

THREE

SUBSTITUTIONAL SOLUTES IN ALPHA IRON

III-1. EFFECTS OF SOLUTES ON LATTICE PARAMETER AND ELASTIC MODULI

To consider properly the several theories of solid solution strengthening of metals, it is essential to know precisely how the lattice parameter and elastic constants of the solvent are affected by various solutes. Table III-1 lists the changes produced in the lattice parameter of alpha iron by elements in binary solid solution.

For Ni, Co, and Si, the dilation of the alpha iron lattice is not linear with solute concentration, but for cobalt the deviation is not significant. In theories of solid solution strengthening, the size misfit parameter

$$\epsilon_a = \frac{1}{a_0}\frac{da}{dc} \tag{III-1}$$

where a_0 = lattice parameter of the pure solvent, is an essential factor, which is complicated when da/dc varies with concentration.

The data in Table III-1 are plotted in Fig. III-1. The dashed lines represent the Hume-Rothery limits ($\Delta a/a = \pm 15\%$) for extended solid solubility.

For pure, polycrystalline iron, free of preferred orientation, the values of the modulus of elasticity E, the shear modulus G, the bulk modulus K, and Poisson's ratio ν are[1]:

[1] G.R. Speich, A.J. Schwoeble, and W.C. Leslie, *Met. Trans.*, 3:2031 (1972).

Table III-1 Dilation of the α iron lattice by elements in dilute solid solution (3 at % or less)[1]

Alloying element	Dilation in Å(10^{-10} m)
Si[2]	$a_{Fe} - 0.0003_5 \times$ (at % Si)
P	$a_{Fe} - 0.0010 \times$ (at % P)
Ti	$a_{Fe} - 0.0031 \times$ (at % Ti)
Cr	$a_{Fe} + 0.0005_5 \times$ (at % Cr)
Mn	$a_{Fe} + 0.0006_7 \times$ (at % Mn)
Co	$a_{Fe} + 0.0000_8 \times$ (at % Co)
Ni[2]	$a_{Fe} + 0.0007 \times$ (at % Ni)
Ru	$a_{Fe} + 0.0027 \times$ (at % Ru)
Rh	$a_{Fe} + 0.0035 \times$ (at % Rh)
Re	$a_{Fe} + 0.0029 \times$ (at % Re)
Ir	$a_{Fe} + 0.0037 \times$ (at % Ir)
Pt	$a_{Fe} + 0.0049_5 \times$ (at % Pt)

[1] L. Zwell, cited in *Met. Trans.*, 3:5 (1972); L. Zwell, G.R. Speich, and W.C. Leslie, *Met. Trans.*, 4:1990 (1973).

[2] Not linear, averaged over 0 to 3 at %.

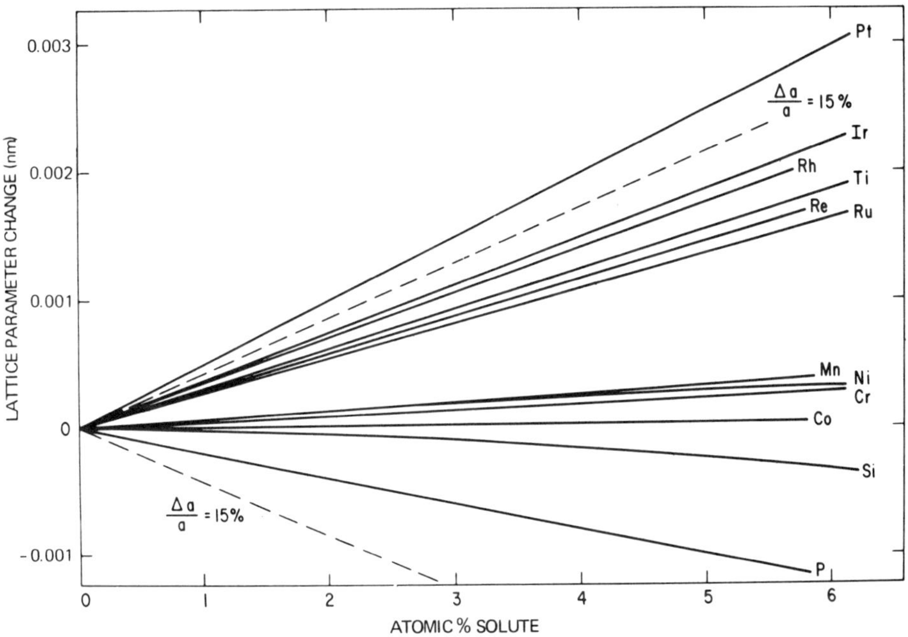

Figure III-1. The effects of elements in solid solution on the lattice parameter of alpha iron at 298 K. (L. Zwell, cited in *Met. Trans.*, 3:5 (1972).)

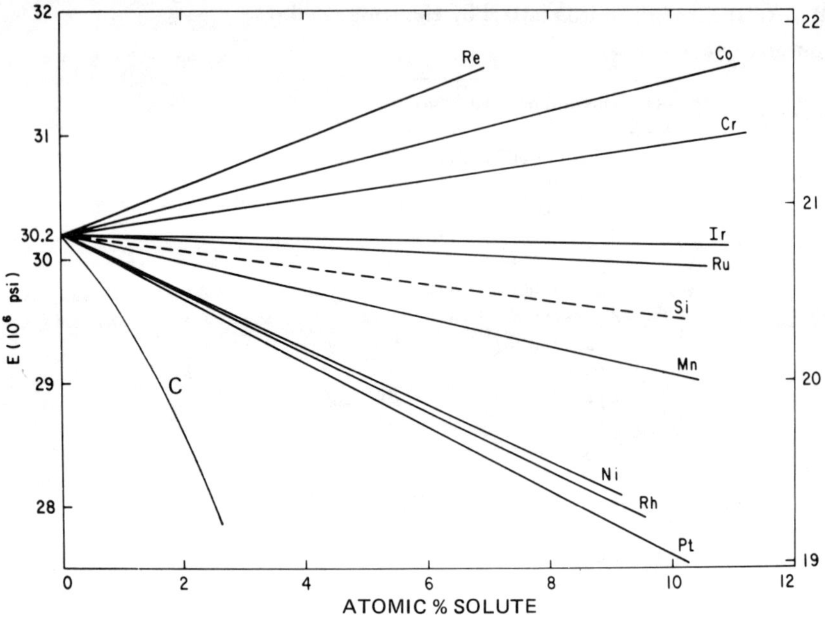

Figure III-2. The effects of elements in solid solution on the modulus of elasticity of alpha iron at 298 K. (G.R. Speich, cited in *Met. Trans.,* 3:5 (1972).)

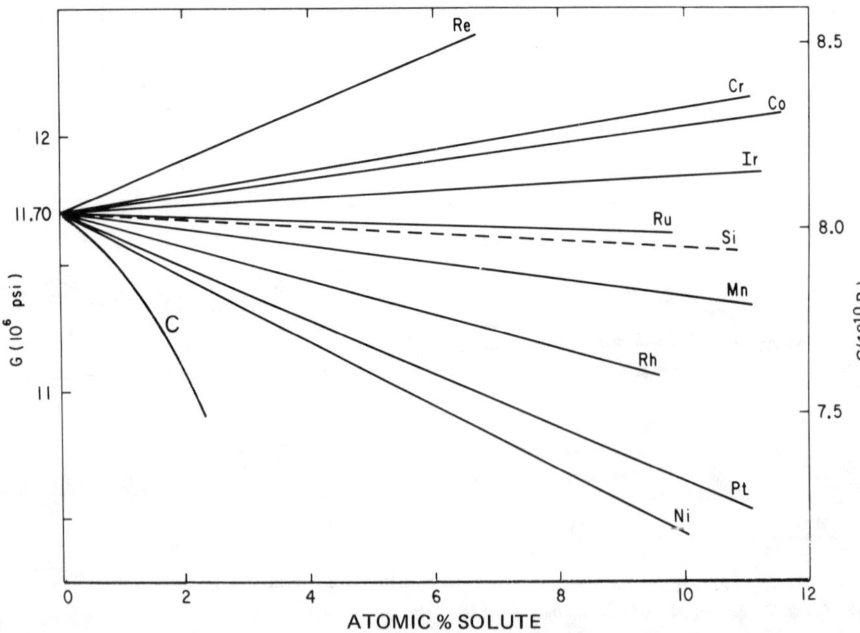

Figure III-3. The effects of elements in solid solution on the shear modulus of alpha iron at 298 K. (G.R. Speich, cited in *Met. Trans.,* 3:5 (1972).)

$$E = 20.82 \times 10^4 \text{ MPa } (30.2 \times 10^6 \text{ psi})$$

$$G = 8.065 \times 10^4 \text{ MPa } (11.70 \times 10^6 \text{ psi})$$

$$K = 16.60 \times 10^4 \text{ MPa } (24.1 \times 10^6 \text{ psi})$$

$$\nu = 0.291$$

The moduli of elasticity of binary alloys as functions of composition are shown in Fig. III-2, and the shear moduli in Fig. III-3. The data for silicon are shown as dashed lines because they were derived from measurements on single crystals.

As can be seen, the elements are about equally divided between those that increase the moduli and those that decrease them. There is no rule of mixtures; elements with greater moduli than iron can decrease the modulus of iron when present in solid solution. The effects of solutes on E and G are roughly parallel.

The effect of each of the solutes shown in Figs. III-2 and III-3 is proportional to concentration, except for carbon and cobalt. Data for the unusual alloying elements Ru, Rh, Re, Ir, and Pt are included because, with the exception of Ni, they are the only elements that increase the toughness of alpha iron.

The values of E and G for iron and its binary alloys decrease nonlinearly as temperature is increased, as shown in Figs. III-4, III-5, and III-6, but the rate of change is almost independent of solute species and concentration, at least up to about 10 at %.

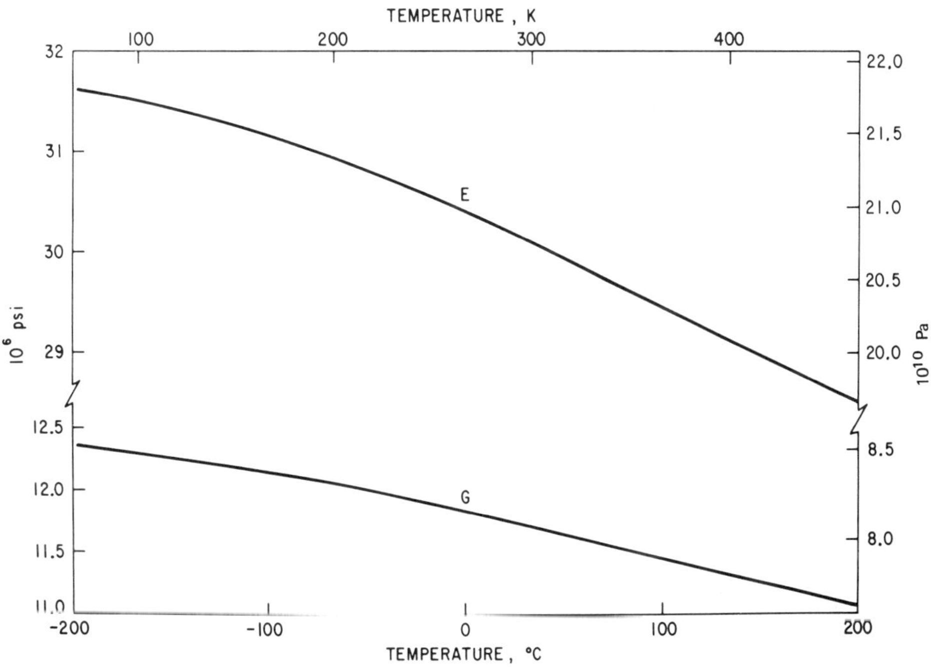

Figure III-4. The temperature dependence of elastic and shear moduli in titanium-gettered alpha iron. (G.R. Speich, cited in *Met. Trans.*, 3:5 (1972).)

Values of K (bulk modulus) and ν (Poisson's ratio) can be derived from E and G by the relations[1]:

$$K = \frac{EG}{3(3G - E)}$$

$$\nu = \frac{E}{2G} - 1$$

Poisson's ratio increased with additions of Ni, Pt, and Co, decreased with additions of Cr, Mn, Re, Rh, Ru, and Ir, and remained about constant with additions of Si. In Fe–Ni alloys, ν increased from 0.291 to 0.331 at 9%Ni. The maximum decrease was observed in Fe–Re alloys in which ν was 0.269 at 6%Re.

[1] H. Kolsky, *Stress Waves in Solids,* Clarendon Press, Oxford, England, 1953.

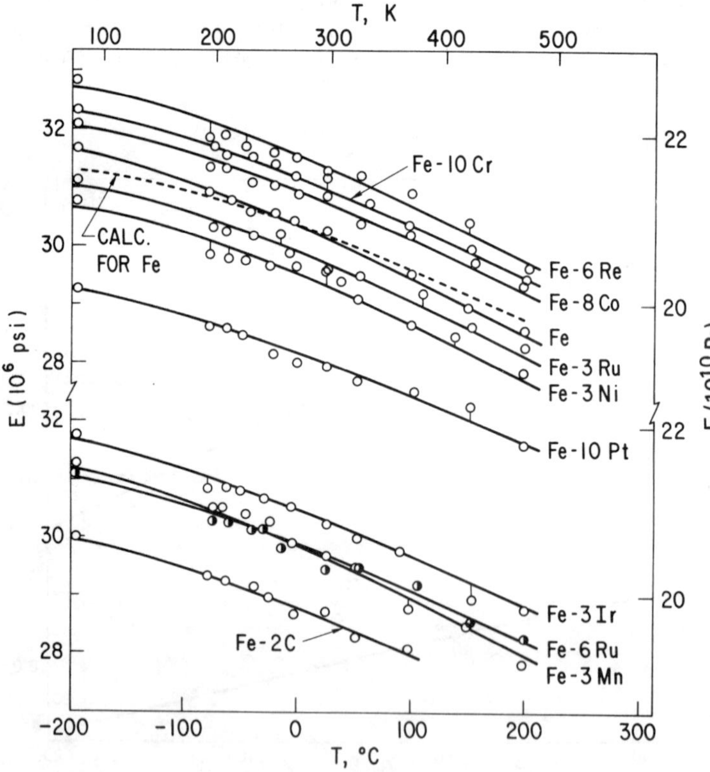

Figure III-5. Effect of temperature on Young's modulus of iron and iron-base solid solutions. (G.R. Speich, A.S. Schwoeble, and W.C. Leslie, *Met. Trans.*, 3:2031 (1972).)

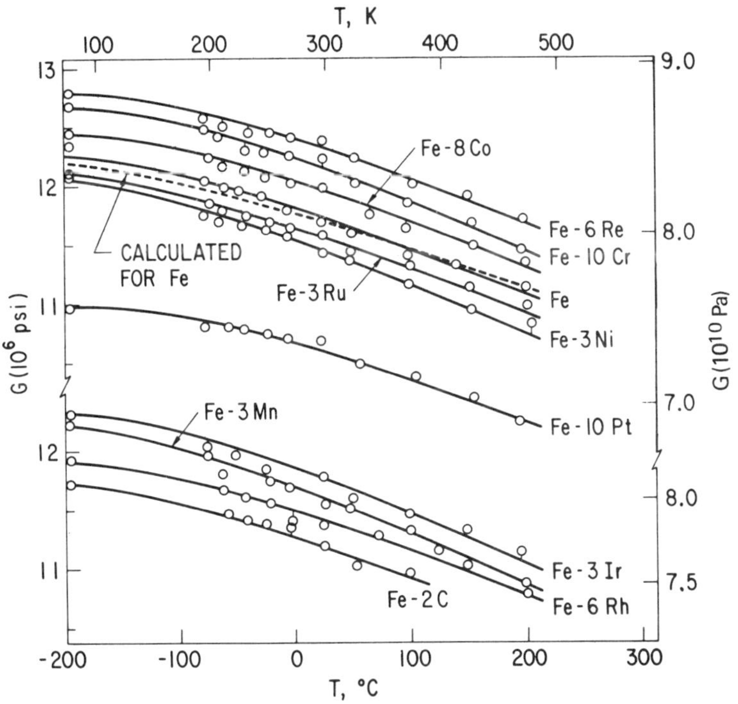

Figure III-6. Effect of temperature on shear modulus of alpha iron and iron-base solid solutions. (G.R. Speich, A.S. Schwoeble, and W.C. Leslie, *Met. Trans.*, 3:2031 (1972).)

III-2. SOLID SOLUTION STRENGTHENING AND SOFTENING

When a substitutional solute is added to alpha iron, one of the consequences is a rounding of the yield stress-temperature curve (compare Figs. I-13 and III-7). In effect, the athermal portion of the plot of Fig. I-13 disappears and there is no region in which the yield stress is independent of temperature. Also, at elevated temperatures ($\simeq$300 to 450°C) dynamic strain aging can occur during the test. The behavior of the Fe–3 at % Ni alloy shown in Fig. III-7 is representative of the behavior of binary alloys. Those containing the ineffective strengthening elements Cr and Co approximated unalloyed iron in behavior; alloys of iron with Si or Pt, which are very effective strengtheners, had even more rounded yield stress–temperature curves than shown in Fig. III-7.

The data for solid solution softening and hardening as a function of temperature can be presented in a condensed form by plotting yield strength of the alloy less the yield strength of iron (Figs. III-8 and III-9). At usual testing-machine strain rates the maximum strengthening occurs near room temperature. The strengthening drops sharply to a minimum near −120°C. The location of these curves along the temperature axis is markedly dependent upon strain rate; increasing the strain rate shifts the

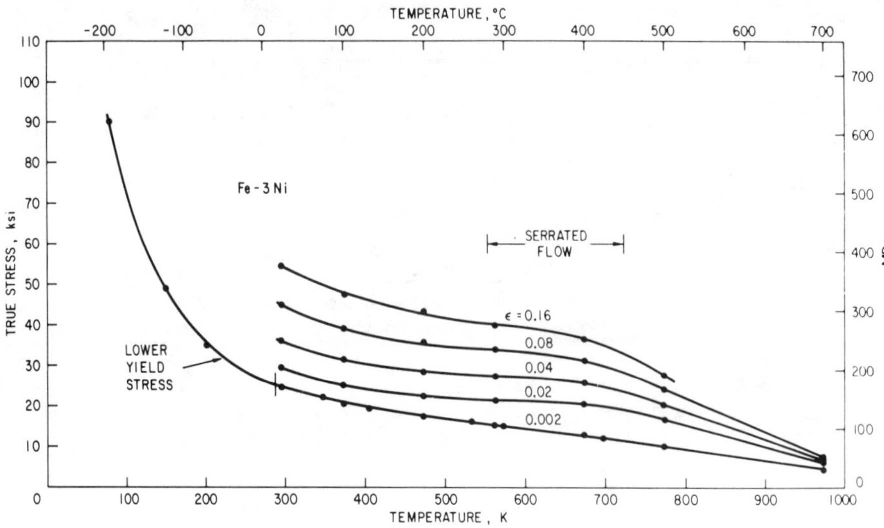

Figure III-7. Temperature dependence of yield and flow stresses in Fe–3 at % Ni, $\dot{\epsilon} \cong 2.5 \times 10^{-4}$ s^{-1}, grain size ASTM 6.5. (W.C. Leslie, *Met. Trans.*, 3:5 (1972).)

curves to higher temperatures.[1] Figure III-10 shows that at strain rates between 10^0 and 10^3 s^{-1} the yield stress of an Fe–1.5Ni alloy is less than that of iron at room temperature at strain rates between 10^0 and 10^3 s^{-1}. The stress at 4% strain for the alloy is less than that for iron at strain rates between 10^1 and 10^3 s^{-1}.

Alloy softening in binary alloys of alpha iron has been reported for at least 22 different solute elements. In a significant experiment, Sato et al.[2] produced softening by knocking iron atoms into interstitial positions by electron irradiation. Solid solution strengthening and softening in bcc metals are both produced by disruption of the lattice by point defects. The greater the disruption by a single defect, the lower the concentration of that defect required to produce detectable softening. However, at higher concentrations solid solutions softening can be replaced by solid solution hardening over the entire temperature range. To illustrate: the minimum concentrations to produce solid solution softening are about 0.09 at % for N, 0.1 at % for P, and 1.5 at % for Ni.

It is regrettably true that no current theory of solid solution strengthening explains all the effects that are seen. The size misfit parameter of Eq. III-1 is a rough guide to the magnitude of strengthening of iron to be expected from any given solute (Fig. III-11), with several notable exceptions. The first of these is that solute atoms smaller than iron atoms (P, Si, Be) are much more effective strengtheners than would be indicated by the magnitude of their size misfits. Apparently the sign of the lattice distortion cannot be ignored. For reasons unknown, Mn and Ni strengthen the lattice out of proportion to their size misfit. As shown in Fig. III-7, a solute addition changes

[1] W.C. Leslie, R.J. Sober, S.G. Babcock, and S.J. Green, *Trans. Am. Soc. Met.*, 62:690 (1969).
[2] A. Sato, J. Mifune, and M. Meshii, *Proc. 2nd Int. Conf. Strength of Metals and Alloys, Am. Soc. Met.*, Metals Park, Ohio, 2:747 (1970).

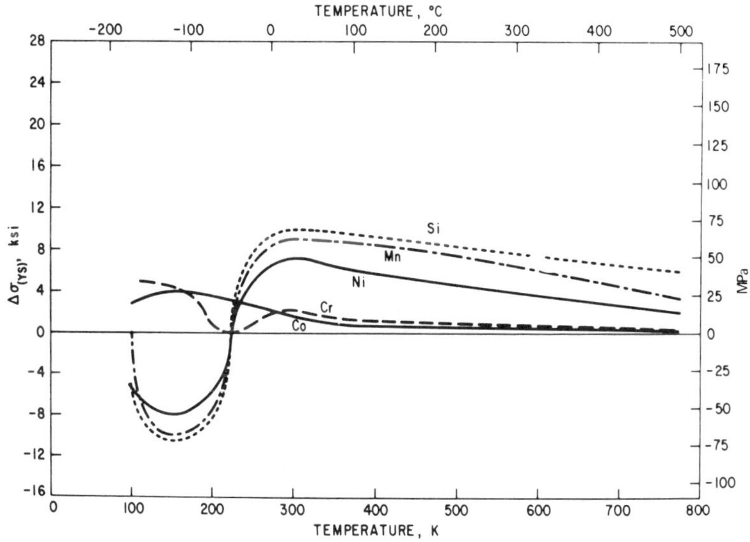

Figure III-8. Solid solution hardening and softening in iron-base alloys (1.5 at %), $\dot{\epsilon} \cong 2.5 \times 10^{-4}\ s^{-1}$. (W.C. Leslie, *Met. Trans.*, 3:5 (1972).)

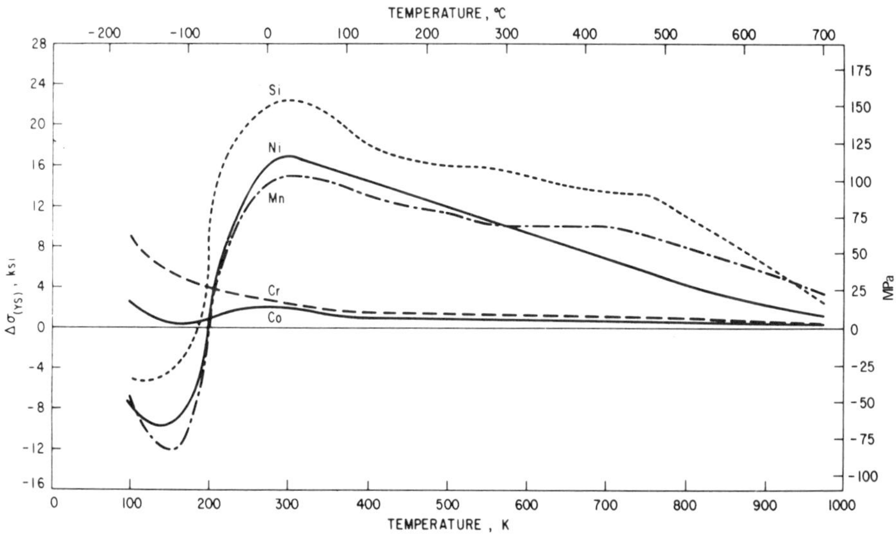

Figure III-9. Solid solution hardening and softening in iron-base alloys (3 at %), $\dot{\epsilon} = 2.5 \times 10^{-4}\ s^{-1}$. (W.C. Leslie, *Met. Trans.*, 3:5 (1972).)

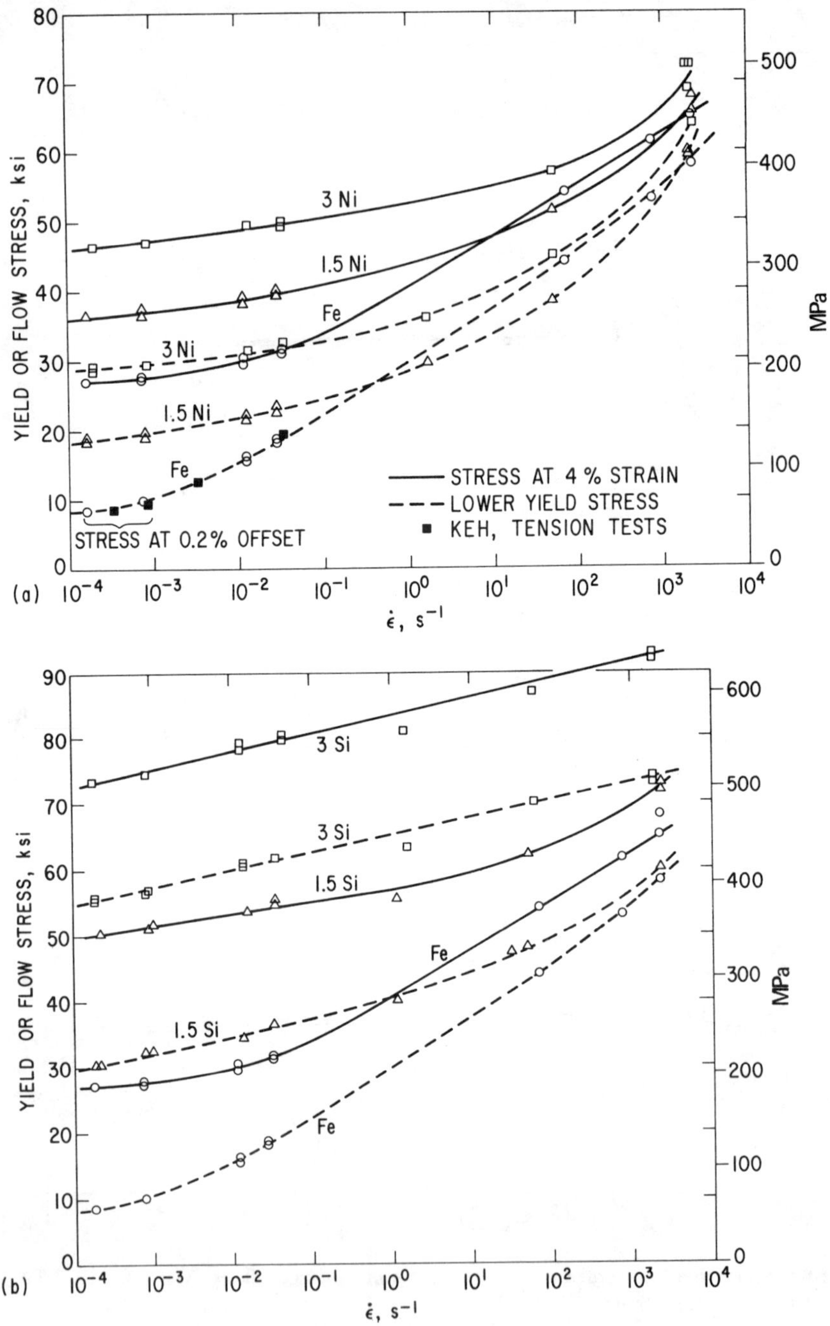

Figure III-10. Effects of alloying elements on the strength and strain-rate sensitivity of iron at 23°C: (*a*) nickel, (*b*) silicon. (W.C. Leslie, R.J. Sober, S.G. Babcock, and S.J. Green, *Trans. Am. Soc. Met.*, 62:690 (1969).)

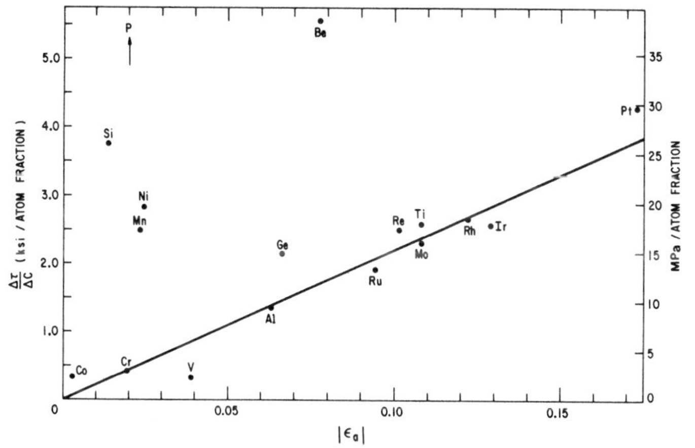

Figure III-11. Correlation of solid solution strengthening of iron-base binary alloys at 298 K with size misfit parameter. (W.C. Leslie, *Met. Trans.*, 3:5 (1972).)

both the thermal and the athermal components of stress, so that solid solution strengthening theories that consider only one or the other of these components are not applicable.

The rate of work-hardening of iron during plastic deformation at room temperature is not affected by dilute solutions of substitutional atoms (Fig. III-12), but this conclusion may not be valid at other temperatures.

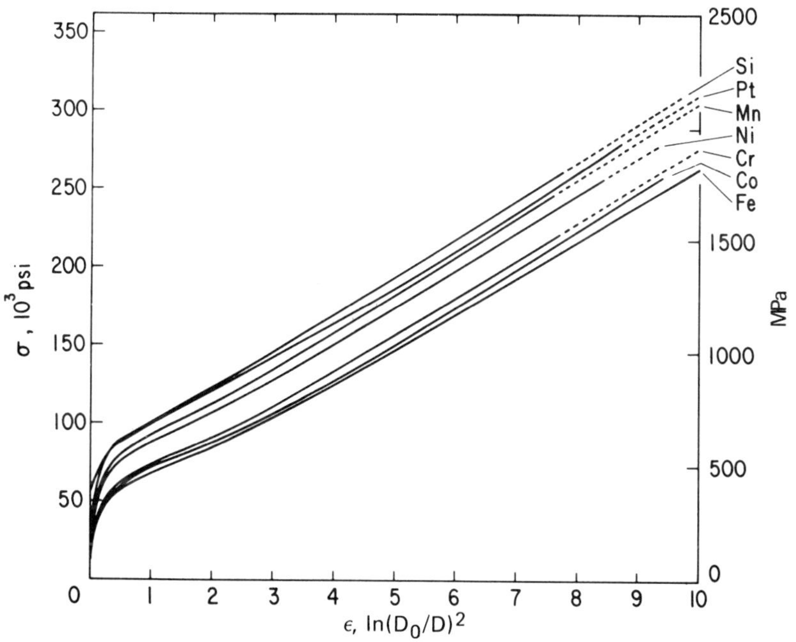

Figure III-12. Tensile strength–prior strain plots for drawn wires of binary iron alloys (3 at %). (G. Langford, P.K. Nagata, R.J. Sober, and W.C. Leslie, *Met. Trans.*, 3:1843 (1972).)

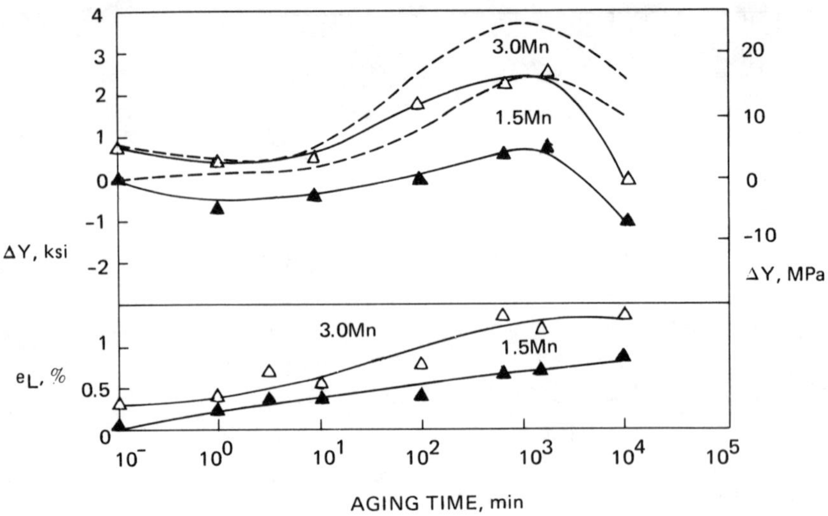

Figure III-13. Static strain aging of Fe–Mn alloys at 350°C (dashed lines represent ΔY after corrections for recovery). (C.C. Li and W.C. Leslie, *Met. Trans. A*, 6A:1987 (1975).)

III-3. STATIC STRAIN AGING BY SUBSTITUTIONAL SOLUTE ATOMS

Under appropriate conditions, atoms in substitutional solid solution can cause static strain aging in iron.[1,2] These conditions are:

1. The absence of interstitial solute atoms (C and N)
2. An aging temperature sufficiently high to allow the short-range migration of substitutional solute atoms

If specimens of a binary solid solution of iron, containing no C or N atoms in solution, are strained in the usual manner, then aged isothermally in the temperature range from 300 to 450°C, they can show an increase in the subsequent yield stress, a return of discontinuous yielding, and an increase in the Lüders strain with increased aging time (Fig. III-13). In this figure ΔY is the difference between the flow stress at 4% strain and the lower yield stress after aging, and ϵ_L is the Lüders strain. ΔY is much smaller than after aging by interstitial atoms. C or N atoms interfere with substitutional strain aging because, with their much greater mobilities and much higher binding energies with dislocations, they saturate the favored sites around dislocations before the substitutional solutes can move.

Note that in static strain aging by substitutional solutes there is no increase in the ultimate tensile strength and no decrease in ductility, which may indicate that because

[1] W.B. Morrison and W.C. Leslie, *J. Iron Steel Inst.*, 211:129 (1973).
[2] C.-C. Li and W.C. Leslie, *Met. Trans. A*, 6A:1987 (1975).

of the low binding energy between dislocations and substitutional atoms the dislocations can be freed by high stresses.

Because of the comparatively high temperatures that must be used for substitutional strain aging, recovery and strain aging are concurrent processes. Iron and Fe–Cr and Fe–Co alloys (dilute) all show recovery rather than strain aging when heated above 300°C after plastic strain.

The activation energies for static strain aging of binary alloys of iron with Si, Ni, Ti, Mn, Mo, or P are equivalent to the activation energies for volume diffusion of each of these solutes in alpha iron.

III-4. DYNAMIC STRAIN AGING BY SUBSTITUTIONAL SOLUTE ATOMS

If substitutional solutes cause static strain aging, they must also produce dynamic strain aging. This process has been observed during plastic deformation in the temperature range 230–500°C.[1] Serrations on the plastic flow curve begin when the number of atomic jumps $N_j = \Gamma\tau \simeq 1$, where Γ is the atomic jump rate and τ is the time between serrations. In a bcc lattice, the number of diffusive jumps N_j made in time t at temperature T by a solute with an activation energy for diffusion Q is[2]:

$$N_j = \frac{8D}{a^2}t = \frac{8D_0}{a^2}t\exp\left(-\frac{Q}{RT}\right) \qquad \text{(III-2)}$$

where a = lattice parameter
$D \simeq 1\text{–}5$

then $$N_j \cong 10^{16}t\exp\left(-\frac{Q}{RT}\right) \qquad \text{(III-3)}$$

In the usual way, the temperature–strain rate regime for serrations is affected by solute concentration (Fig. III-14). Increasing solute concentration shifts downward the temperature at which serrations are first observed at a given strain rate but has little effect on the temperature at which serrations disappear. In contrast to the situation with interstitial solutes, the activation energy for the initiation of serrations is about 110 kJ/mol, half the activation energy for volume diffusion of substitutional solutes in alpha iron. This, together with the fact that at the concentrations required for dynamic strain aging there are many solute atoms within one Burgers vector of a dislocation, indicates that dynamic strain aging is caused by reorientation of solute atoms in the strain fields of dislocations. The observation that the activation energy for the initiation of serrations in Fe–(C, N) alloys is equal to that for volume diffusion may indicate that diffusion of interstitial solutes is little affected by dislocation strain fields. The mechanism for the termination of serrations remains uncertain.

[1] W.C. Leslie, L.J. Cuddy, and R.J. Sober, *Proc. 3rd Int. Conf. on Strength of Metals and Alloys*, Inst. of Metals, London, 1:11 (1973).

[2] Y. Adda and J. Philibert, *La Diffusion Dans Les Solids*, Presses Univ. de France, 1966, p. 376.

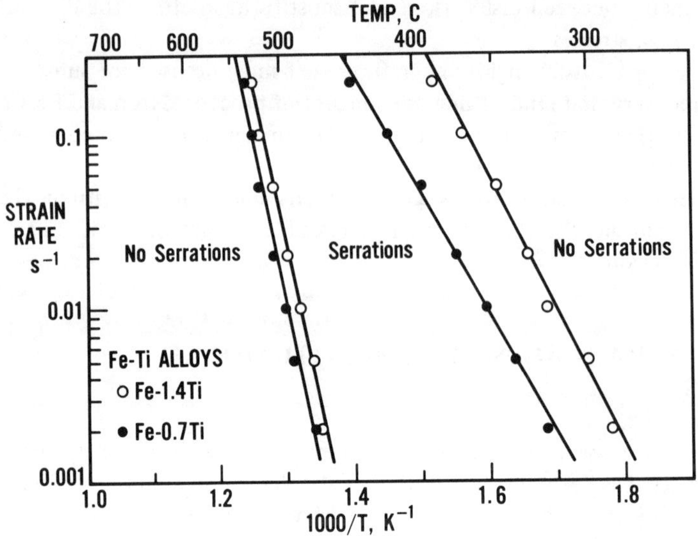

Figure III-14. Temperature and strain-rate dependence of serrated flow in Fe–Ti alloys. (W. C. Leslie, L.J. Cuddy, and R.J. Sober, *Proc. 3d Intern. Conf. Strength of Metals and Alloys,* Inst. of Metals, London, 1:11 (1973).)

The minimum concentrations of substitutional solutes required to produce serrations on the flow curves of binary alloys of iron are:

Ti:	between 0.15 and 0.7 wt %
Ni:	between 3.0 and 5.0 wt %
Mn:	between 1.5 and 4.0 wt %
Mo:	between 0.6 and 1.8 wt %
Re:	between 3.0 and 9.0 wt %
Pt:	between 3.0 and 6.0 wt %
Rh:	no serrations up to 6 wt %
Co:	no serrations
Cr:	no serrations

In these binary alloys of iron, no critical amount of prestrain is needed before serrated flow can begin.

We will return to the topic of serrated flow when interactions between solute atoms of different species are discussed.

III-5. EFFECTS OF SUBSTITUTIONAL SOLUTES ON TOUGHNESS OF IRON

An economical means of increasing the resistance of bcc iron-base alloys to brittle cleavage fracture would be of intense interest. Although a great deal of improvement can be made by controlling complex microstructures, as in quenched and tempered alloy steels, the fact remains that at low temperatures fcc alloys such as austenitic stainless steels or aluminum alloys must be used when brittle cleavage fractures cannot be tolerated. This tendency towards brittle cleavage fracture is an inherent characteristic of bcc metals[1] that can be ameliorated by alloying with a few substitutional solutes. In order of decreasing effectiveness in improving the toughness of iron, these elements are Pt, Ni, Ru, Rh, Ir, and Re.[2] From a practical standpoint, therefore, the list reduces to one—Ni. All other alloy additions *decrease* the toughness of alpha iron. The effects produced on low-carbon iron by alloying with Ni are shown in Fig. III-15. The effect of Pt is particularly striking because of its large strengthening effect on iron (Fig. III-16) and the retention of strengthening to high temperatures. There is no theory pertaining to this toughening effect of solutes; we can speculate that the addi-

[1] V. Vitek, *Cryst. Lattice Defects,* 5:1 (1974).
[2] W.C. Leslie, *Met. Trans.,* 3:5 (1972).

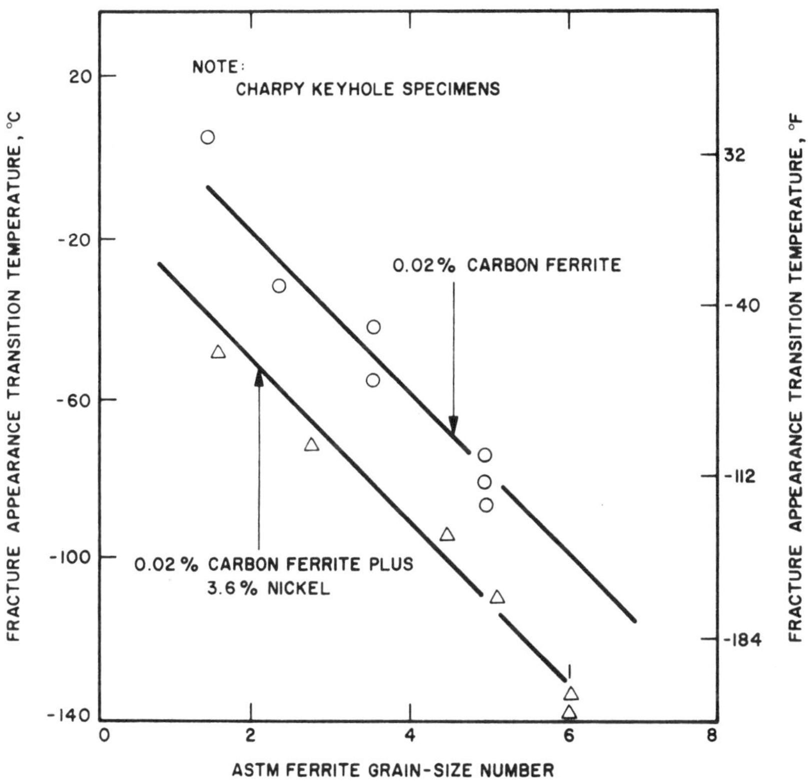

Figure III-15. Effects of grain size and nickel content on the fracture appearance transition temperature of ferrite. (J.M. Hodge, R.D. Manning, and H.M. Reichhold, *Trans. AIME,* 185:233 (1949), by permission.)

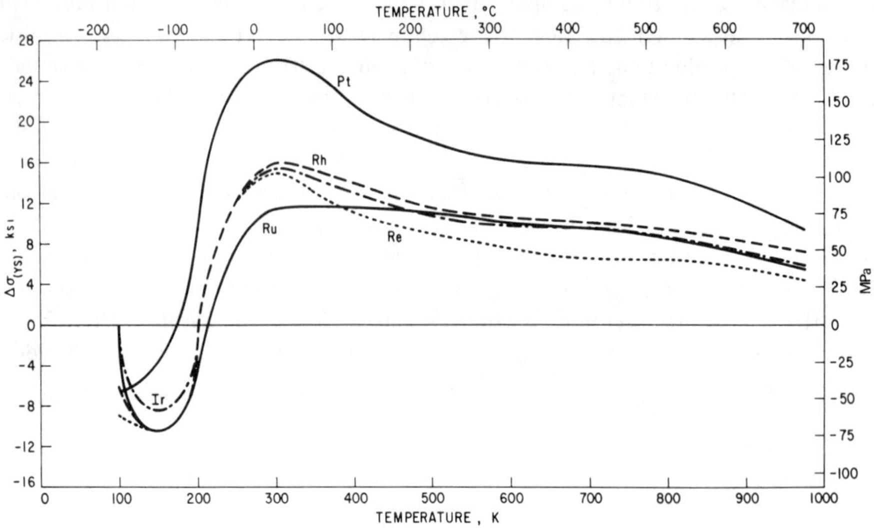

Figure III-16. Solid solution hardening and softening in iron-base alloys (3 at %), $\dot{\epsilon} \cong 2.5 \times 10^{-4}\ s^{-1}$. (W.C. Leslie, *Met. Trans.*, 3:5 (1972).)

tion of one of these few metals to iron increases the cross slip of screw dislocations by reducing their dissociation, or perhaps by spreading of the shear misfit of $a\langle 100\rangle$ edge dislocations, making them less effective as crack nuclei.

III-6. SEGREGATION OF SOLUTES IN IRON

Elements in solid solution in iron are not distributed uniformly on a microscopic scale; they are redistributed thermally between the matrix and exterior surfaces, grain boundaries, and precipitate-matrix interfaces. The tendency for such segregation is a strong function of solid solubility, as indicated in Fig. III-17. The lower the solid solubility, the greater the enrichment of the solute at grain boundaries. Therefore, S, P, B, C, Te, N, Sb, and Sn can be highly concentrated at grain boundaries, in equilibrium with a random solid solution. Such redistribution is reversible with temperature change. We are more interested in grain boundary segregation than in exterior surface segregation because of the greater effect of the former on mechanical properties.

Segregation is a time-dependent process, as shown in Fig. III-18, calculated from an expression developed by McLean[1] for the grain boundary concentration after a specimen has been suddenly cooled from a high to a low temperature:

$$\frac{C_{gbt} - C_{gb0}}{C_{gb\infty} - C_{gb0}} = 1 - \exp\left(\frac{4Dt}{\alpha^2 d^2}\right) \operatorname{erfc}\left(2\sqrt{\frac{Dt}{\alpha d^2}}\right) \qquad \text{(III-4)}$$

[1] D. McLean, *Grain Boundaries*

where C_{gb0} = initial grain boundary concentration
C_{gbt} = grain boundary concentration after time t at the lower temperature
$C_{gb\infty}$ = equilibrium grain boundary concentration after infinite time at the lower temperature
D = solute diffusion coefficient

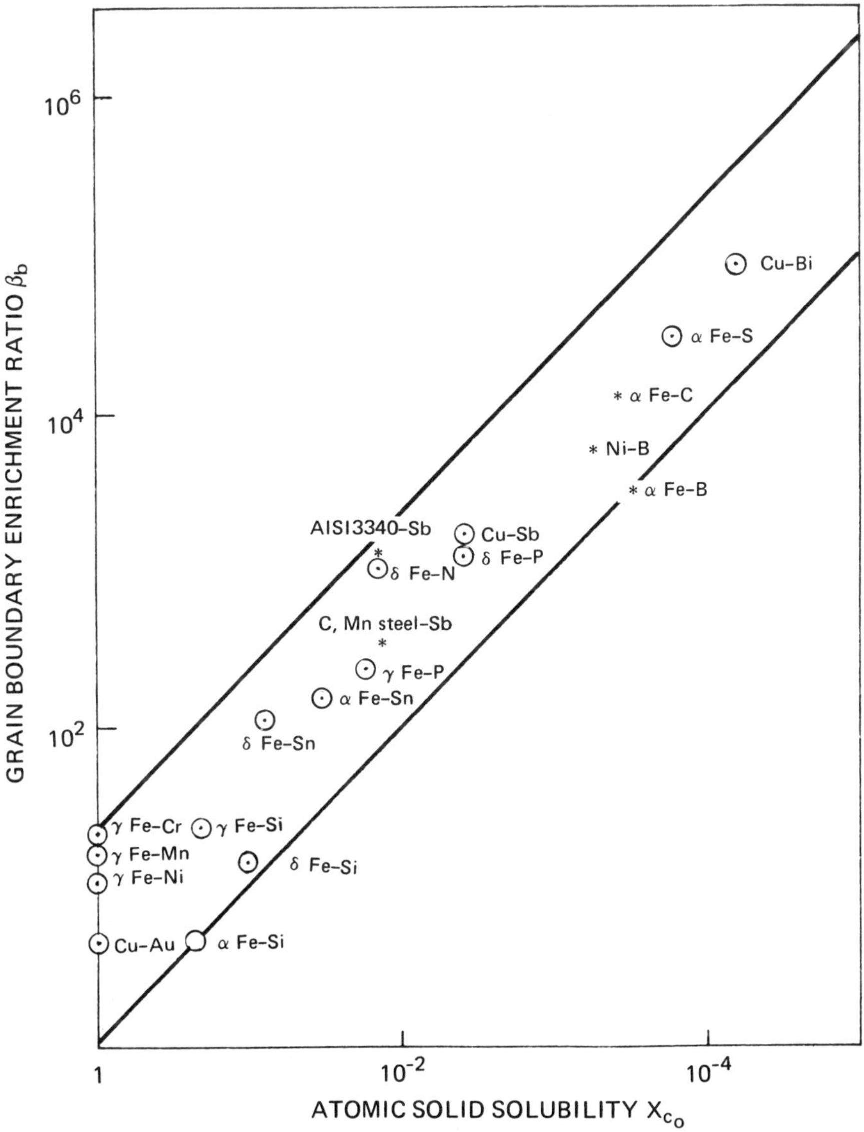

Figure III-17. Correlation of grain-boundary enrichment ratio β_b with the inverse of solid solubility for several systems. (E.D. Hondros and M.P. Seah, *Int. Met. Rev.*, 22:262 (1977), by permission.)

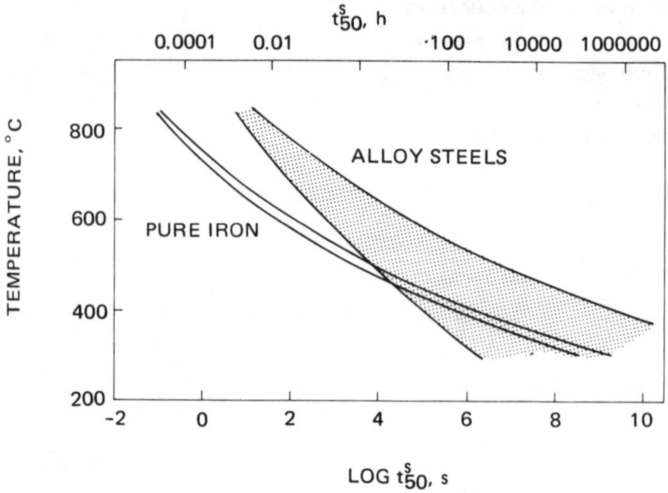

Figure III-18. The time t^s_{50} for segregation of P and Sn in pure iron ($\alpha = 300$) and alloy steels ($\alpha = 3000$), as a function of temperature; α is the ratio of solute concentration in the boundary to that in the matrix at equilibrium. (E.D. Hondros and M.P. Seah, *Int. Met. Rev.*, 22:262 (1977), by permission.)

α = ratio of solute concentration in the boundary to that in the matrix at equilibrium
d = grain boundary thickness

From this expression, the time for 50% completion of segregation is

$$t^s_{50} = 0.591 \frac{\alpha^2 d^2}{4D} \tag{III-5}$$

We will return to the subject of solute segregation when we consider temper embrittlement of steels; in that instance, segregation of more than one element is involved.

Equilibrium segregation at grain boundaries can complicate determination of solid solution strengthening in polycrystalline metals. This is indicated by the variation of k in the Hall–Petch equation,

$$\sigma = \sigma_0 + kd^{-1/2} \tag{III-6}$$

when alloying elements are present in solid solution in iron.[1] Ni and Si increase k, but Cr has no effect (Fig. III-19). Small amounts of C or N can greatly increase k. The data in Fig. III-19 are summarized in Table III-2, with the addition of the effects of P.

The tentative conclusion reached is that solutes segregated to grain boundaries alter the capacity of the boundaries to act as dislocation sources.

Interdendritic segregation carried over from incongruent freezing of alloys is very difficult to eradicate, as demonstrated later. Although such segregation can be reduced by homogenization anneals, it is best to perform such anneals after mechanical working to reduce the segregate spacing.

[1] W.B. Morrison and W.C. Leslie, *Met. Trans.*, 4:379 (1973).

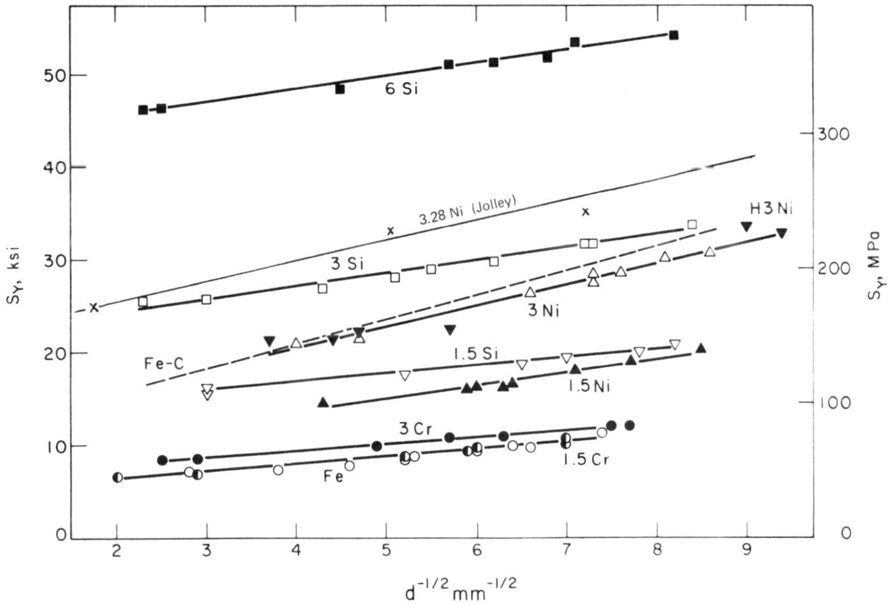

Figure III-19. Relation between yield stress and grain size in binary alloys of iron at room temperature; H3Ni = Wet-H_2 purified 3Ni alloy. (W.B. Morrison and W.C. Leslie, *Met. Trans.*, 4:379 (1973); W. Jolley, *Trans. AIME*, 242:306 (1968), by permission.)

Table III-2 Values of *k* in the Hall–Petch equation, at 295 K

Alloy	k (Mpa/mm$^{-1/2}$)
Fe	5.45 ± 0.83
1.5%Cr	5.45 ± 0.83
3%Cr	4.96 ± 0.62
1.5%Ni	9.65 ± 2.48
3%Ni	15.6 ± 1.56
1.5%Si	5.75 ± 5.72
3%Si	9.45 ± 1.45
6%Si	9.52 ± 1.52
Ferrovac E	
(C & N present)	18.0 ± 1.24
0.09%P[1]	26.9
0.18%P[1]	27.5
1.08%P[1]	3.79

[1] W.A. Spitzig, *Met. Trans.*, 3:1183 (1972).

III-7. IMPURITY EFFECTS ON GRAIN BOUNDARY MIGRATION

It is known that the rate of boundary migration is very sensitively dependent upon the purity of the metal. For example, Aust and Rutter[1] found that the addition of only 0.006 wt % tin to zone-refined lead reduced the rate of boundary migration by four orders of magnitude, but only for so-called "random" boundaries. For grain boundaries at or near a coincidence site relationship, in which atoms spaced at regular intervals along the boundary are held in common by both grains, the effect was much less, only about one order of magnitude. The retardation of grain boundary migration in zone-refined lead was specific for different impurities.[2] Silver and gold in lead reduced the rate of migration about 15 times as much as an equal concentration of tin. Both atom size and valence are important in determining grain boundary–solute interactions.

Similar effects exist in iron, but they are much more difficult to study than in an easily-purified metal such as lead. The effects of manganese and molybdenum on rates of grain growth in polycrystalline, cold-worked iron have been examined.[3,4] The driving force for grain growth was not constant but decreased with time because of recovery. The retardation by solutes was not as great as those demonstrated by Aust and Rutter (Figs. III-20 and III-21) for at least three reasons: the starting material was not as pure, the size differences between solute and solvent atoms was not as great, and last, the valence difference was probably slight. However, this work brought out another factor not obvious from the work of Aust and Rutter, a saturation effect. As more solute is added a concentration is reached at which the retarding effect is reversed (Figs. III-22 and III-23). The concentration at saturation differs for each solute element, some being much more effective than others.

The observation of a "critical" concentration of solute for inhibition of recrystallization of cold-worked iron was made by Abrahamson and Blakeney[5] by very different techniques. Beginning with bars cold-swaged to 75% reduction, they defined "recrystallization temperature" as the temperature at which the first recrystallized grain could be seen at 200 diameters magnification following a 1-h anneal in a furnace with a known temperature gradient. The results for iron are given in Fig. III-24. "Critical" concentrations of Mn (~0.20 at %) and Mo (~0.05 at %) are indicated. These are very close to the "saturation" concentrations, shown in Figs. III-22 and III-23, of 0.30 at % for Mn and 0.04 at % for Mo. Abrahamson and Blakeney[5] show an apparent correlation between the rate of change of recrystallization temperature per atom percentage solute and the number of *d*-shell electrons in the solute atom. This intriguing correlation has not been studied subsequently.

There have been several theories proposed to account for the effects of solute atoms in inhibiting grain boundary migration. These have been summarized by

[1] K.T. Aust and J.W. Rutter, *Trans. AIME*, 215:820 (1959).
[2] J.W. Rutter and K.T. Aust, *Trans. AIME*, 218:682 (1960).
[3] W.C. Leslie, F.J. Plecity, and J.T. Michalak, *Trans. AIME*, 221:691 (1961).
[4] W.C. Leslie, F.J. Plecity, and F.W. Aul, *Trans. AIME*, 221:982 (1961).
[5] E.P. Abrahamson II and B.S. Blakeney, Jr., *Trans. AIME*, 218:1101 (1960).

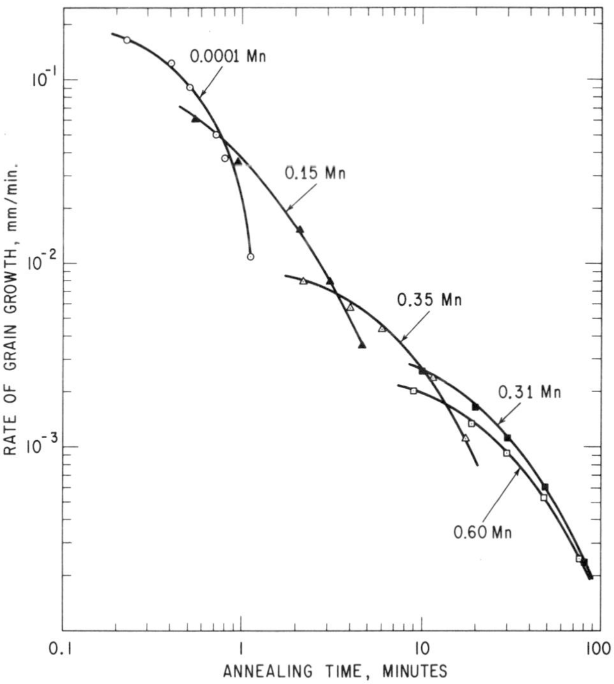

Figure III-20. Rates of grain growth in iron and Fe–Mn alloys at 540°C, width dimension. (W.C. Leslie, F.J. Plecity, and J.T. Michalak, *Trans. AIME,* 221:691 (1961).)

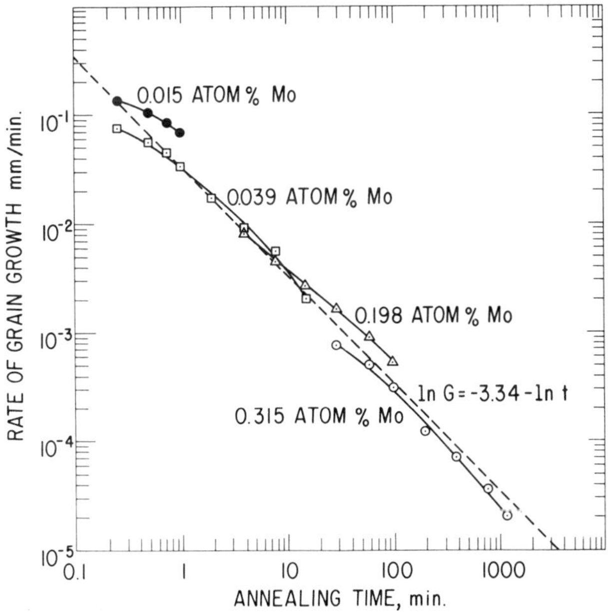

Figure III-21. Rates of grain growth in Fe–Mo alloys at 595°C, width dimension. (W.C. Leslie, F.J. Plecity, and F.W. Aul, *Trans. AIME,* 221:982 (1961).)

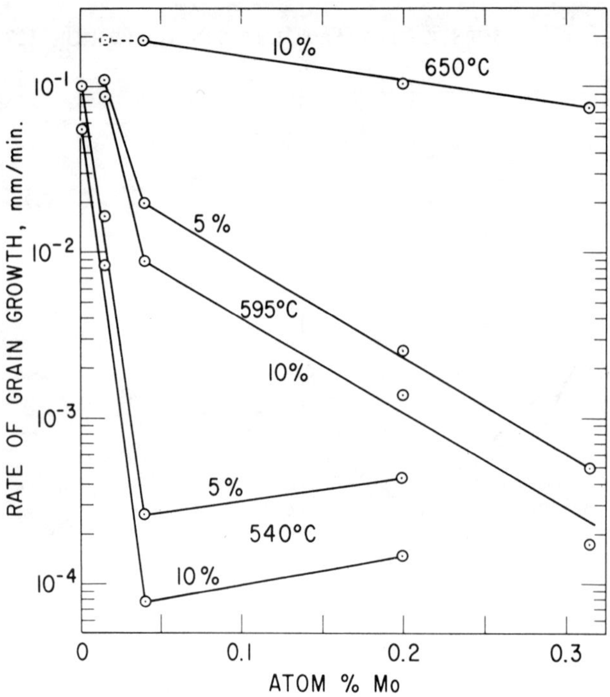

Figure III-22. Effect of manganese on rate of growth of recrystallized grains in iron. (W.C. Leslie, F.J. Plecity, and J.T. Michalak, *Trans. AIME,* 221:691 (1961).)

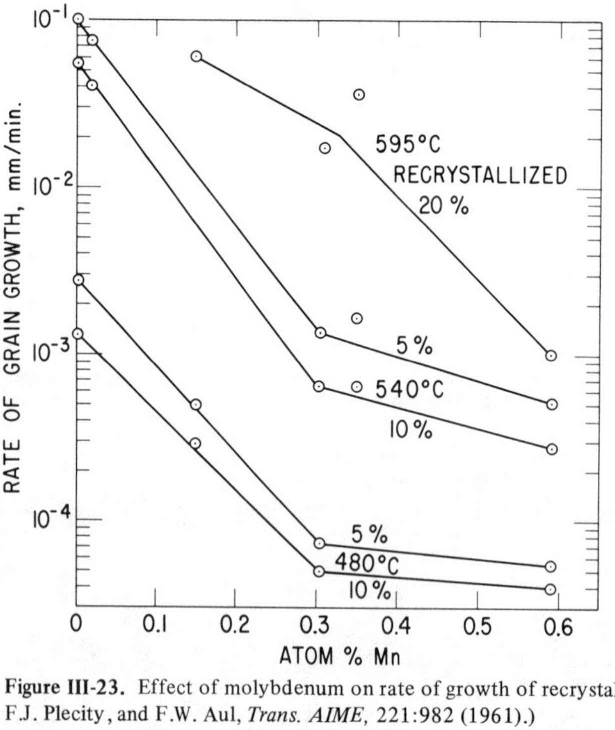

Figure III-23. Effect of molybdenum on rate of growth of recrystallized grains in iron. (W.C. Leslie, F.J. Plecity, and F.W. Aul, *Trans. AIME,* 221:982 (1961).)

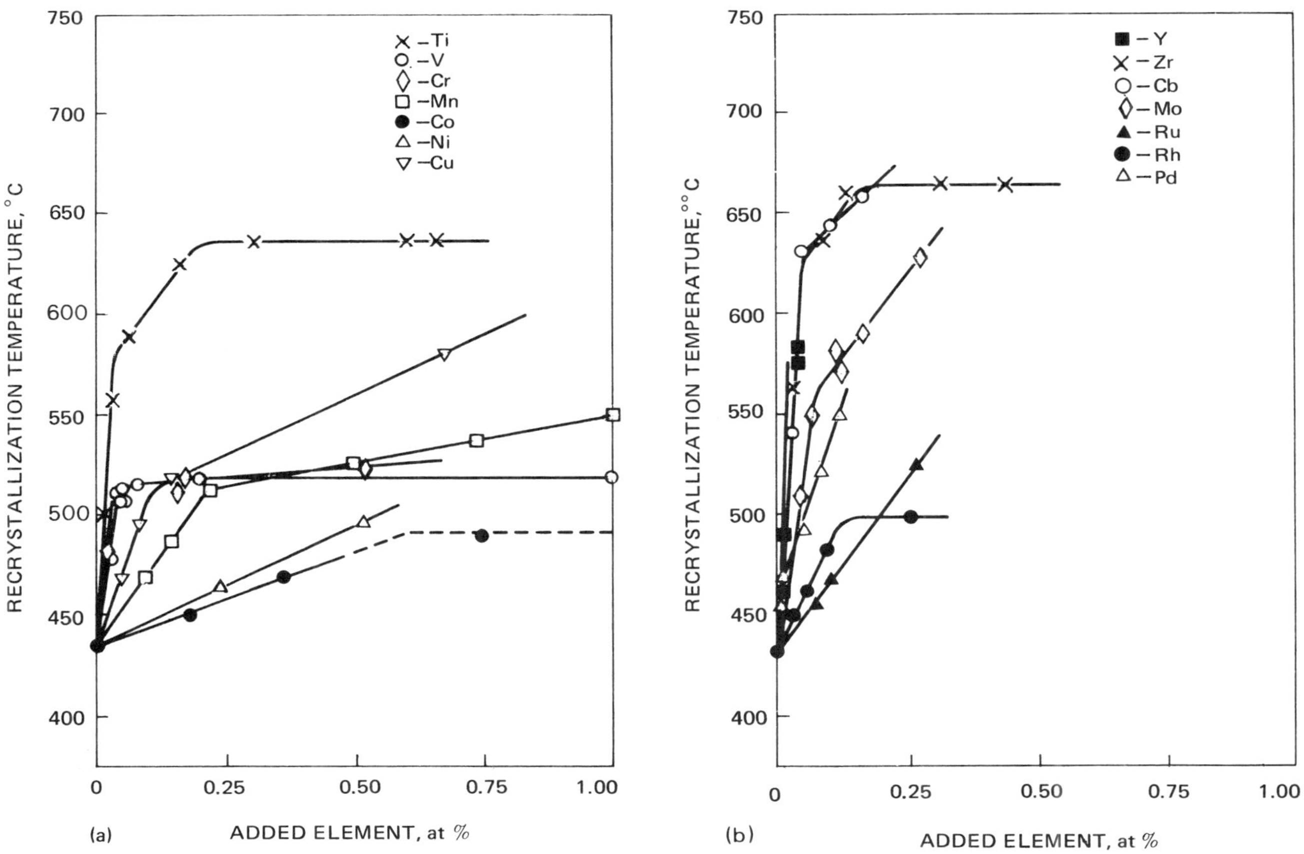

Figure III-24. "Recrystallization temperatures" of iron binary alloys: (*a*) first transition series, (*b*) second transition series. (E.P. Abrahamson II and B.S. Blakeney Jr., *Trans AIME*, 218:1101 (1960), by permission.)

Simpson et al.[1] For a "pure" metal the velocity of a moving grain boundary is proportional to the driving force, as previously discussed:

$$V = M(\Delta G) \tag{III-7}$$

with M depending upon temperature by the relation

$$M = M_0 \exp\left(-\frac{Q}{RT}\right) \tag{III-8}$$

But when solute atoms are present Eq. III-7 no longer holds; there is an interaction between a moving boundary and the solute atoms. There have been several theories dealing with this interaction. The first, by Lücke and Detert,[2] was based on an elastic interaction of solute atoms with grain boundaries. The energy gained by putting a solute atom into a grain boundary was assumed to be due only to the elimination of the stress field surrounding the foreign atom in the lattice. When a boundary moves, the solute atoms associated with it will tend to move with it, so the rate of migration of the boundary will be reduced, at any given temperature, by the necessity of diffusing solute atoms along with the moving boundary. This leads to two distinctive types of boundary migration. At low temperatures and with small driving forces the rate of migration will be determined by the rate at which solute atoms can move with the boundary. They will tend to make their diffusive jumps toward the boundary rather than away from it, because of the attraction between the solute atoms and the boundary. The second type of boundary motion occurs at high temperatures and large driving forces, when the boundary can break away and move independent of the solutes.

This theory has had to be modified because of disagreement with experiment, but all subsequent theories have been based upon it.[3,4,5] In general, a differential equation is formulated to describe the composition profile across a boundary in steady-state motion. The interaction energy and the diffusion coefficient of the solute are both considered to vary with distance from the boundary. Because both of these functions are unknown, solutions for solute distribution are obtained for the limiting cases of high- and low-velocity migration. For high migration rates the drag is inversely proportional to the velocity. For low-velocity migration rates the solute drag is proportional to the velocity. Hillert has made the most recent contribution to this topic.[6] He relates the drag on a moving boundary to work done by this boundary—that is, the loss in Gibbs free energy due to the irreversibility of diffusion of solute atoms.

These theories are hardly amenable to experimental verification, and although

[1] C.J. Simpson, W.C. Winegard, and K.T. Aust, in *Grain Boundary Structure and Properties*, Academic Press, New York, 1976, p. 201.

[2] K. Lücke and K. Detert, *Acta Metall.*, 5:628 (1957).

[3] J.W. Cahn, *Acta Metall.*, 10:789 (1962).

[4] K. Lücke and H.-P. Stüwe, in *Recovery and Recrystallization of Metals*, Interscience, New York, 1963, p. 171.

[5] K. Lücke and H.-P. Stüwe, *Acta Metall.*, 19:1087 (1971).

[6] M. Hillert, *Met. Sci.*, 13:118 (1979).

recent experimental data (not on iron-base systems) are in qualitative agreement with them, we cannot yet state that any particular theory is an accurate representation of reality.

III-8. PRECIPITATION OF SUBSTITUTIONAL SOLUTES FROM ALPHA IRON

In precipitation from binary substitutional solid solutions of iron, most solutes form intermetallic compounds with iron, creating complex situations that are outside the scope of this text. The topic has been reviewed by Hornbogen.[1] In two instances, however, the solute elements Cu and Au do not form intermetallic compounds; their precipitation is well understood and for Cu the process is technologically important. Both Cu and Au precipitate as solute-rich coherent bcc zones in the iron lattice.[2,3] This stage has been followed in detail by a field ion microscope atom probe. The maximum solubility of Cu in alpha iron is 2.1 wt % at 851°C. When an Fe–Cu alloy is quenched from a high temperature in the α field, then aged at a lower temperature (500°C, for example), the copper atoms form spherical clusters in the iron lattice. These clusters can be detected when they are only about 0.8 nm in diameter. At this stage, in a 1.4 wt % Cu alloy, their number density is about 10^{18}/cc. The clusters grow to a diameter of about 2.4 nm, at which point the strength of the alloy is maximized. These clusters are not pure Cu; the earliest clusters contain about 50% iron. As the particles grow, they become richer in copper. At a size of about 5.0 nm, they transform in situ from bcc to fcc, becoming noncoherent. At that point the alloy is overaged with respect to yield strength. Only in the overaged condition do the particles approach the equilibrium composition of the ϵ phase, which is essentially pure copper. The rate of growth is controlled by rate of volume diffusion of copper.

The copper precipitates as spheres because the strain energy of the coherent particle is very small. The atomic radii of Cu and Fe differ by only about 0.3%. This allows an interesting comparison between the forms of the precipitate in Fe–Cu and Fe–Au. In the latter, the difference in atomic radii is about 16%, so the strain energy associated with cluster formation is large. Therefore, the clusters and the subsequent equilibrium ϵ phase (nearly pure gold) form as disks on $\{100\}_\alpha$. The noncoherent gold particles strongly favor dislocation sites for nucleation, whereas the copper particles nucleate on dislocations or in the matrix with equal facility.

The precipitation of Cu from solid solution in ferrite continues to attract interest because of the possibility of controlling texture and strength by this process.[4,5]

[1] E. Hornbogen, in *Precipitation from Iron-Base Alloys,* Gordon and Breach, New York, 1965, p.1.

[2] E. Hornbogen and R.C. Glenn, *Trans. AIME,* 218:1064 (1960).

[3] S.R. Goodman, S.S. Brenner, and J.R. Low Jr., *Met. Trans.,* 4:2363, 2371 (1973).

[4] R.L. Rickett and W.C. Leslie, *Trans. Am. Soc. Met.,* 51:310 (1959).

[5] W.B. Hutchinson and B.J. Dugan, *Met. Sci.,* 12:372 (1978).

CHAPTER

FOUR

INTERACTIONS BETWEEN SOLUTE ATOMS IN IRON

IV-1. COMPOUND FORMATION

Interactions between solute atoms in iron take several forms: formation of second phases composed of interstitial and substitutional solutes, clustering in solid solution between interstitial and substitutional atoms, clustering in solid solution between substitutional atoms of different species, and precipitation of second phases composed of two or more substitutional solutes. The first of these is the best known.

When two solutes form a compound of very limited solubility in iron, it is customary to report this solubility, as a function of temperature, as the solubility product. This quantity is identical with an equilibrium constant for the reaction of dissolving the compound in a solid matrix when the activity of the compound is unity and the activity coefficients of the solute species are independent of the solute concentrations. For example, when aluminum nitride dissolves in austenite to form a solid solution of Al and N,[1]

$$AlN_{(solid)} \rightleftarrows \underline{Al} + \underline{N}$$

the solubility product can be represented by

$$K_s = [\%Al][\%N]$$

The solubilities in austenite of several frequently encountered compounds are listed in Table IV-1. With respect to MnS, it is necessary to include f_s^{Mn}, the activity coefficient of sulfur in austenite. This activity coefficient varies with the concentration of manganese. The activity of MnS can be considered to be unity (as in the equation shown) only when the manganese content of steel exceeds 0.3%. Figure IV-1 was constructed by plotting points from the equation for Nb[C, N].

[1] Dissolved solids are indicated by underlining.

Table IV-1 Temperature dependent solubility of several compounds in austenite

Compound	Solubility limit
AlN[1]	$\log [\%Al][\%N] = (-7400/T) + 1.95$
MnS[2]	$K_S = [\%Mn][\%S] f_S^{Mn}$
	$\log K_S = (-9020/T) + 2.929$
	$\log f_S^{Mn} = [(-215/T) + 0.097][\%Mn]$ in γ)
NbC[3]	$\log [\%Nb][\%C]^{0.87} = (-7530/T) + 3.11$
NbN[4]	$\log [\%Nb][\%N] = (-10{,}230/T) + 4.04$
Nb[C,N][5]	$\log [\%Nb][\%C]^{0.24}[\%N]^{0.65} = (-10{,}400/T) + 4.09$
V_4C_3[4]	$\log [\%V]^{4/3}[\%C] = (-10{,}800/T) + 7.06$
VN	$\log [\%V][\%N] = (-7{,}733/T) + 2.99$
TiC[6]	$\log [\%Ti][\%C] = (-7000/T) + 2.75$

[1] L.S. Darken, R.P. Smith, and E.W. Filer, *Trans. AIME,* 191:1174 (1951).

[2] E.T. Turkdogan, S. Ignatowicz, and J. Pearson, *J. Iron Steel Inst.,* 180: 349 (1955).

[3] H. Nordberg and B. Aronsson, *J. Iron Steel Inst.,* 206:1263 (1968).

[4] H. Chino and K. Wada, *Yawata Tech. Report 251,* June 1965, p.75.

[5] T. Mori, M. Tokizane, Y. Nakazima, and T. Saheki, *Tetsu to Hagane,* 54:3, no. 7, (1968).

[6] K.J. Irvine, F.B. Pickering, and T. Gladman, *J. Iron Steel Inst.,* 205:161 (1967).

The solution and precipitation of these compounds in commercial steels produce many important effects. Aluminum nitride particles inhibit austenite grain growth in Al-killed steels. Precipitation of this compound, or preprecipitation clustering of Al and N during annealing after cold work, determines the texture, microstructure, and drawability of Al-killed low-carbon sheet steels. Solution of MnS during the reheating

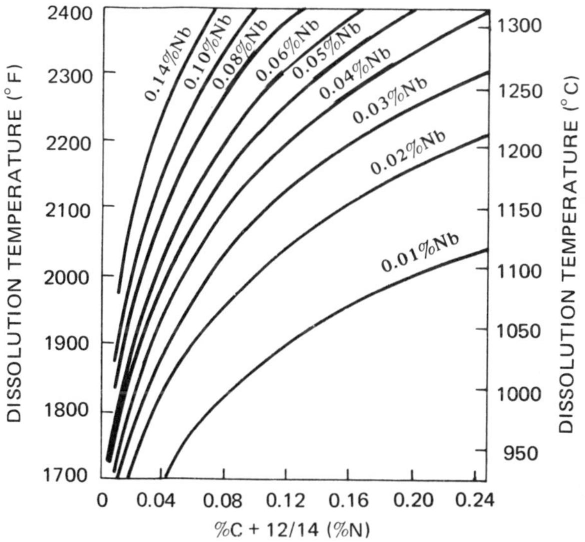

Figure IV-1. The temperature dependence of the solubility of niobium carbonitride in austenite. (M. Lafrance, C.A. Caron, G.R. Lamant, and J. Leclerc, in *Microalloying 75,* Union Carbide Corp., New York, 1975, by permission.)

of austenite—in preparation for forging, for example—and subsequent precipitation during cooling produces the phenomenon called overheating. Precipitation or pre-precipitation stages of MnS during annealing of cold-rolled, low-Mn steels for porcelain enameling can prevent recrystallization at all temperatures below that required for transformation to austenite. Solution and precipitation of the carbides and nitrides (or carbonitrides) of Nb, V, or Ti in austenite during hot-working can inhibit recrystallization of the austenite, minimize grain growth of austenite, and add a component of precipitation hardening to the strength of the as-rolled steel at room temperature. These processes have made possible the development of a variety of "micro-alloyed" high-strength, low-alloy steels, as discussed later.

The determination of the temperature dependence of the solubility of sparingly soluble compounds in austenite is a difficult task, subject to error, so the data in Table IV-1 may be revised in the future. The compounds listed are all less soluble in ferrite than in austenite and these differences can lead to nucleation of precipitate particles on the α–γ interface when a steel transforms from austenite to ferrite during cooling.

IV-2. INTERACTIONS BETWEEN SUBSTITUTIONAL AND INTERSTITIAL SOLUTE ATOMS IN FERRITE

This discussion is *not* concerned with formation of precipitate particles of a second phase, but with interactions between substitutional atoms, interstitial atoms, and dislocations in ferrite. In commercial steels both substitutional and interstitial solutes are present. The substitutional atoms are most commonly Mn, Cr, Si, and Mo, although V, Ti, and Al can produce similar effects. These elements hold C and N atoms in close association, as indicated by internal friction.[1,2,3] Figure IV-2 shows interstitial positions around a Mn atom, calculated on the basis of elastic strains in an isotropic medium, which is probably an oversimplification. The resulting substitutional–interstitial dipoles in turn can interact strongly with dislocations over a greatly expanded temperature range as compared with interstitial atoms acting alone. This is well illustrated by Fig. IV-3, which shows the flow stress of iron increasing with increasing test temperature when Mo and N atoms are present together in solid solution. Baird and Jamieson[4] call this "interaction solid solution strengthening." Similar but smaller effects are obtained with Mn or Cr in combination with N or C. These carbide- and nitride-forming elements interact with dislocations via their association with interstitial atoms.

These substitutional–interstitial associations in ferrite can be surprisingly stable. For example, Cuddy et al.[5] studied the hardening of Fe–Ti alloys in the temperature

[1] L.J. Dijkstra and R.J. Sladek, *Trans. AIME*, 197:69 (1953).

[2] D.T. Hasson and R.J. Arsenault, in, *Treatise on Materials Science and Technology*, Academic Press, New York, 1972, p. 179.

[3] J. Gouzou, J. Wégria, and L. Habraken, *Metall. Rep. CRM*, 33:65 (Dec. 1972).

[4] J.D. Baird and A. Jamieson, *J. Iron Steel Inst.*, 210:841 (1972).

[5] L.J. Cuddy, H.E. Knechtel, and W.C. Leslie, *Met. Trans.*, 5:1999 (1974).

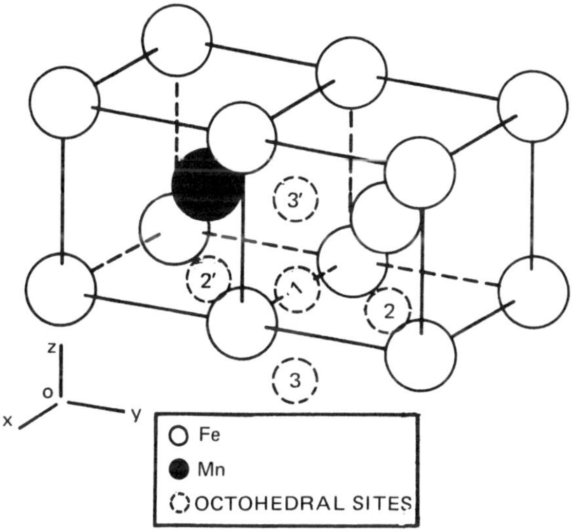

Figure IV-2. Nonequivalent interstitial positions about a manganese atom in alpha iron, assuming elastic strains in an isotropic continuum. Site 1—most favored. Sites 2 and 3—equivalent and less favored. Sites 2′ and 3′—least favored. (J. Gouzou, J. Wegria, and L. Habraken, *Met. Rep. CRM,* 3:65 (Dec. 1972), by permission.)

range from 23 to 600°C, with the results shown in Fig. IV-4. The ordinate in this figure is the increase of hardness over that of a hydrogen-purified, zone-melted iron at the same temperature. A hardening peak occurs at about 500°C. This peak, in an alloy containing 0.15% Ti and 0.015% C, remained at 500°C through three successive temperature cycles from 23 to 900°C, 900 to 23°C, and 23 to 900°C. This sort of hardness peak cannot be due to dispersion hardening by Ti[C, N] particles, nor to solid solution hardening by Ti. It can only result from diffusional processes. The strengthening comes from dynamic interaction between dislocations and clusters of C or N atoms

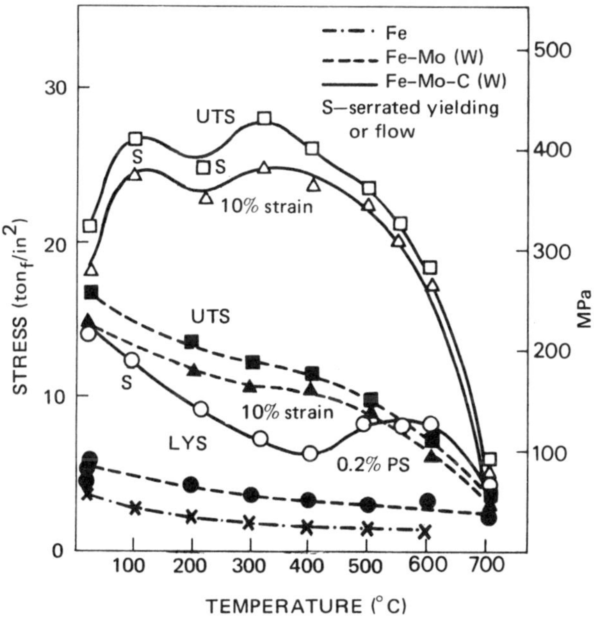

Figure IV-3. Tensile properties of Fe–1%Mo and Fe–1%Mo–0.01%C alloys; also shown is the 0.2% offset yield stress of high-purity iron. (J.D. Baird and A. Jamieson, *J. Iron Steel Inst.,* 210:841 (1972), by permission.)

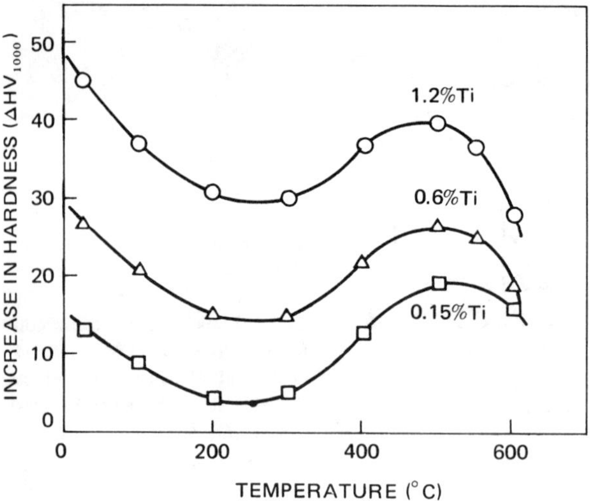

Figure IV-4. Hardening of iron between 23 and 600°C by titanium additions. Compositions (wt %): 1.2 Ti, 0.027 C, 0.003 N; 0.6 Ti, 0.011 C, 0.003 N; 0.15 Ti, 0.005 C, 0.002 N. (L.J. Cuddy, H.E. Knechtel, and W.C. Leslie, *Met. Trans.*, 5:1999 (1974).)

about Ti atoms. These clusters have been observed in an Fe–0.15%Ti alloy by a field ion microscope atom probe.[1]

If a noncarbide- or nonnitride-forming element (Ni, for example) is present in addition to C or N, the presence of the interstitial solute eliminates high-temperature jerky flow (serrations) caused by the substitutional atoms. The introduction of an interstitial atom into the strain field of a dislocation prevents interaction of the substitutional solute atom with that strain field.

The formation of substitutional–interstitial clusters causes profound changes in the creep resistance of iron alloys,[2,3] as indicated in Fig. IV-5. The addition of 0.15% Ti to an Fe-3%Ni alloy containing 0.004% C and 0.004% N reduced the steady-state creep rate by a factor of 100, in the range of stresses from 27 to 150 MPa. The strengthening remained after 200 h at 500°C. Much of the creep resistance of Cr–Mo steels commonly used at high temperatures is due to the formation of substitutional–interstitial clusters.[3]

IV-3. INTERACTIONS BETWEEN UNLIKE SUBSTITUTIONAL ATOMS IN FERRITE

If two species of substitutional atoms are associated in solid solution in ferrite, such association, by creating anisotropic distortion of the lattice, can increase the inter-

[1] H.W. Pickering, Y. Kuk, and T. Sakurai, *Appl. Phys. Lett.*, 36:902 (1980).
[2] L.J. Cuddy, H.E. Knechtel, and W.C. Leslie, *Met. Trans.*, 5:1999 (1974).
[3] J.D. Baird and A. Jamieson, *J. Iron Steel Inst.*, 210:847 (1972).

action energy between substitutional atoms and dislocations.[1] Such interactions have been observed between titanium and several other elements with which titanium forms stable intermetallic compounds in binary alloys. Interactions of this type do not appear to be significant in the common grades of steel, but they may occur in some more highly alloyed types, such as the maraging steels.

IV-4. EFFECTS OF SUBSTITUTIONAL SOLUTES ON PRECIPITATION OF IRON CARBIDES

The effects of substitutional solutes on precipitation of iron carbides from solution in ferrite or martensite are of considerable interest in the control of tempering of martensite and in the magnetic aging of silicon steels. Such effects have been studied for three alloying elements; Mn, Si, and Al.[2] An alloying element having a higher solubility in the carbide than in ferrite is not likely to have a large effect on the rate of growth of the carbide, although it may control the structure of the carbide precipitated, the nucleation sites, and the rate of nucleation. If the alloying element is

[1] W.C. Leslie, L.J. Cuddy, and R.J. Sober, *Proc. 3rd Intern. Conf. Strength of Metals and Alloys,* 1:11 (1973).

[2] W.C. Leslie and G.C. Rauch, *Met. Trans. A,* 9A:343 (1978).

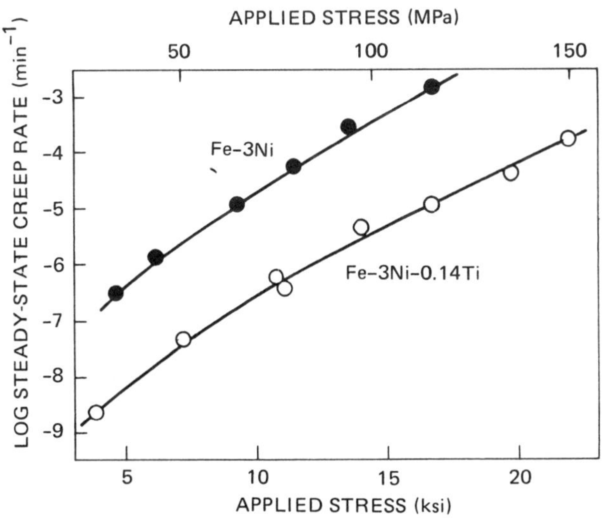

Figure IV-5. The effect of stress on steady-state creep rates of two alloys at 500°C. Compositions (wt %): 3 Ni, 0.005 C, 0.001 N; 3 Ni, 0.14 Ti, 0.004 C, 0.004 N. (L.J. Cuddy, H.E. Knechtel, and W.C. Leslie, *Met. Trans.,* 5:1999 (1974).)

nearly insoluble in the carbide, it can have a very large effect on the rate of growth of precipitated carbides.

Of the three elements studied, Mn distributes preferentially to cementite and lowers the temperature at which Fe_3C forms in preference to ϵ carbide. Al and Si have negligible solubility in cementite and retard cementite precipitation. Also, Si has a pronounced effect in decreasing the rate of growth of ϵ carbide particles (Fig. IV-6). Owen[1] explained this as being caused by a necessity for Si to diffuse away from a growing carbide particle. As Si is rejected from the carbide, its concentration in the surrounding matrix rises. Because Si increases the activity of C in ferrite, the gradient in C activity causing the C atoms to diffuse toward the growing carbide particle is decreased by the accumulation of Si, so growth of the carbide stops.

This mechanism, however, does not explain the retarding effect of Al on the rate of carbide growth. Aluminum functions by decreasing the activity of C in ferrite and increasing it in cementite, both effects acting to retard cementite precipitation. However, growth of the carbides continues; thus Al is not as effective as Si in inhibiting magnetic aging. The effects of these three elements are summarized in Table IV-2. In this table, "carbide transition temperature" refers to the temperature at which ϵ carbide is replaced by cementite. The conclusion reached is that the effect of substitutional solutes on cementite stability is the predominant influence on carbide precipitation during quench-aging of ferrite and the tempering of martensite.

[1] W.S. Owen, *Trans. Am. Soc. Met.*, 46:812 (1954).

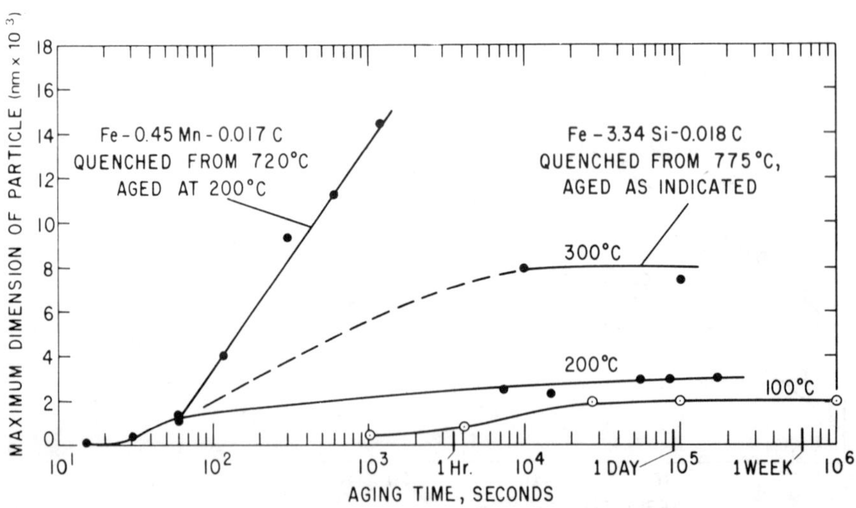

Figure IV-6. Effect of silicon on rate of growth of carbide particles during the quench aging of ferrite. (A.S. Keh and W.C. Leslie, in *Materials Science Research*, vol. 1, Plenum Publishing Corp., N.Y., 1963, p. 208.)

Table IV-2 Influence of substitutional solutes on carbon activity and carbide precipitation

Solute	Mn	Si	Al
Activity coefficient of carbon in ferrite	Deceased	Increased	Decreased
Carbon activity in cementite	Decreased	Increased	Increased
Carbon activity in transition carbide	Unknown	Unknown	Unknown
Carbide transition temperature	Decreased	Increased	Increased
Rate of carbide growth	Unchanged	Decreased	Decreased

CHAPTER

FIVE

CARBON STEELS

V-1. LOW-CARBON SHEET STEELS

The American Iron and Steel Institute[1] defines carbon steel as follows:

> Steel is considered to be carbon steel when no minimum content is specified or required for Al, Cr, Co, Cb, Mo, Ni, Ti, W, V, or Zr, or any other element added to obtain a desired alloying effect; when the specified minimum for Cu does not exceed 0.40 percent; or when the maximum content specified for any of the following elements does not exceed the percentages noted: Mn 1.65, Si 0.60, Cu 0.60.
>
> In all carbon steels small quantities of certain residual elements, such as Cu, Ni, Mo, Cr, etc., are unavoidably retained from raw materials. These elements are considered as incidental.

That definition, although it smacks more of the marketplace than of technology, is certainly precise and it will serve our purpose.

By far the largest fraction of carbon steels goes into flat-rolled products of low carbon content. The remainder is made as bars, rods, wire, plates, structural shapes, rails, and railroad wheels. No sharp division can be made between these two groups of products on the basis of physical metallurgy. They are commonly used without heat treatment subsequent to forming, but of course carbon steels can be heat treated by the user. The compositions of the standard carbon steels currently produced in the United States are listed in Table V-1. The column headed "UNS No." gives the Unified Numbering System designation of the steels. This system employs 15 series of numbers for metals and alloys, each series consisting of a letter followed by five digits. It is a system for identifying *commercial* alloys for which chemical composition limits have been set. For additional information on UNS the reader should refer to ASTM E527.

[1] *AISI Steel Products Manual,* Section 1, May 1972.

Table V-I AISI-SAE standard carbon steels[1]

AISI-SAE No.	UNS No.	Composition (%)[2] C	Mn	P (max)	S (max)
Nonresulfurized grades					
Manganese 1.00% maximum					
1005[3]	G10050	0.06 max	0.35 max	0.040	0.050
1006[3]	G10060	0.08 max	0.25–0.40	0.040	0.050
1008	G10080	0.10 max	0.30–0.50	0.040	0.050
1010	G10100	0.08–0.13	0.30–0.60	0.040	0.050
1012	G10120	0.10–0.15	0.30–0.60	0.040	0.050
1015	G10150	0.13–0.18	0.30–0.60	0.040	0.050
1016	G10160	0.13–0.18	0.60–0.90	0.040	0.050
1017	G10170	0.15–0.20	0.30–0.60	0.040	0.050
1018	G10180	0.15–0.20	0.60–0.90	0.040	0.050
1019	G10190	0.15–0.20	0.70–1.00	0.040	0.050
1020	G10200	0.18–0.23	0.30–0.60	0.040	0.050
1021	G10210	0.18–0.23	0.60–0.90	0.040	0.050
1022	G10220	0.18–0.23	0.70–1.00	0.040	0.050
1023	G10230	0.20–0.25	0.30–0.60	0.040	0.050
1025	G10250	0.22–0.28	0.30–0.60	0.040	0.050
1026	G10260	0.22–0.28	0.60–0.90	0.040	0.050
1029	G10290	0.25–0.31	0.60–0.90	0.040	0.050
1030	G10300	0.28–0.34	0.60–0.90	0.040	0.050
1035	G10350	0.32–0.38	0.60–0.90	0.040	0.050
1037	G10370	0.32–0.38	0.70–1.00	0.040	0.050
1038	G10380	0.35–0.42	0.60–0.90	0.040	0.050
1039	G10390	0.37–0.44	0.70–1.00	0.040	0.050
1040	G10400	0.37–0.44	0.60–0.90	0.040	0.050
1042	G10420	0.40–0.47	0.60–0.90	0.040	0.050
1043	G10430	0.40–0.47	0.70–1.00	0.040	0.050
1044	G10440	0.43–0.50	0.30–0.60	0.040	0.050
1045	G10450	0.43–0.50	0.60–0.90	0.040	0.050
1046	G10460	0.43–0.50	0.70–1.00	0.040	0.050

[1] Sources: *Steel Products Manual,* American Iron and Steel Institute, August 1977 (revised January 1979) and *ASM Metal Progress Databook* June 1979, Vol. 116, p.17.

[2] The following notes refer to boron, copper, lead, and silicon additions:

Boron: Standard killed carbon steels, which are generally fine grain, may be produced with a boron treatment addition to improve hardenability. Such steels can be expected to contain 0.0005–0.003%B. These steels are identified by inserting the letter "B" between the second and third numerals of the AISI or SAE number, e.g., 10B46.

Copper: When copper is required, 0.20% minimum is generally specified.

Lead: Standard carbon steels can be produced with a lead range of 0.15–0.35% to improve machinability. Such steels are identified by inserting the letter "L" between the second and third numerals of the AISI or SAE number, e.g., 12L15 and 10L45.

Silicon: It is not common practice to produce the 12XX series of resulfurized and rephosphorized steels to specified limits for silicon because of its adverse effect on machinability. When silicon ranges or limits are required for resulfurized or nonresulfurized steels, however, these values apply: a range of 0.08% Si for Si max up to 0.15% inclusive, a range of 0.10%Si for Si max over 0.15 to 0.20% inclusive, a range of 0.15%Si for Si max over 0.20 to 0.30% inclusive, and a range of 0.20%Si for Si max over 0.30 to 0.60% inclusive. Example: Si max is 0.25%, range is 0.10–0.25%.

[3] Standard grades for wire rods and wire only.

Table V-I AISI–SAE standard carbon steels *(Continued)*

AISI–SAE No.	UNS No.	Composition (%) C	Mn	P	S
1049	G10490	0.46–0.53	0.60–0.90	0.040	0.050
1050	G10500	0.48–0.55	0.60–0.90	0.040	0.050
1053	G10530	0.48–0.55	0.70–1.00	0.040	0.050
1055	G10550	0.50–0.60	0.60–0.90	0.040	0.050
1059[3]	G10590	0.55–0.65	0.50–0.80	0.040	0.050
1060	G10600	0.55–0.65	0.60–0.90	0.040	0.050
1070	G10700	0.65–0.75	0.60–0.90	0.040	0.050
1078	G10780	0.72–0.85	0.30–0.60	0.040	0.050
1080	G10800	0.75–0.88	0.60–0.90	0.040	0.050
1084	G10840	0.80–0.93	0.60–0.90	0.040	0.050
1086[3]	G10860	0.80–0.93	0.30–0.50	0.040	0.050
1090	G10900	0.85–0.98	0.60–0.90	0.040	0.050
1095	G10950	0.90–1.03	0.30–0.50	0.040	0.050
Manganese maximum over 1.00%					
1513	G15130	0.10–0.16	1.10–1.40	0.040	0.050
1522	G15220	0.18–0.24	1.10–1.40	0.040	0.050
1524	G15240	0.19–0.25	1.35–1.65	0.040	0.050
1526	G15260	0.22–0.29	1.10–1.40	0.040	0.050
1527	G15270	0.22–0.29	1.20–1.50	0.040	0.050
1541	G15410	0.36–0.44	1.35–1.65	0.040	0.050
1548	G15480	0.44–0.52	1.10–1.40	0.040	0.050
1551	G15510	0.45–0.56	0.86–1.15	0.040	0.050
1552	G15520	0.47–0.55	1.20–1.50	0.040	0.050
1561	G15610	0.55–0.65	0.75–1.05	0.040	0.050
1566	G15660	0.60–0.71	0.85–1.15	0.040	0.050
Free-machining grades					
Resulfurized					
1110	G11100	0.08–0.13	0.30–0.60	0.040 max	0.08–0.13
1117	G11170	0.14–0.20	1.00–1.30	0.040 max	0.08–0.13
1118	G11180	0.14–0.20	1.30–1.60	0.040 max	0.08–0.13
1137	G11370	0.32–0.39	1.35–1.65	0.040 max	0.08–0.13
1139	G11390	0.35–0.43	1.35–1.65	0.040 max	0.13–0.20
1140	G11400	0.37–0.44	0.70–1.00	0.040 max	0.08–0.13
1141	G11410	0.37–0.45	1.35–1.65	0.040 max	0.08–0.13
1144	G11440	0.40–0.48	1.35–1.65	0.040 max	0.24–0.33
1146	G11460	0.42–0.49	0.70–1.00	0.040 max	0.08–0.13
1151	G11510	0.48–0.55	0.70–1.00	0.040 max	0.08–0.13
Resulfurized and rephosphorized					
1211	G12110	0.13 max	0.60–0.90	0.07–0.12	0.10–0.15
1212	G12120	0.13 max	0.70–1.00	0.07–0.12	0.16–0.23
1213	G12130	0.13 max	0.70–1.00	0.07–0.12	0.24–0.33
1215	G12150	0.09 max	0.75–1.05	0.04–0.09	0.26–0.35
12L14[4]	G12144	0.15 max	0.85–1.15	0.04–0.09	0.26–0.35

[4] 0.15–0.35%Pb.

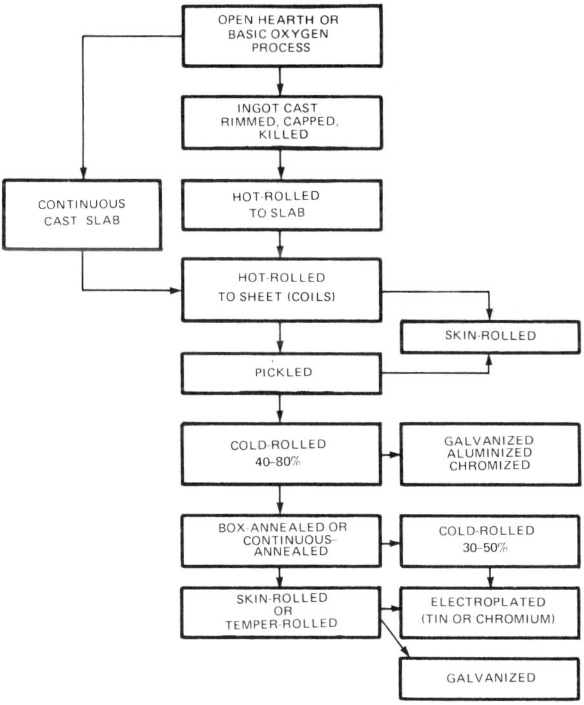

Figure V-1. Simplified flow-chart of the production of low-carbon sheet steel. (D.J. Blickwede, *Trans. Am. Soc. Met.*, 61:653 (1968), by permission.)

The physical metallurgy of flat-rolled sheet products has been described by Rickett and Kristufek[1] and by Blickwede.[2] The processing of such products is summarized in Fig. V-1.

Types of Steel, Based on Deoxidation Practice

Carbon steels, when cast into ingots, can be classified into four types by the deoxidation practice employed, or alternatively, by the amount of gas evolved during solidification. These types are rimmed, capped, semikilled, and killed. In the making of a rimmed steel no deoxidizing agents are added in the furnace. Ferromanganese can be added to the ladle and small additions of aluminum to the ingot molds to control the rimming action. Capping is a variation of rimmed steel practice. After the ingot has rimmed for a short period a cast-iron cap is placed on the top of the mold to freeze the top of the ingot and stop the rimming action. This kind of ingot has a thin zone of relatively pure iron and less segregation in the core than has a rimmed ingot.

Semikilling is often used for carbon steels with carbon contents in the range 0.15 to 0.30%. Ferrosilicon and some aluminum are added to the ladle, but less than required fully to kill the steel.

[1] R.L. Rickett and F.C. Kristufek, *Trans. Am. Soc. Met.*, 41:1113 (1949).
[2] D.J. Blickwede, *Trans. Am. Soc. Met.*, 61:653 (1968).

A killed steel is one that has been sufficiently deoxidized to prevent any gas evolution during solidification. It lies quietly in the mold, and when it solidifies a shrinkage cavity (pipe) is left at the top of the ingot. Killed steels are nearly always cast in big-end-up molds with hot tops, so that the shrinkage cavity is contained within the hot top which is later cut off and discarded. All steels containing more than about 0.30% carbon are killed. Ingot structures typical of these four types of steel are illustrated in Fig. V-2.

Because of difficulties in producing either rimmed or Al-killed steels, most continuously cast steel has been semikilled or Si-killed, but this situation may change as experience is gained in the techniques of making Al-killed steels by continuous casting.

To an increasing extent, liquid steels are being deoxidized by vacuum before casting. This process occurs by the reaction:

$$\underline{C} + \underline{O} \rightarrow CO \uparrow$$

It has the advantages of eliminating solid deoxidation products, thus giving a "cleaner" steel, and of reducing the hydrogen content. As a result of the latter effect the internal cracks (flaking) that can occur in heavy sections during cooling after rolling or forging can be eliminated.

All steels cast into ingot molds suffer to a greater or lesser extent from segregation of solute elements. The current trend in steel production is toward larger ingots, which increases the degree of segregation because of slower rates of solidification. The nature of solidification is such that the last steel to freeze in an ingot contains the highest concentration of solutes. Segregation can be positive (the concentration of an element is greater than the average) or negative (concentration less than the average).

Any disturbance of the liquid steel during freezing, such as by gas evolution, increases the tendency towards segregation; therefore, segregation increases in the order killed steel, semikilled steel, rimmed or capped steel. In a rimmed steel ingot

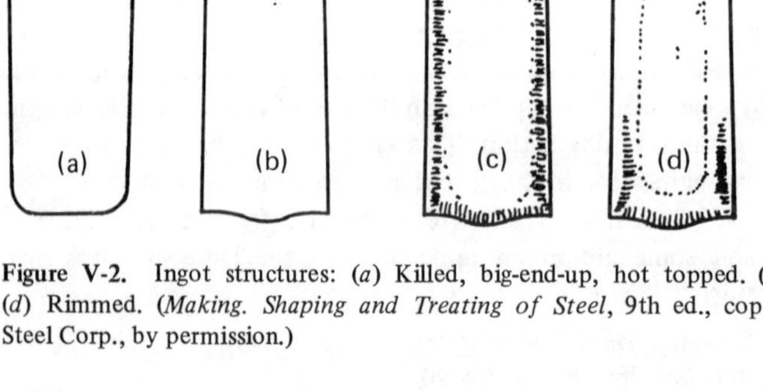

Figure V-2. Ingot structures: (*a*) Killed, big-end-up, hot topped. (*b*) Semikilled. (*c*) Capped. (*d*) Rimmed. (*Making. Shaping and Treating of Steel*, 9th ed., copyright 1971, United States Steel Corp., by permission.)

there is very high negative segregation in the first metal to freeze at the ingot wall. The rim zone is very low in carbon. Segregation also occurs along the center line of the ingot, from bottom to top. An example is cited by Hayes.[1] In rods of AISI 1038 steel the carbon content *at the center of the rod* can vary ±0.15% from the average, that is, the carbon content at the center of a rod rolled from the bottom of the ingot can be 0.23% whereas in a rod rolled from the top of the ingot the carbon content can be 0.53%.

The single most widely used steel is low-carbon rimmed sheet, containing less than 0.10%C and about 0.30–0.40%Mn. The Mn content adds some solid solution strengthening, but this effect is incidental to its principal roles as a deoxidizer and as a getter for sulfur. The Mn content is determined by the minimum necessary to avoid hot shortness—that is, to allow hot rolling without cracking. Hot shortness is caused by the formation in the steel of FeS, which has a low melting point. Rimmed steels are used because the rimming action in the ingot mold—the formation of carbon oxide bubbles in the liquid through interaction of carbon and oxygen—leaves a skin of very low carbon, relatively clean metal at the ingot surface and the closed bubbles compensate for shrinkage during ingot solidification, leading to a high yield of product from each ingot. The continuous casting process has not entirely displaced ingot casting because rimmed steels cannot be produced consistently by continuous casting.

Low-carbon sheet and strip are sold as hot-rolled or after cold-rolling and annealing. The hot-rolled product is the least expensive, but because of the difficulty of controlling thickness accurately during hot-rolling, it is not produced in thicknesses less than 1.8 mm. It cannot have the smooth surface typical of cold-rolled and annealed sheet. This hot-rolled sheet is used extensively for forming chassis members for automobiles and trucks, where appearance is secondary to cost and formability.

The microstructure of low-carbon sheet and strip is composed of ferrite and carbide, the latter being in the form of pearlite or individual carbide particles, as determined by composition and thermal history.[2]

Factors Affecting the Yield and Tensile Strengths of Carbon Steels

The stress in a uniaxial tension test at which macroscopic plastic flow begins in annealed carbon steels containing less than about 0.30% carbon depends primarily upon the ferrite grain size and upon the content of Mn, Si, and N in solid solution. An empirical equation developed by Pickering[3] gives the relative importance of each of these factors:

$$\text{Yield stress (MPa)} = 15.4[3.5 + 2.1(\%\text{Mn}) + 5.4(\%\text{Si}) + 23(\sqrt{\%N_f}) + 1.13d^{-1/2}] \qquad \text{(V-1)}$$

[1] D. Hayes, in *Hardenability Concepts with Applications to Steel*, AIME, Warrendale, Pennsylvania, 1978, p. 617.

[2] R.L. Rickett and F.C. Kristufek *Trans. Am. Soc. Met.*, 41:1113 (1949).

[3] F.B. Pickering, in *Towards Improved Toughness and Ductility*, Climax Molybdenum Co., Greenwich, Connecticut, 1971, p. 9.

Here $\%N_f$ is the weight percent nitrogen in solid solution in ferrite and d is grain diameter in millimeters. Certainly, the content of "tramp" elements (Cu, Sn, Cr, Mo, Ni, P) will affect the yield stress and they must be minimized if a consistent low yield stress is desired. At these low carbon contents the fraction of pearlite in the microstructure does not affect the yield stress.

The ultimate tensile strength of low-carbon steels is affected by the amount of pearlite present, as indicated by this equation, also from Pickering:

$$\text{UTS (MPa)} = 15.4[19.1 + 1.8(\%\text{Mn}) + 5.4(\%\text{Si}) + 0.25(\%\text{ pearlite}) + 0.5d^{-1/2}] \quad \text{(V-2)}$$

Similar equations for yield and tensile strengths are listed by Baird and Preston.[1]

It is obvious that grain size must be controlled if the mechanical properties of the steel are to be consistent. As mentioned previously, surface roughening (orange-peel effect) during cold forming places an upper limit on the ferrite grain size of about ASTM 5 to 6 (d = 57 to 40 μm). Below this size, the grain size in rimmed, capped, or semikilled steels is dependent upon the amount of cold reduction prior to annealing and upon the temperature of batch annealing. The greater the cold reduction, the finer will be the grain size after annealing.

Temper Rolling and Return of Discontinuous Yielding

Discontinuous yielding in a sheet steel is highly undesirable because by concentrating the strain in a small region it produces unsightly markings on a smooth surface after forming. These "stretcher strains" can be eliminated by temper rolling or roller leveling the sheet after annealing. In addition to eliminating discontinuous yielding, temper rolling also improves surface finish and reduces variation in thickness. Therefore, cold-rolled and annealed sheet steels are usually temper rolled before shipment to a user.

Extension during temper rolling is only between 0.25 and 1.0%. This eliminates the yield point elongation, as shown in Fig. V-3, by introducing a pattern of localized high residual stresses.[2] Residual stresses are the product of nonhomogeneous plastic deformation, and in industrial temper rolling, plastic deformation is highly inhomogeneous. After 1% extension by cold-rolling, only about half of the volume of the sheet has undergone plastic deformation; about 5% extension is required before deformation is reasonably uniform. In a highly simplified representation (Fig. V-4), the deformation pattern takes the form of thin lamellae of alternating deformed and undeformed regions, lying normal to the rolling direction. Because of this pattern of residual stresses, the yield stress after temper rolling is less than in the annealed condition and yielding is continuous, a consequence of easy nucleation of yielding at many points in the sheet.

To eliminate the yield point by uniaxial tensile straining requires a strain of about

[1] J.D. Baird and R.R. Preston, *Processing and Properties of Low Carbon Steel, AIME,* New York, 1973, p. 1.

[2] R.D. Butler and D.V. Wilson, *J. Iron Steel Inst.*, 201:16 (1963).

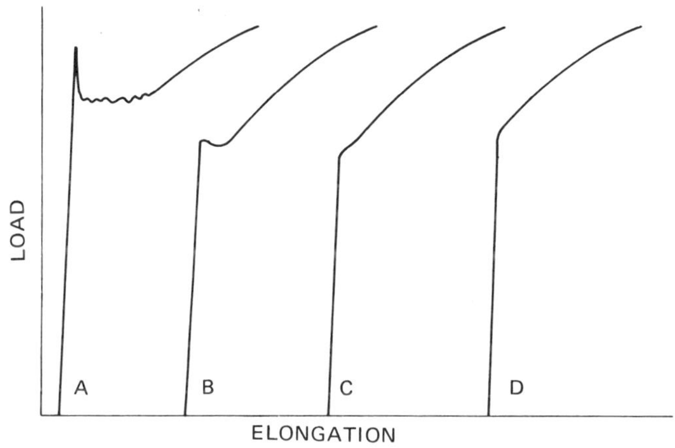

Figure V-3. Load-elongation curves of annealed, low-carbon steel sheet varying from (*a*) no to (*d*) heavy temper rolling or roller levelling. (Courtesy of J.S.H. Lake, John Lysaght (Australia) Pty. Ltd.)

5%. This would seriously reduce ductility in subsequent forming. Also, the yield point returns more rapidly during ambient temperature aging if the sheet has been deformed in tension than if it has been deformed by temper rolling, another consequence of the high residual stresses imparted by temper rolling.

To counteract the effects of aging after temper rolling, sheets are often roller-leveled in the user's plant. The sheets are passed between two rows of small-diameter rolls, the upper row being offset from the lower. The rolls are adjusted so that the amount of flexure decreases as the sheet nears the exit end of the leveller.

Koistinen[1] has developed an interesting thermal analog to temper rolling. The yield point in steel sheets is removed by rapid heating to a temperature in the range

[1] D.J. Bailey and D.P. Koistinen, *Proc. 3rd Intern. Conf. Strength of Metals and Alloys,* 1:586 (1973).

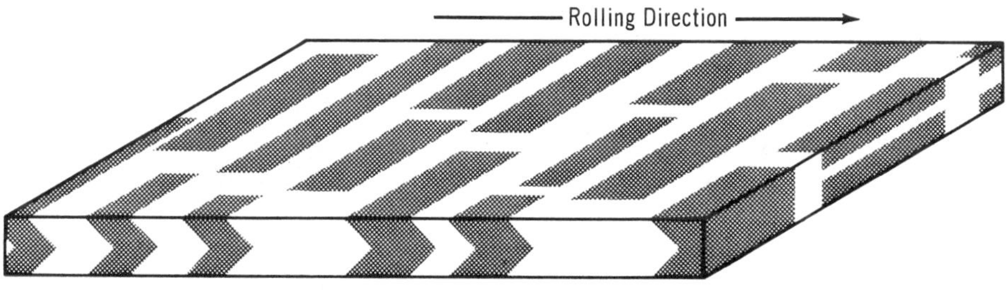

Figure V-4. Idealized pattern of deformed (black) and undeformed (white) regions in annealed, temper-rolled sheet steel. (R.D. Butler and D.V. Wilson, *J. Iron Steel Inst.*, 201:16 (1963), by permission.)

400–600°C, followed by a water quench. The rapid temperature change during quenching produces inhomogeneous plastic deformation,[1] especially around carbide particles. The sheet must be formed soon after this process, because the increased concentration of C and N atoms in solution after the quench causes a more rapid return of the yield point than after temper rolling.

Processing of Al-Killed, Low-Carbon Steel Sheets

The processing of Al-killed low-carbon steel sheets is a specific example of the general case of controlling simultaneous precipitation and annealing. The technique consists of cold-working a supersaturated solid solution, then adjusting the subsequent thermal history so that precipitation begins in the cold-worked matrix, before recrystallization. This precipitation in the cold-worked structure has several important consequences:

1. The rate of recrystallization is retarded much more than by elements in solid solution.
2. The size and shape of recrystallized grains are changed markedly.
3. Most important, the texture of the fully recrystallized alloy is altered.

These phenomena are quite general, occurring in the precipitation of AlN, MnS, and Cu in Fe; Fe in Al; and TaO from Ni. In contrast with various effects produced by precipitates existing before cold work, where a considerable volume fraction is required, these processes occur with only traces of the precipitating elements. While this is an economic blessing, it creates problems for the investigator because it is very difficult to locate and identify precipitates or even to prove that they exist. In the most important instance, the precipitation of AlN in Al-killed, low-carbon steels, the concentration of nitrogen involved is only about 20 to 40 ppm and the concentration of aluminum is about 300 ppm by weight.

As is often the case in metallurgy, the critical steps in the control of the production of Al-killed deep-drawing steel sheets were determined long before the agent responsible was identified. The critical factors were found by experience to be:

1. The temperature of the strip at the last stand of the continuous hot strip mill (Fig. V-1), which must be high (~890°C)
2. The temperature at which the hot strip was coiled, which must be low, ~580°C—this means that the strip must be cooled quickly by water sprays after passing through the last stand of the mill
3. The aluminum content, which must be held between 0.015 and 0.08 wt %

When these requirements were met, the batch annealed sheet had pancake-shaped grains, elongated in the direction of rolling. It was also noted that when processed in this manner the sheet recrystallized very sluggishly as compared to rimmed steel,[2] or as

[1] O. Richmond, W.C. Leslie, and R.J. Sober, *Met. Trans.*, 3:2593 (1972).
[2] R.L. Rickett, S.H. Kalin, and J.T. MacKenzie, *Trans. AIME*, 185:242 (1949).

compared to Al-killed steel that had been coiled at too high a temperature. After the development of a method for determining the amount of nitrogen combined as AlN[1] and after the determination of the solubility product of AlN in austenite (Table IV-1) it was possible to study the effects of the solution and precipitation of AlN on the structure of low-carbon steels.[2] It became obvious that the objective of all prior processing should be to bring the product to the cold-rolling operation as a solid solution supersaturated with respect to Al and N. Later, transmission electron microscopy was used to show that the retardation of recrystallization, the pancake-shaped grains, and the pronounced (111) [110] cube-on-corner texture were consequences of the formation of preprecipitation clusters of Al and N in cell and grain boundaries of the cold-rolled steel.[3] By this means, the number of nucleation sites for recrystallized grains was greatly reduced, a barrier effect was created in former boundaries, and rate of grain growth was strongly anisotropic in a textured matrix. Formation of a (111) $[1\bar{1}0]$ texture component was enhanced during recrystallization; formation of the (001) [110] component was inhibited.

Al-killed steel with the proper cube-on-corner orientation can be produced by box (batch) annealing or by open-coil annealing after cold rolling. In open-coil annealing the sheet is coiled with a spacer between each wrap of the coil, so that the furnace gases have access to all the surfaces of the sheet. Most other flat-rolled products are now softened by the much more rapid continuous annealing process, but in that process the heating is so rapid that recrystallization is complete before the AlN can precipitate. The much slower heating rates used in box annealing or in open coil annealing allow the formation of Al–N clusters at the proper time for maximum effect. Richards[4] has shown that pre-heating to 450–500°C before continuous annealing allows this process to be used for production of Al-killed steels.

Al-killed steels are expensive to produce because of their lower ingot-to-slab yield and higher surface conditioning costs. It would be highly desirable to have a steel that would combine the lower cost and excellent surface of rimmed steel with the nonaging characteristics and good drawing properties of Al-killed steel. At least three approaches are being tried. If Al-killed steels of consistent quality can be continuously cast with good surface the cost should be considerably reduced. Alternatively, rimmed steels can be cast into ingots, allowed to rim for about 2 min, then killed with an injection of liquid Al.[5] Lastly, a precipitation process or compositional control applicable to rimmed steel may be developed to produce the desirable cube-on-corner texture. Hu[6] has summarized the recent developments in this highly complex research. Texture can be controlled by reducing the Mn content of low-carbon steels to less than 0.1%. When this is done, the r_m value can exceed 2.0, where r_m is the average

[1] H.F. Beeghly, *Annal. Chem.*, 21:1513 (1949).

[2] W.C. Leslie, R.L. Rickett, C.L. Dotson, and C.S. Walton, *Trans. Am. Soc. Met.*, 46:1970 (1954).

[3] J.T. Michalak and R.D. Schoone, *Trans. AIME*, 242:1149 (1968).

[4] P.N. Richards, *J. Aust. Inst. Met.*, 12:2 (1967).

[5] M.A. Orehoski, J.W. Bales Jr., and D.W. Venseret, *Open Hearth Proc.*, 57:386 (1974).

[6] H. Hu, *Textures of Materials*, vol. II, Springer-Verlag, Berlin, 1978, p. 3.

plastic strain ratio, defined in next section. A second approach is to add P, in amounts up to 0.12%, to low-carbon steels containing about 0.1%Mn. The r_m value increases linearly with P content, approaching 4 at 0.12%P. In this instance the principal component of the texture of the cold-reduced and annealed sheet is (111)[112]. Much remains to be learned of the mechanisms by which preferential nucleation and growth are affected in these two types of experimental steels. Phosphorus in amounts up to 0.1 wt % does not embrittle iron[1] and it segregates to both grain boundaries and dislocations.[2]

r Values and Texture in Steel Sheets for Deep Drawing

As mentioned earlier, the most important effect of simultaneous precipitation and recrystallization is the alteration of the primary annealing texture. When sheet steel fails in drawing, it generally fails through the onset of plastic instability—that is, it becomes rapidly thinner in a small region until it breaks. One factor that controls the beginning of necking is the plastic strain ratio, or r value.[3,4] This is defined as the ratio of width true strain to thickness true strain in a tension test of a specimen cut from sheet:

$$r = \frac{\ln (W_i/W_f)}{\ln (T_i/T_f)} \tag{V-3}$$

where W_i = initial width
W_f = final width
T_i = initial thickness
T_f = final thickness

Because measurement of the thickness of a thin sheet is subject to considerable error, it is commonly assumed that the volume of the specimen is constant; therefore

$$r = \frac{\ln (W_i/W_f)}{\ln (L_f W_f/L_i W_i)} \tag{V-4}$$

where L_i and L_f are the initial and final lengths. The r value can be measured at any point in the tension test between the end of the Lüders strain and the maximum load.

Values of r are conventionally reported as an average, r_m,

$$r_m = \tfrac{1}{4}(r_0 + 2r_{45°} + r_{90°}) \tag{V-5}$$

where r_0 = value determined in the rolling direction (longitudinal)
$r_{45°}$ = value determined at 45° to the rolling direction
$r_{90°}$ = value determined in the transverse direction

[1] W.A. Spitzig, *Met. Trans.*, 3:1189 (1972).
[2] C.-C. Li, E.J. Dziura, and W.C. Leslie, *Met. Trans. A*, 8A:705 (1977).
[3] W.J. Lankford, S.C. Snyder, and J.A. Bauscher, *Trans. Am. Soc. Met.*, 42:1197 (1950).
[4] J.A. Elias, R.H. Heyer, and J.H. Smith, *Trans. AIME*, 224:678 (1962).

The variation of r with direction in the sheet is an indication of the planar anisotropy

$$\Delta r = \frac{r_0 - 2r_{45°} + r_{90°}}{2} \tag{V-6}$$

An isotropic material would have an r value of 1 and a Δr value of 0. For deep drawing a high value of r is desired, for that means that the sheet is resistant to thinning. High r values for sheet steel means 1.5 or more. A minimum value of 1.5 in the longitudinal direction is usually required for good performance in difficult draws. The relation between r and the limiting drawing ratio is shown in Fig. V-5. The limiting drawing ratio is the ratio of blank diameter to punch diameter for the largest blank that can be drawn successfully into a flat-bottom cup. Values of r_m in rimmed steels can vary between about 0.8 and 1.2. In Al-killed steels, properly processed, r_m varies between 1.5 and 1.8. In some experimental low-carbon steels, r_m values as high as 3.0 have been obtained. In the opposite direction, cube-textured copper or austenitic stainless steel can have r_m as low as 0.1.

To obtain consistently high r_m values, the hot rolling of the steel must be completed in the γ range. The coiling temperature after hot rolling has little effect on the r_m values in rimmed steel, but, as previously discussed, it is essential that the coiling temperature of Al-killed steel be low. High r values are produced when the final cold reduction of the sheet is in the range 60–75%. Such high r values are consistently correlated with the extent of {111} ⟨110⟩ texture in the annealed sheet; if cold reduction exceeds 75%, components of the {111} ⟨112⟩ type appear. Values of r are not altered by temper rolling or by aging after annealing.

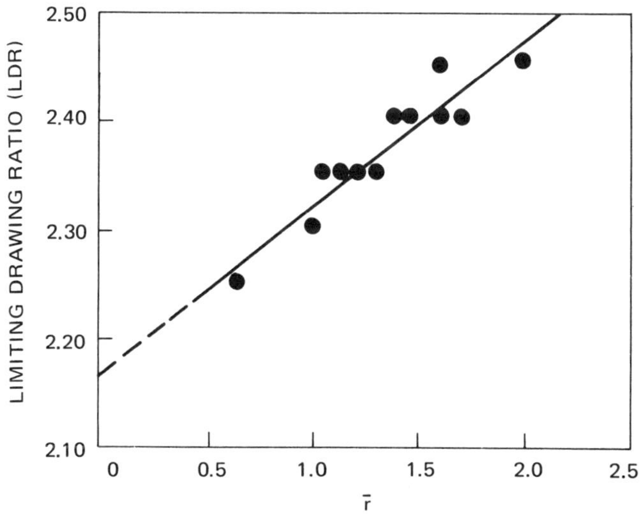

Figure V-5. Correspondence between limiting drawing ratio and the average plastic-strain ratio. (C.A. Stickels and P.R. Mould, *Met. Trans.*, 1:1303 (1970), by permission.)

To obtain high r values, it is necessary to minimize the {100} texture and maximize {111}, and this can be accomplished by precipitation of AlN during annealing after cold work, as previously mentioned. In cold-rolled alpha iron the energy stored in grains having a {111} orientation is much greater than in those having a {100} orientation.[1] Therefore, during annealing the {111} grains recover and recrystallize first. In properly processed Al-killed, low-carbon steels, the recrystallization of {111}-oriented grains is retarded, but once begun it proceeds rapidly before AlN forms as a distinct second phase.[2] In {100} grains, both recovery and nucleation of new grains are inhibited by preprecipitation clustering of Al and N atoms. This enhancement of the relative rates of growth leads to a greater volume of {111}-oriented grains in Al-killed than in rimmed steel.

The tendency to form "ears" in the drawing of cups is a function of Δr. When Δr is positive, ears tend to form in the rolling direction and at 90° from the rolling direction. When Δr is negative, ears tend to form at 45° from the rolling direction.

A later development[3,4] in the measurement of r takes advantage of the anisotropy of elastic constants in alpha iron:

$$E[100] = 13.1 \times 10^4 \text{ MPa}$$

$$E[110] = 21.7 \times 10^4 \text{ MPa}$$

$$E[111] = 27.7 \times 10^4 \text{ MPa}$$

The Young's moduli of small specimens cut from sheet are determined ultrasonically. A high value of $\bar{E}$ is associated with a high value of r_m, as indicated in Fig. V-6*a*; but the relation between the two is not linear. Values of Δr and ΔE are linearly related (Fig. V-6*b*).

The search for better ways to develop texture in cold-rolled and annealed sheets continues. Hu and Goodman[5] found that r_m decreased with increasing Mn content between 0.05 to 0.55 wt %, with excellent r_m values being obtained at Mn contents of 0.1% or less. These promising results are probably associated with the action of sulfur atoms at low Mn contents. Blickwede[6] obtained r_m values as high as 2.8 in steels with a Ti/C + N ratio of 15 at a grain size of ASTM 7. Messien and Greday[7] obtained r_m values over 2 and Δr near zero with

$$12 < \frac{\text{Ti}}{\text{C} + \text{N}} < 15 \qquad \text{and} \qquad 9 < \frac{\text{Nb}}{\text{C} + \text{N}} < 12$$

[1] I.L. Dillamore, G.J.E. Smith, and T.W. Watson, *Met. Sci. J.*, 1:49 (1967).
[2] W. Jolley, *Met. Trans.*, 3:245 (1972).
[3] C.A. Stickels and P.R. Mould, *Met. Trans.*, 1:1303 (1970).
[4] P.R. Mould and T.E. Johnson Jr., *Sheet Metal Ind.*, 50:328 (1973).
[5] H. Hu and S.R. Goodman, *Met. Trans.*, 1:3057 (1970).
[6] D. J. Blickwede, *Trans. Am. Soc. Met.*, 61:653 (1968).
[7] P. Messien and T. Greday, *Metall. Rep. CRM*, no. 49, Dec. 1976.

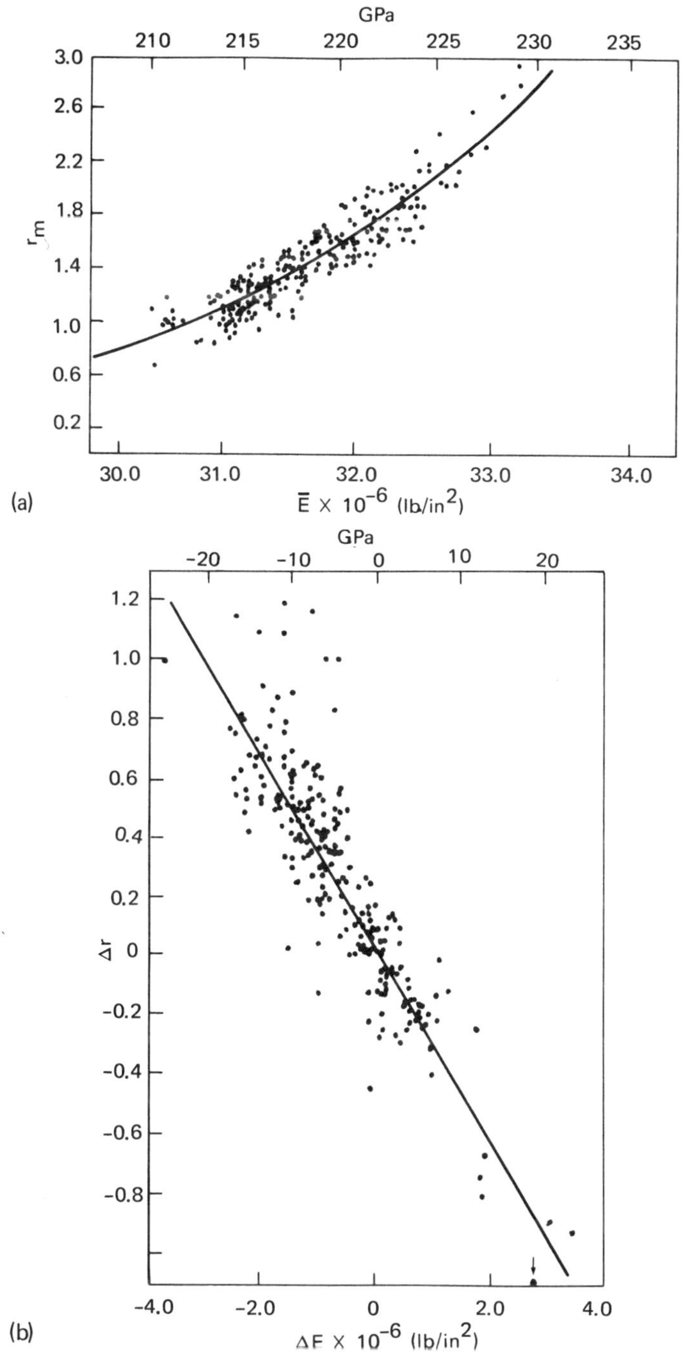

Figure V-6. (*a*) Relation between the average plastic-strain ratio r_m and the average Young's modulus $\bar{E}$. (P.R. Mould and T.E. Johnson Jr., *Sheet Met. Ind.*, 50:328 (1973), by permission.) (*b*) Relation between the planar variation in plastic-strain ratio Δr and the planar variation in Young's modulus ΔE. (P.R. Mould and T.R. Johnson Jr., *Sheet Met. Ind.*, 50:328 (1973), by permission.)

The Strain-Hardening Exponent *n* and the Stretchability of Sheet Steels

On page 8 it was shown that n, the strain-hardening exponent, is numerically equal to the uniform strain in a standard tension test. An experimentally determined correlation is shown in Fig. V-7. Therefore, for optimum formability a high value of n is required in addition to a high value of r. Unfortunately, n is not as susceptible to improvement as r. The strain-hardening exponent decreases with increasing concentrations of substitutional solutes,[1] with decreasing grain size (Eq. I-39),[2] and with increasing volume fractions of second phases.[3] The improvement of n by increasing the grain size is limited by surface roughening, as previously noted.

A distinction must be made between n, the strain-hardening exponent, and $d\sigma/d\epsilon$, the rate of work hardening at any given strain. There is no simple relation between the two.

Quench Aging of Low-Carbon Sheet Steels

The quench aging of Fe–C and Fe–N alloys has been discussed at some length in Chap. II-9. In commercial steels the process is complicated by the presence of man-

[1] D.J. Blickwede, *Trans. Am. Soc. Met.*, 61:653 (1968).
[2] W.B. Morrison, *Trans. Am. Soc. Met.*, 59:824 (1966).
[3] J.F. Enrietto, M.G.H. Wells, and E.R. Morgan, in *Precipitation from Iron-Base Alloys,* Gordon and Breach, New York, 1965, p. 141.

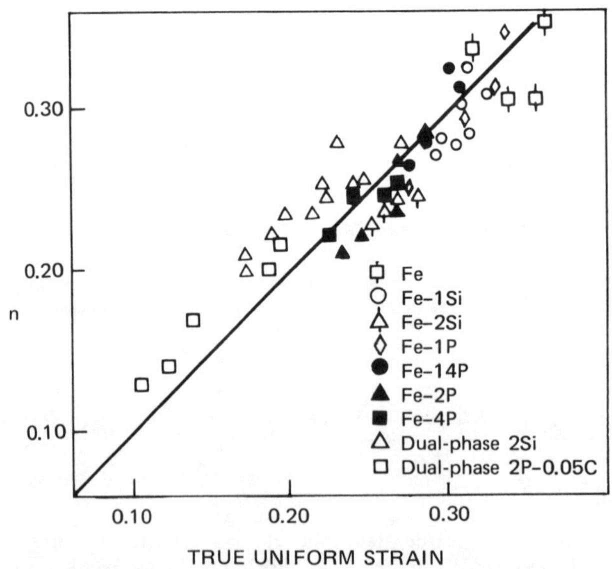

Figure V-7. Correlation between n and true uniform strain; the solid line represents perfect correlation. (R.G. Davies, *Met. Trans. A*, 10A:113 (1979), by permission.)

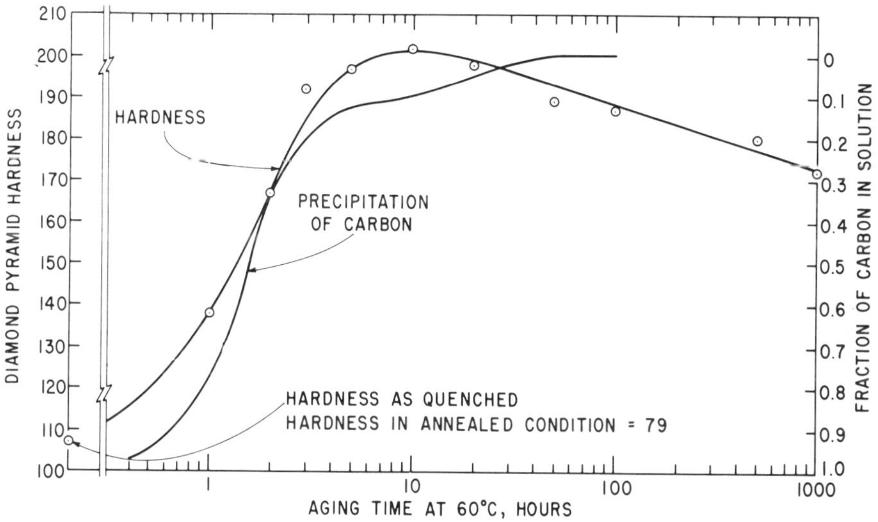

Figure V-8. Relation between hardness changes and the precipitation of carbon during aging at 60°C of 0.03%C rimmed steel. (A.S. Keh, W.C. Leslie, and D.L. Sponseller, in *Precipitation from Iron-Base Alloys*, AIME, 1965.)

ganese and other solutes. This has the effects of promoting the formation of Fe_3C at low aging temperatures and of increasing the number of nucleation sites for the precipitation of carbides. The changes in hardness and tensile properties of a low-carbon rimmed steel during aging at 60°C after quenching from 725°C are shown in Figs. V-8 and 9. The changes of hardness, yield strength, and tensile strength give no indication of the complexity of the process, but this is hinted at by the presence of a shoulder in the plot of precipitation of carbon, determined by internal friction. Transmission electron microscopy revealed that two forms of iron carbide were involved. The size of the metastable carbide increased to a maximum after about 5 h at 60°C (Fig. V-10). Nucleation of Fe_3C at other sites began at about that time; then the particles of cementite grew at the expense of the metastable carbide. As a result, the interparticle spacing increased during the early stages of aging, then decreased before the attainment of maximum hardness. Particle size increased to a maximum, then decreased, then increased again during overaging. During overaging, the tensile strength decreased much more rapidly than did the yield strength, indicating a marked decrease in the rate of work hardening. The change in ductility during aging was opposite to the change in strength, in the usual manner.

The quench aging of carbon steel can affect the mechanical properties of hot-dipped galvanized sheet. The annealed sheet is dipped into liquid zinc held at about 450°C. The sheet is in the liquid zinc for about 10 s. This is sufficient to dissolve about 25 ppm (C + N),[1] which can subsequently precipitate at room temperature,

[1] P.N. Richards and K.V. Barratt, *Trans. Am. Soc. Met.*, 58:611 (1965).

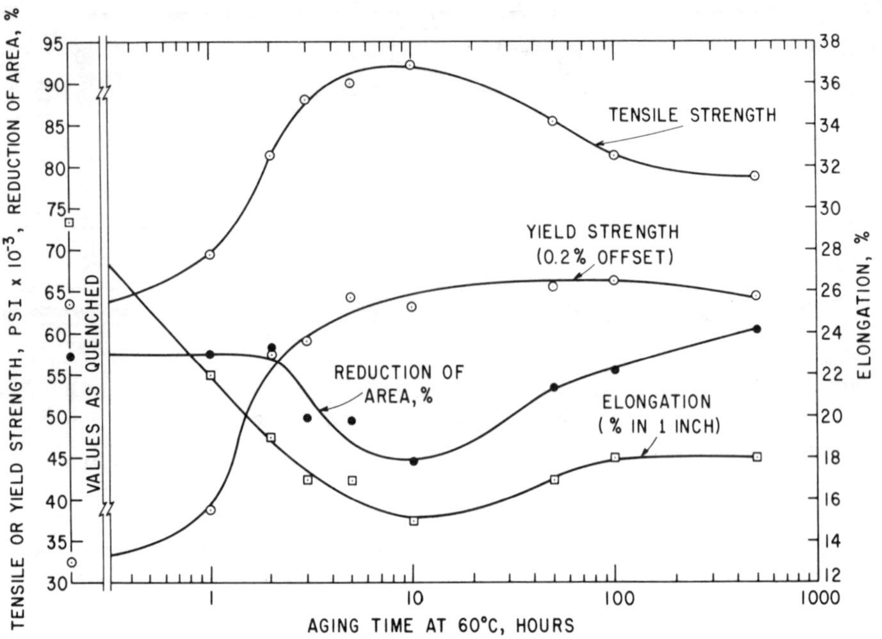

Figure V-9. Changes in tensile properties during aging of 0.03%C rimmed steel, after quenching from 725°C. (A.S. Keh, W.C. Leslie, and D.L. Sponseller, in *Precipitation from Iron-Base Alloys,* AIME, 1965.)

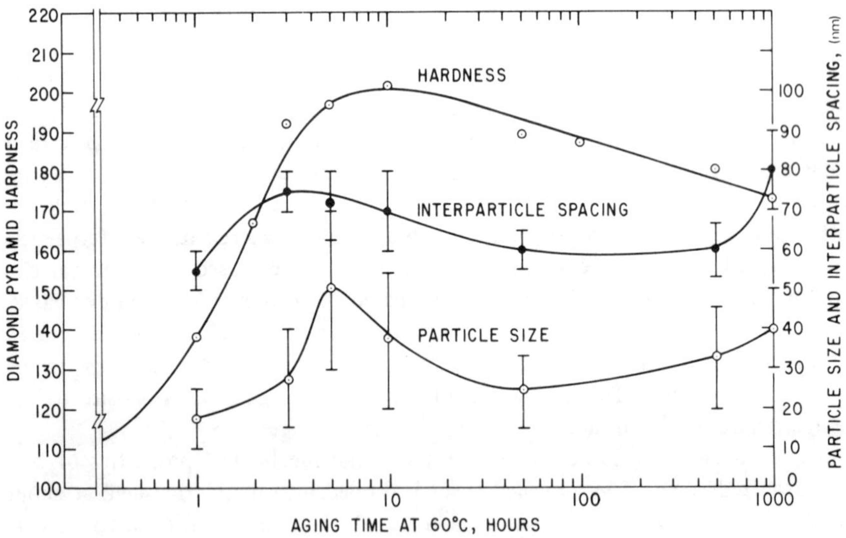

Figure V-10. Change of hardness, interparticle spacing, and particle size during aging of 0.03%C rimmed steel at 60°C. (W.C. Leslie and A.S. Keh, in *Mechanical Working of Steel*, vol. 2, AIME, 1965.)

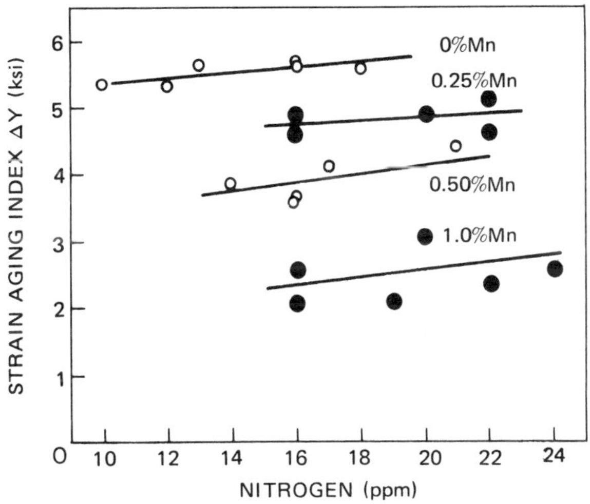

Figure V-11. Effect of manganese and nitrogen on the strain-aging index, ΔY: ΔY is the lower yield stress of prestrained and aged specimens minus prestraining stress. Specimens prestrained 10%, then aged 5 min at 100°C. (D.J. Blickwede, *Trans. Am. Soc. Met.*, 61:653 (1968), by permission.)

strengthening the sheet and reducing its ductility.[1] Reheating the galvanized sheet to 300°C for 1 h can reduce the (C + N) in solution to about 10 ppm and eliminate any subsequent quench aging.

Strain Aging of Carbon Steels

The static strain aging of low-carbon steels resembles that of Fe–N alloys, discussed in Chap. II, with one notable exception. The presence of Mn reduces considerably the extent of aging, as indicated in Fig. V-11. The presence of Mn reduces the mobility of the interstitial solutes.

Aluminum-killed steels are usually termed "non-aging" because they do not strain age at ambient temperatures. Aluminum and other nitride-forming elements, such as Ti, V, or B, produce the same effect by combining with N. Aluminum-killed steels will strain age if the aging temperature is raised to the vicinity of 100°C, at which point C can cause aging.

If commercial carbon steels are strained at temperatures between 200 and 400°C, their subsequent strength at room temperature is greatly increased, the fatigue life is slightly improved, the Bauschinger effect is greatly decreased, and the notch-impact transition temperature is raised.[2] This process of dynamic strain aging is particularly effective in the AISI 1500 series of carbon steels with Mn contents in excess of 1%. Figure V-12 shows the tensile properties of an AISI 1522 steel after strains of 3 or 5% at temperatures from 100 to 600°C. The optimum properties are obtained by straining less than 5% at temperatures between 200 and 400°C. After such treatment, this steel has properties comparable to HSLA (high-strength, low-alloy) steels:

[1] A.S. Keh, W.C. Leslie, and D.L. Sponseller, in *Precipitation from Iron-Base Alloys*, Gordon and Breach, New York, 1965, p. 281.

[2] C.-C. Li and W.C. Leslie, *Met. Trans. A*, 9A:1765 (1978).

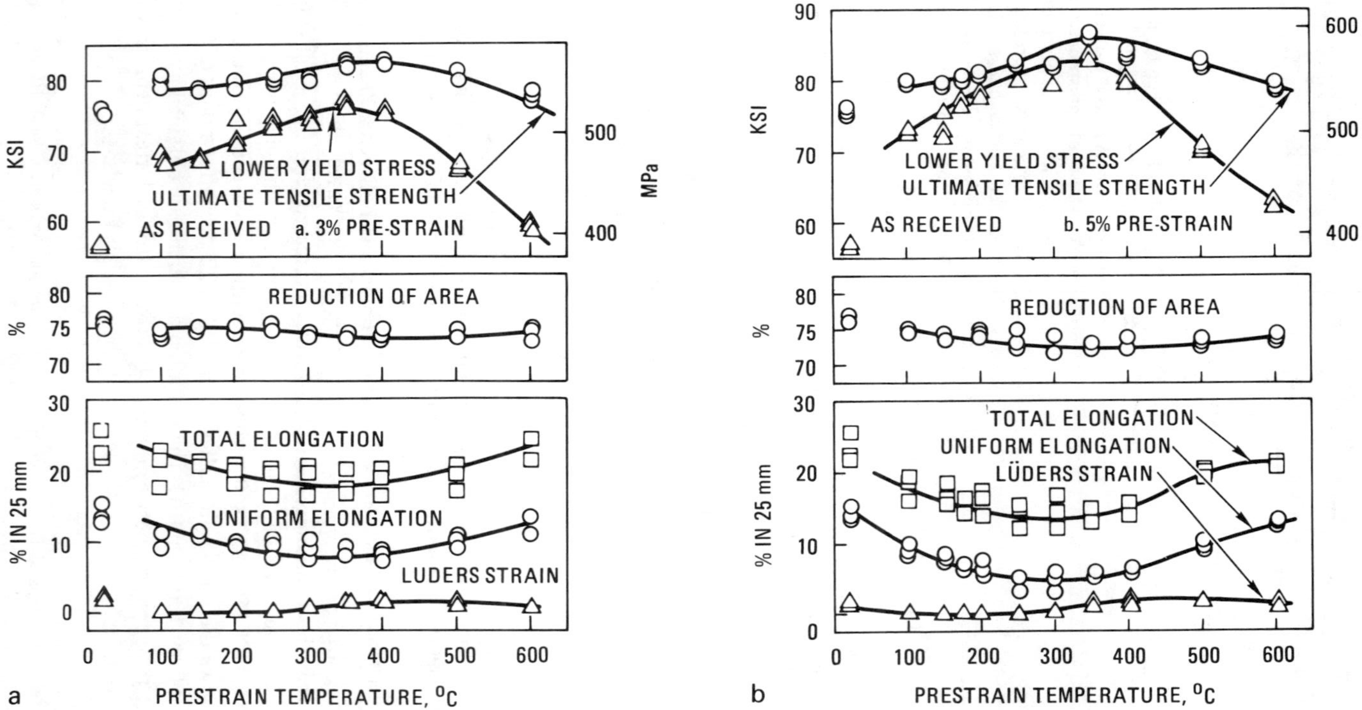

Figure V-12. Ambient temperature properties of AISI 1522 steel after prestraining at elevated temperatures: $\dot{\epsilon}$ at ambient temperature = 1.75×10^{-4} s^{-1}. (*a*) 3% Prestrain. (*b*) 5% Prestrain. (C.-C. Li and W.C. Leslie, *Met. Trans. A*, 9A:1765 (1978).)

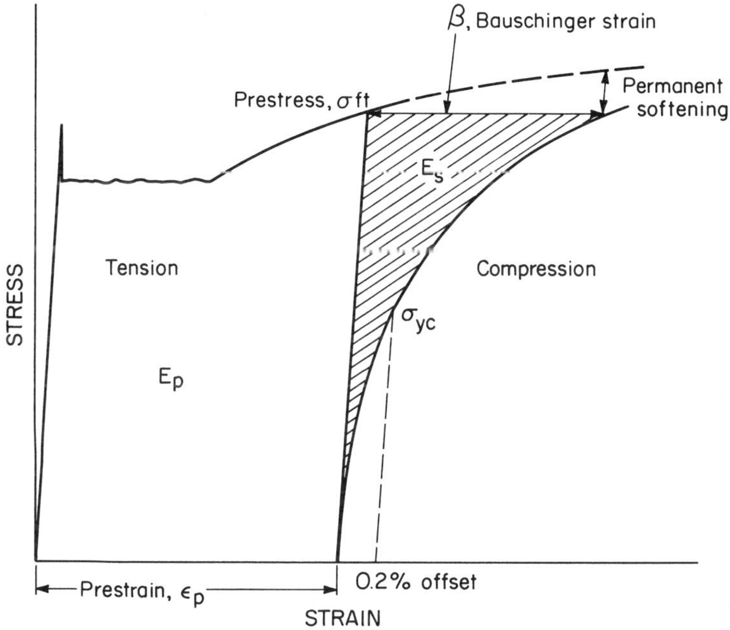

Figure V-13. Schematic diagram of the Bauschinger effect. (C.-C. Li, J.D. Flasck, J.A. Yaker, and W.C. Leslie, *Met. Trans. A*, 9A:85 (1978).)

Lower yield stress, 490 to 580 MPa
Ultimate tensile stress, 545 to 600 MPa
Elongation in 25 mm, 15 to 20%

The decrease in the Bauschinger effect by dynamic strain aging[1] is of particular interest. The Bauschinger effect, in which prior plastic deformation in one direction can greatly reduce the flow stress in subsequent straining in the opposite direction, arises from three causes:

1. The presence of long-range internal stresses (back stresses)
2. The presence of mobile dislocations with directionality in their resistance to movement
3. Annihilation of dislocations during reverse straining (work softening)

These are several measures of the Bauschinger effect, as summarized in Fig. V-13. These are:

1. The Bauschinger strain, β, the strain developed by fully reversed loading[2]
2. The Bauschinger strain parameter, defined by[3]

[1] C.-C. Li, J.D. Flasck, J.A. Yaker, and W.C. Leslie, *Met. Trans. A*, 9A:85 (1978).
[2] S.N. Buckley and K.M. Entwistle, *Acta Metall.*, 4:352 (1956).
[3] A. Abel and H. Muir, *Phil. Mag.*, 26:489 (1972).

$$\beta_\epsilon = \frac{\text{Bauschinger strain } \beta}{\text{plastic prestrain } \epsilon_p}$$

3. The Bauschinger effect factor[1]

$$\text{BEF} = \frac{\text{flow stress in compression } (0.2\%\ \epsilon)}{\text{maximum flow stress in tensile prestraining}}$$

$$= \frac{\sigma_{yc}}{\sigma_{ft}}$$

4. The Bauschinger energy parameter

$$\text{BEP} = \frac{E_s}{E_p}$$

where E_p is the energy expended in prestrain and E_s (as in Fig. V-13) is the energy saved in reverse loading, sometimes called the Bauschinger energy

From these definitions, the Bauschinger strain, the Bauschinger strain parameter, and the Bauschinger energy parameter all *increase* with increasing Bauschinger effect at a fixed prestrain. The Bauschinger effect factor *decreases* with increasing Bauschinger effect.

Strictly speaking, measurements of the Bauschinger effect should be made at the same temperature for the prestrain and subsequent strain, but by prestraining steels at elevated temperatures the subsequent "Bauschinger effect" at room temperature can be greatly reduced, between 50 and 90%.[2] The reduction is due primarily to the pinning of mobile dislocations and secondarily to the relief of back stresses. The conclusion is reached that the straightening of steel parts after heat treatment, or the straightening of test specimens cut from tubing, should be done at the most effective temperature for dynamic strain aging of the steel. This minimizes deterioration of mechanical properties by the Bauschinger effect.

Dual-Phase Carbon Steels

The great interest in low-cost, formable, high-strength, low-carbon steel sheets has led to studies of the properties and structures of dual-phase carbon steels.[3,4] Dual-phase steels, which will be discussed in greater detail later, have structures consisting of islands of martensite in a ferrite matrix. The structure in carbon steels is developed by reheating into the $\alpha + \gamma$ region, quenching to form martensite, then tempering at a low temperature to improve ductility. The carbon steels so treated are characterized by a

[1] R.V. Milligan, W.H. Koo, and T.E. Davidson, *J. Basic Eng.*, 88D:480 (1966).
[2] C.-C. Li, J.D. Flasck, J.A. Yaker, and W.C. Leslie, *Met. Trans. A*, 9A:1765 (1978).
[3] D.J. Bailey and R. Stevenson, *Met. Trans. A*, 10A:47 (1979).
[4] R. Stevenson, D.J. Bailey, and G. Thomas, *Met. Trans. A*, 10A:57 (1979).

low yield stress and a rapid rate of work hardening. The low yield stress is a consequence of the tesselated residual stresses arising from the transformation of the austenite pools to martensite; the high rate of work hardening is typical of alloys containing a dispersion of relatively hard particles in a softer matrix.

Enameling Sheet

Steel sheet for porcelain enameling must have a low carbon content to prevent formation of defects in the enamel arising from carbon oxide gases created by reactions between Fe_3C in the steel and the oxides in the liquid enamel during enamel firing. Also, low carbon and manganese contents are desirable to minimize sagging of the sheet during enamel firing. This sagging is exaggerated by the formation of austenite. The temperature at which the first austenite appears can be raised above the enamel firing temperature by reducing both manganese and carbon contents. The total of both is often less than 0.1%. This leads to two problems. The first is that the steel is now hot-short; that is, it is susceptible to edge cracking during hot rolling in the temperature range between 1040 and 900°C. This difficulty can be evaded by rolling the steel through the roughing stands at temperatures above 1040°C, holding the steel between the roughing and the finishing stands until it cools below 900°C, then completing the hot rolling. The hot rolling is finished while the steel is ferritic.

The second problem is that these very low Mn steels cannot be recrystallized consistently after cold rolling by annealing at any temperature in the ferrite region. To obtain a product with a consistent hardness the sheets must be normalized—that is, heated into the austenite range and air cooled. This is an expensive operation.

The sluggish recrystallization depends upon the prior thermal history of the sheet. If hot-rolled sheets are reheated in the range 815-900°C before cold rolling, the resistance to recrystallization is removed. Reheating to temperatures above 1000°C before cold rolling restores the sluggish recrystallization behavior.

The recrystallization of these low-Mn steels is inhibited by the precipitation of "MnS," or the clustering of Mn and S atoms that precedes such precipitation.[1,2] Because of the low carbon content of enameling steel, the oxygen content tends to be high, so a portion of the Mn content is combined with oxygen. By reducing the already low concentration of Mn available to combine with S, this enables all or nearly all of the MnS to be dissolved in austenite (Table IV-1) during heating for hot rolling. Both elements can remain in solution during hot rolling and cold rolling, then form clusters or MnS precipitate particles during batch annealing after cold work. The MnS is even more effective than AlN in inhibiting recrystallization.

By enabling the C content of sheet steel to be reduced to a very low level, <0.005%, the open-coil annealing process opened another avenue for the production of enameling sheet. A steel can be produced with C content of about 0.06% and Mn

[1] J.D. Baird and J.M. Arrowsmith, *J. Iron Steel Inst.*, 204:240 (1966).

[2] D.A. Witmer and G. Krauss, *Trans. Am. Soc. Met.*, 62:447 (1969).

content of 0.15% or more, sufficient to minimize hot-shortness, then be decarburized during open coil annealing. Unfortunately, this is not the end of the difficulties, for when such a very low carbon sheet is cold-formed, then heated during enamel firing, in certain regions the conditions may match those required for growth of very large crystals by the strain-anneal method. When this occurs, the yield strength of the enameled part can be too low to resist plastic deformation during ordinary use and the enamel can flake off. To avoid this problem elements such as V, which can inhibit grain growth by forming small, stable carbide particles, must be added to the steel.

Coated Flat-Rolled Products

For the production of hot-dipped galvanized sheet or terne-coated sheet (terne is an alloy of lead and tin), the steel used can be rimmed, capped, semikilled, Al-killed, or continuously cast. Up to 0.30% Cu can be added to improve atmospheric corrosion resistance. Steels for tin plate or electrolytic chrome plate (tin-free steel) are rimmed, capped, or continuously cast. Sheet steels for the manufacture of tin plate are cold-rolled to between 80 and 90% reduction in thickness. After annealing they are temper-rolled, or the temper-rolling step can be replaced by a second cold reduction, to between 30 and 40% reduction in thickness, and the material is used in this condition without annealing. If a higher strength tin-plate steel is required, phosphorus can be added in amounts up to 0.15%, or nitrogen up to about 0.015%.

V-2. FERRITE–PEARLITE STEELS

To this point in Chap. V, the discussion has been limited to low-carbon sheet steels in which Fe_3C, whether present as a constituent of pearlite or as individual particles, has little effect on the yield strength. As the carbon content is increased, the pearlite plays a large role in determining the strength, ductility, toughness, and wear resistance of the steel.

Tensile Properties

For steels with pearlite content ranging up to 100% (eutectoid composition), Gladman et al.[1] developed the empirical equation

$$\sigma_{ys}\ (\text{MPa}) = 15.4\{f_\alpha^{1/3}[2.3 + 3.8(\%\text{Mn}) + 1.13d^{-1/2}] + (1 - f_\alpha^{1/3})[11.6 + 0.25S_p^{-1/2}] + 4.1(\%\text{Si}) + 27.6(\%\text{N})\} \qquad \text{(V-7)}$$

where f_α = volume fraction of ferrite

S_p = interlamellar spacing of pearlite, the distance in mm from the center of one cementite lamella to the center of the next, measured normal to the plane of the lamellae

d = mean linear intercept grain diameter, in mm

[1] T. Gladman, I.D. McIvor, and F.B. Pickering, *J. Iron Steel Inst.*, 210:916 (1972).

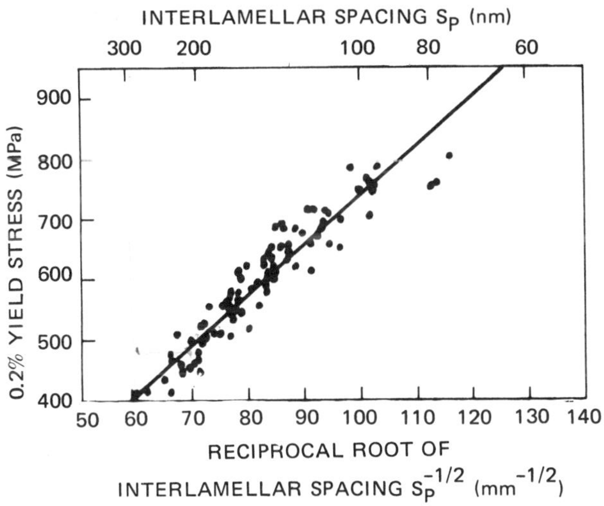

Figure V-14. Relationship between 0.2% offset yield stress and interlamellar spacing of pearlite in eutectoid carbon steel. (W. Heller, in *Rail Steels*, STP 644, ASTM, Philadelphia, 1979, by permission.)

The term $f_\alpha^{1/3}$ was included so that the yield strength could vary nonlinearly with pearlite content. As the pearlite content is decreased, the yield strength of the steel becomes quite insensitive to the pearlite. As the pearlite content is increased, the effect of ferrite grain size on yield strength becomes progressively less important.

The yield strength of high-carbon steels is controlled by the pearlite interlamellar spacing. Therefore, the yield strength can be increased by increasing the cooling rate from the austenite range, so that transformation to pearlite occurs at a lower temperature, leading to smaller interlamellar spacing. Heller[1] claims that aside from their effect on interlamellar spacing, alloying elements in eutectoid carbon steels have no effect on the yield stress. He proposes the simple equation

$$\sigma_{ys(0.2\%)}\,(\text{MPa}) = -85.9 + 8.3S_p^{-1/2}(\text{mm}^{-1/2}) \qquad \text{(V-8)}$$

as illustrated in Fig. V-14.

The ultimate tensile strength of ferrite-pearlite steels is given by the equation[2]

$$\text{UTS (MPa)} = 15.4\{f_\alpha^{1/3}[16.0 + 74.2(\sqrt{\%\text{N}}) + 1.18d^{-1/2}] + (1 - f_\alpha^{1/3})(46.7 + 0.23S_p^{-1/2}) + 6.3(\%\text{Si})\} \qquad \text{(V-9)}$$

The pearlite content, because of its effect on work hardening, has a greater influence on the tensile strength than on the yield strength.

Toughness

Because toughness depends upon such factors as grain boundary segregation and distribution of solute elements between ferrite and carbide phases,[3] it is inherently

[1] W. Heller, *Rail Steels,* STP 644, Am. Soc. Test. Mater., Philadelphia, 1978 p. 162.
[2] T. Gladman, I.D. McIvor, and F.B. Pickering, *J. Iron Steel Inst.,* 210:916 (1972).
[3] W.C. Leslie, R.L. Rickett, and W.D. Lafferty, *Trans. AIME,* 218:699 (1960).

more difficult to relate to gross composition and to microstructure. The impact transition temperature depends upon the temperature dependences of the stresses required for crack initiation and crack propagation. For low-carbon ferrite–pearlite steels, Petch[1] suggested the following for the fracture transition (change from fibrous to cleavage fracture):

$$\beta T_c = \ln B - \ln\left(\frac{4qG\gamma'}{K^*} - K^*\right) - \ln d^{-1/2} \tag{V-10}$$

where T_c = fracture transition temperature
β, B = constants
q = triaxiality factor, $\frac{1}{3}$ for a Charpy V-notch
G = shear modulus
γ' = effective surface energy
K^* = constant
d = mean linear intercept grain diameter

This has been shortened to

$$\beta T_c = \ln B - \ln C - \ln d^{-1/2} \tag{V-11}$$

β is a constant related to the resistance of the lattice to deformation and C is a measure of the resistance to crack propagation. Equation V-11 predicts a linear relation between notch-impact transition temperature and $d^{-1/2}$. Duckworth and Baird[2] suggest that

$$\underset{(85\%\ \text{cleavage})}{\text{FATT}\ (^\circ\text{C})} = 36 + \frac{690 - 190d^{-1/2}}{3.5 + (\%\ \text{pearlite})} \tag{V-12}$$

where FATT is fracture appearance transition temperature.

Because $d^{-1/2}$ is in the range 4 to 10 mm$^{-1/2}$ in most commercial steels, the numerator is always negative and so is the total fraction. On the bases of Eqs. V-9 and 12, increasing the carbon content increases both the tensile strength and the notch-impact transition temperature. The tensile strength increases about 4 MPa per % pearlite, and the transition temperature increases about 2.2°C per % pearlite. Increasing the carbon content from 0.1% (15% pearlite) to 0.2% (30% pearlite) raises the yield strength only slightly, raises the tensile strength about 60 MPa, and raises the notch-impact transition temperature about 33°C. This embrittlement, added to the adverse effects of the higher carbon content on weldability, emphasizes the undesirability of strengthening ferrite–pearlite steels by adding C and the improvement to both strength and toughness to be gained by decreasing the grain size.

Pickering[3] has described the notch-impact transition temperature in ferrite-pearlite steels by the equation

[1] N.J. Petch, *Phil. Mag.*, 3:1089 (1958).
[2] W.E. Duckworth and J.D. Baird, *J. Iron Steel Inst.*, 207:854 (1969).
[3] F.B. Pickering, in *Towards Improved Toughness and Ductility*, Climax Molybdenum Co., Greenwich, Connecticut, 1971, p. 9.

$$\mathrm{ITT}(^{\circ}\mathrm{C}) = -19 + 44(\%\mathrm{Si}) + 700(\sqrt{N_f}) + 2.2(\%\ \mathrm{pearlite}) - 11.5d^{-1/2} \quad \text{(V-13)}$$

This points out the same important features as Eq. V-12 and in addition shows the desirability of reducing the nitrogen in solution in ferrite. The factor for Si, however, is highly suspect in view of studies[1,2] that indicate that additions of Si up to about 0.5% *lower* the notch impact transition temperature of low-carbon steels.

Higher carbon steels, essentially 100% pearlite in the as-hot-rolled condition, are used for rails, railroad car wheels and axles, drill rods, and rods for wire drawing. As we have seen, toughness decreases with increased pearlite content, so it is desirable to use the lowest carbon content consistent with the desired strength. For eutectoid carbon steels, Hyzak and Bernstein[3] have shown that strength depends upon the interlamellar spacing of the pearlite, but that toughness is largely dependent upon the prior austenite grain size. For yield strength they propose the relation

$$\sigma_{ys}\ (\mathrm{MPa}) = 6.90[3.16 \times 10^{-1} Sp^{-1/2} - 5.79 \times 10^{-2} P^{-1/2} - 4.17 \times 10^{-1}(d^{-1/2}) + 7.58] \quad \text{(V-14)}$$

where Sp = pearlite interlamellar spacing in cm
P = pearlite colony size in cm
d = austenite grain size in cm

and for notch-impact transition temperature, they propose

$$\mathrm{TT}(^{\circ}\mathrm{F}) = -8.25 \times 10^{-2}(Sp^{-1/2}) - 1.22P^{-1/2} - 5.55(d^{-1/2}) + 4.35 \times 10^{2} \quad \text{(V-15)}$$

To convert Eq. V-15 to °C, multiply (TT − 32) by 5/9. The relation between transition temperature and prior austenite grain size is shown in Fig. V-15. Pearlite colonies are groups of pearlite nodules with ferrite and cementite lamellae in nearly the same orientation. Colony boundaries serve as obstacles to the propagation of brittle cracks by forcing the crack to change direction in the manner of ferrite grain boundaries.

Grain Size Control

The control of ferrite grain size in ferrite-pearlite steels, or of pearlite colony size in eutectoid carbon steels, is obtained by controlling the austenite grain size. This can be done in at least four ways:

1. Normalizing
2. Repetitive rapid austenitizing
3. Controlled rolling
4. Additions of aluminum to the steel

[1] K.D. Sibley and N.N. Breyer, *Met. Trans. A*, 7A:1602 (1976).
[2] B. Mintz and P.J. Turner, *Met. Trans. A*, 9A:1611 (1978).
[3] J.M. Hyzak and I.M. Bernstein, *Met. Trans. A*, 7A:1217 (1976).

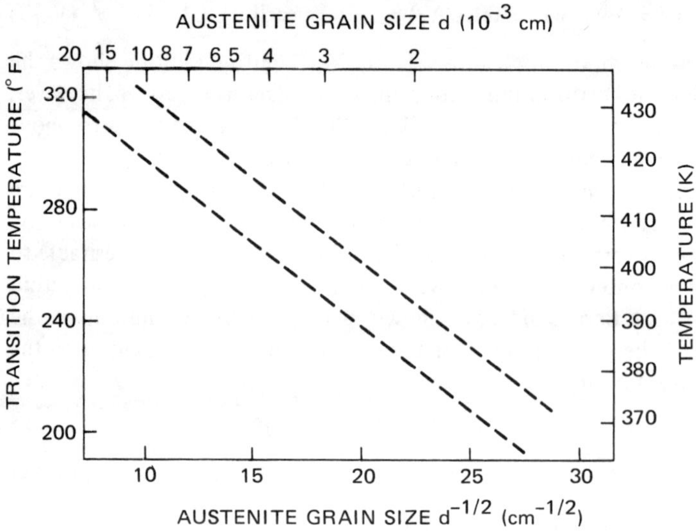

Figure V-15. Relationship between austenite grain size and notch-impact transition temperature in pearlitic, eutectoid carbon steels. (J.M. Hyzak and I.M. Bernstein, *Met. Trans. A*, 7A:1217 (1976), by permission.)

Normalizing is accomplished by heating the steel to a low temperature in the austenite region, holding for a period long enough to attain temperature uniformity, then air cooling. This process forms relatively fine austenite grains that then transform to fine ferrite grains or fine pearlite colonies. In as-hot-rolled steels the ferrite grain size is generally determined by the finishing temperature in hot rolling, which is subject to wide variation. Normalizing is also employed to reduce carbide size in alloy steels.

A variation of normalizing that can produce ultrafine ferrite grains is short-cycle repetitive austenitizing.[1] The important factors are

1. The reheating (austenitizing) temperature
2. Time of reheating
3. Initial microstructure

The reheating temperature should be barely above the Ac_3 temperature of the steel. For low-carbon, unalloyed steels this is about 845°C or lower. Time at temperature should be a matter of seconds. The best results are obtained when the starting structure is very fine—that is, tempered martensite. The ferrite grains become smaller with each cycle, up to four or five cycles, beyond which there is little further refinement. Grains can be made as fine as ASTM 15 (1.7 μm) in carbon steels. The process does not depend upon deoxidation practice; Si-killed steels perform as well as Al-killed.

[1] R.A. Grange, *Met. Trans.*, 2:65 (1971).

The procedure has been used to increase the toughness of high-strength plate steels.[1] A plate is raised to a high temperature in the austenite + ferrite + carbide region in a conventional gas-fired furnace. Because of the presence of the three phases, neither the austenite nor the ferrite grains can grow. The plate is run out of the furnace and through an induction coil which very quickly raises the temperature above Ac_3; then the plate is rapidly cooled by a water spray and the cycle is repeated.

Although the technique is applicable to carbon steels, its cost is such that it has been applied only to expensive alloy steels. Nevertheless, the fact that ultrafine grains can be developed in carbon steels should be remembered.

Controlled rolling, in which the last pass is made at a low temperature in the austenite region, can be used to control grain size in sheet, plates, or bars, but it cannot be applied to the rolling of sections because of the different finishing temperatures associated with different parts of the same section.

Grain refinement by Al is essential in alloy steels that are to be carburized. Also, Al, Nb, or Ti can be added to inhibit austenite grain growth in steels that are to be used in the as-hot-rolled condition, or to limit grain growth and thereby improve toughness in the heat-affected zones of welds. Inhibition of austenite grain growth by AlN is the classic example of grain boundary pinning by small particles of a second phase.

When a semikilled steel, or a steel that has been killed with Si, is heated above Ac_3, austenite grain growth is continuous with time and temperature. When an Al-killed steel is reheated, the austenite grains grow very little during the periods customarily employed, up to a temperature of about 1100°C; but if this temperature is exceeded, grain growth is rapid and the final grain size at a given temperature can be greater than in the Si-killed steel. The pinning agent is AlN, but all of the AlN need not be dissolved before the grains can grow rapidly. If some grains can break away while others are still pinned, the resulting grain size can be larger than if no inhibiting agent had been used. Austenite grain growth characteristics are shown in Figs V-16 and V-17.

The problem of the pinning of grain boundaries by second-phase particles has been treated by Zener (see Smith[2]) and by Paranjpe.[3] When a second-phase particle lies in the plane of a grain boundary, the missing segment of the boundary must be created when the boundary moves away from the particle. Energy is required to create this boundary segment (see Fig. V-18). The particles are assumed to be spherical, of equal size, and uniformly distributed in the matrix. The grains, of radius R, were assumed to be spherical and isolated. The total length of the line of contact between the particle and the boundary is $2\pi r \cos\theta$, when the particle radius is r and θ is the angle between the *average* surface of the grain boundary and the surface where it meets the particle. There is a component of surface tension normal to the average boundary surface, equal to $\sigma \sin\theta$, where σ is the grain boundary surface tension. The

[1] L.F. Porter, S.J. Manganello, D.S. Dabkowski, and J.H. Gross, *Met. Eng. Q.*, 6:17 (Aug. 1966).

[2] C.S. Smith, *Trans. AIME*, 175:15 (1948).

[3] V.P. Paranjpe, *Trans. Indian Inst. Met.*, 6:203 (1952).

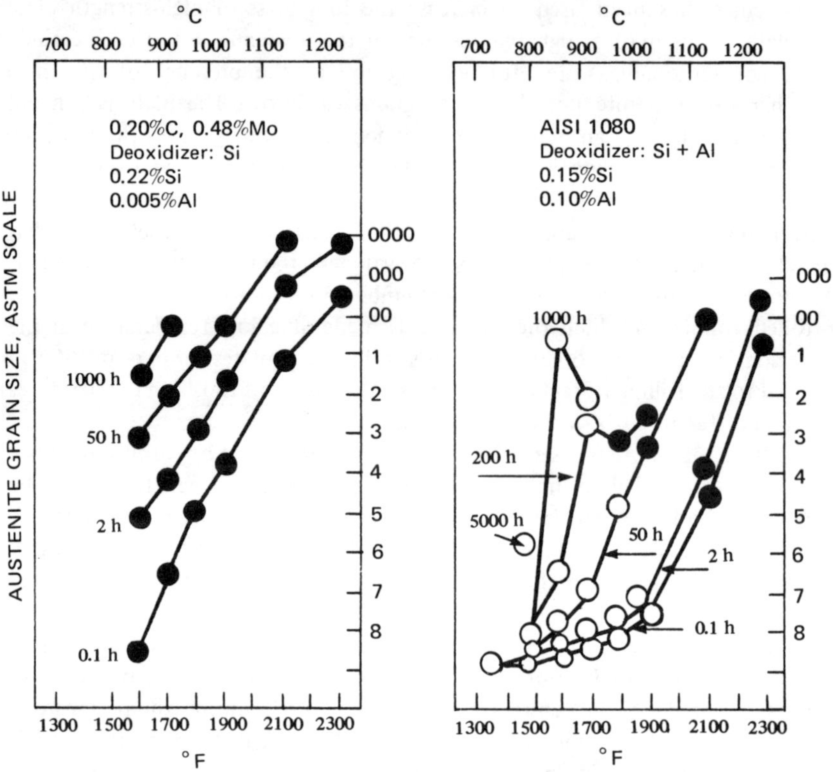

Figure V-16. Contrast between austenite grain growth in Si-killed and Al-killed carbon steels. (O.O. Miller, *Trans. Am. Soc. Met.*, 43:260 (1951), by permission.)

product of this component and the length of the line of contact gives the restraining force on the boundary

$$f = 2\pi r\sigma \sin\theta \cos\theta \tag{V-16}$$

This is at a maximum when $\theta = 45°$; then $\sin\theta \cos\theta = \frac{1}{2}$, so

$$f_{\max} = \pi r\sigma \tag{V-17}$$

The total restraining force on the boundary is approximately $N_s\pi r\sigma$, when N_s is the number of inclusions per unit area of boundary. Surface tension provides the driving force for grain growth. For a shrinking, isolated spherical surface this is equal to $2\sigma/R$ per unit area. Equating this to the restraining force due to particles,

$$\frac{2\sigma}{R} = N_s\pi r\sigma \tag{V-18}$$

Pinning then requires

$$\pi N_s R r = 2 \tag{V-19}$$

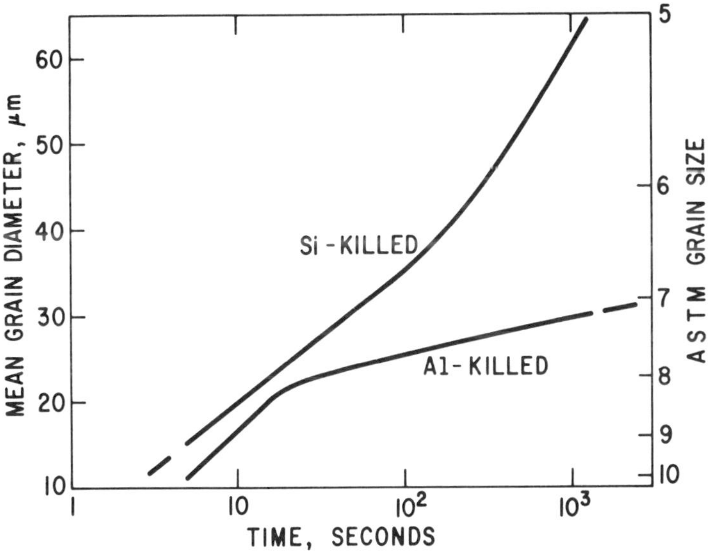

Figure V-17. Austenite grain growth with time at 870°C, AISI 1020 steels. (R.A. Grange, *Met. Trans.*, 2:65 (1971), by permission.)

It was assumed initially that the particles were uniformly distributed. The number of particles holding back a grain boundary of area A, then, is the number of particles present in a volume equal to $A \times 2r$. If the number of particles per unit volume is N_V, then this will be $2N_V Ar$. Also,

$$N_V = \frac{F}{(4/3)\pi r^3} \qquad \text{(V-20)}$$

where F = volume fraction of the second phase

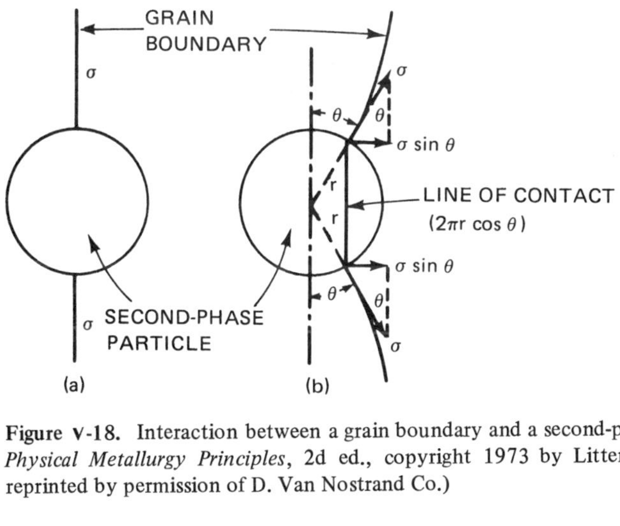

Figure V-18. Interaction between a grain boundary and a second-phase inclusion. (R.E. Reed-Hill, *Physical Metallurgy Principles*, 2d ed., copyright 1973 by Litten Educational Publishing, Inc., reprinted by permission of D. Van Nostrand Co.)

The volume of a single particle is $(4/3)\pi r^3$ and N_S, the number of particles per unit area, equals $2N_V r$. Substituting in Eq. V-18, we obtain

$$\frac{r}{R} = \frac{3}{4}F \tag{V-21}$$

or

$$R = \frac{4r}{3F} \tag{V-22}$$

The equilibrium grain size becomes smaller as the particle size is decreased and as the volume fraction of particles is increased.

Gladman[1] pointed out that Zener's assumption of an isolated spherical grain leads to an overestimation of the driving force for grain growth. In a polycrystalline aggregate, as one grain shrinks the grain boundary area of others must increase. On the assumption of space filled by tetrakaidecahedra, there will be a net reduction of grain boundary area if the ratio of the radius of the growing grain to that of its neighbors is greater than 1.33. Heterogeneity in grain size is essential for grain growth, and such heterogeneity always exists in commercial alloys.

Gladman concluded that the maximum particle size for restricting grain growth is given by

$$r = \frac{6R_0F}{\pi}\left(\frac{3}{2} - \frac{2}{Z}\right)^{-1}$$

where R_0 is the matrix grain size and Z is the ratio of the radii of growing grains and matrix grains.

It is obvious from this analysis, and from experience, that complete solution of pinning particles is not necessary for grain growth to occur; if the temperature is such that particles can coalesce to greater than the critical size, grains will be freed to grow.

At the moment, however, we are not considering the grain growth characteristics of heat-treatable alloy steels, but grain size control in as-hot-rolled steels that are customarily reheated to about 1250°C for rolling. In most instances this temperature is sufficient to dissolve all AlN, making this compound ineffective in controlling grain growth. However, if a carbon steel is made with about 0.1%Al and 0.02%N, and if the slab reheating temperature is kept below about 1200°C, then grain sizes in hot-rolled plates can be in the range ASTM 8 to 11.[2] A lower slab-heating temperature means a low finishing temperature, which also aids in producing a fine-grained steel. More will be said later about the effects of Nb, V, and Ti on austenite grain growth.

Banding in Carbon Steels

Banding is the name given to a microstructure in as-rolled carbon steels that consists of alternate layers of ferrite and pearlite. It is caused by the interdendritic segregation

[1] T. Gladman, *Proc. Roy. Soc. (London)*, 294A:298 (1966).

[2] H. Nakamura and Y. Kuriyama, Document No. IX–370–63, 1963 Meeting, Intern. Inst. of Welding.

of Mn during solidification, which is difficult to avoid in cast steel. Banding is most pronounced in steels containing ferrite and pearlite in about equal proportions, for example, in a steel containing 0.25%C and 1.5%Mn. In hot-rolled plates, the interdendritic segregation of Mn has been converted to knife-edged sheets of above-average Mn content. These are easily revealed by microradiography, as shown in Fig. V-19. Because this technique is neither well known nor widely used, a brief digression to describe it may be appropriate. An X-ray absorption discontinuity called the absorption edge (λ_k) is characteristic of every element. At a certain wavelength there is an abrupt change in the mass absorption coefficient. Wavelengths shorter than λ_k are absorbed to a much greater extent than wavelengths longer than λ_k. In our case, knowing the absorption edges of Mn and Fe, we can reveal the segregation of Mn in Fe by using Co K_α radiation, as shown in Fig. V-20. At this wavelength, the mass absorption coefficient of Fe is 59.5 and of Mn is 431 cm^2/gm.

A section of steel about 0.04 mm thick is placed in direct contact with a photographic film having a very fine grained emulsion. It is then irradiated by X-rays from the selected target. After development, the photographic image is magnified. The useful magnification is determined by the grain size of the emulsion and is probably never more than 200X. Of course, Mn segregation can also be detected by electron microprobe.

To return to banding, the sheets of pearlite coincide with sheets of Mn segregation. Mn reduces the activity of C in austenite, so C segregates with the Mn. During the transformation of austenite after hot rolling, pearlite forms at the concentrations of C and Mn, as shown in Fig. V-19*b*.

For most applications of plate steels, banding is not detrimental and in some instances it may be beneficial. Banding causes anisotropy in tensile ductility and notch-impact energy, but it must be emphasized that banding is *not* the primary cause of anisotropy of mechanical properties in plate steels. It is easier to propagate a crack parallel to bands than normal to them, so banded steels should be avoided in applications where lamellar tearing is possible.

Efforts are often made to homogenize steel by soaking at a high temperature, but this operation should always be considered carefully in advance, lest it be a waste of money. The removal of banding is a classical problem in diffusion.[1,2] We can assume, quite reasonably, that the variation of Mn concentration in a rolled plate occurs in only one direction. For simplicity, we will assume that the variation of composition along a straight line normal to the bands is sinusoidal. The concentration after hot rolling would then be (Fig. V-21):

$$C_x = C_0 + C_m \sin\left(\pi \frac{X}{l}\right) \qquad \text{(V-23)}$$

where C_x = the concentration of Mn at any point X through the thickness of the plate
C_0 = average concentration of Mn

[1] J.D. Lavender and T.W. Jones, *J. Iron Steel Inst.*, 163:14 (1949).

[2] P.G. Shewmon, in *Physical Metallurgy*, R.W. Cahn, ed., North Holland, Amsterdam, 1965, p. 370.

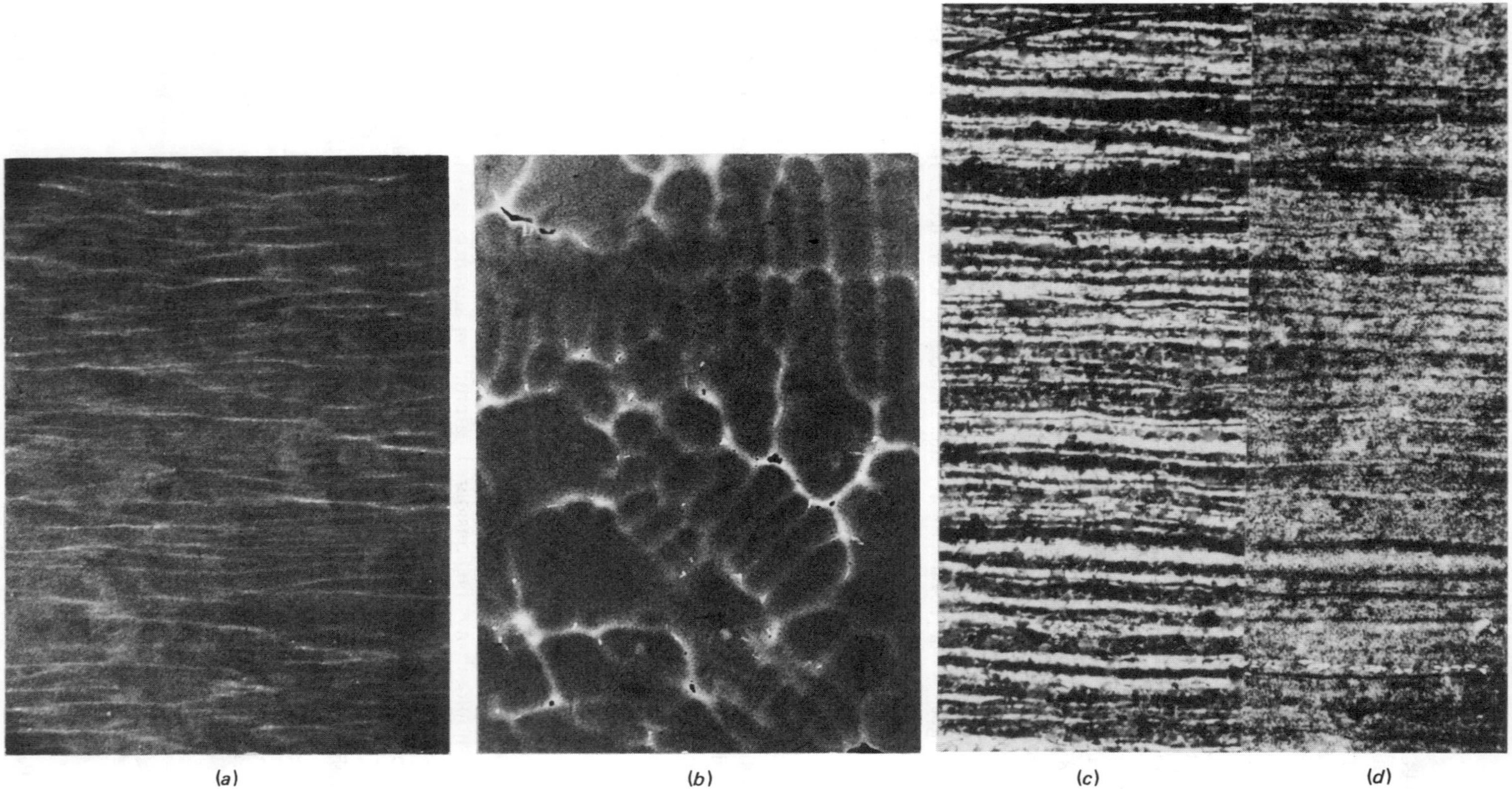

Figure V-19. Manganese segregation and its effect on microstructure. (*a*) Manganese segregation in an Fe–1.5%Mn–0.2%C alloy cast into a 75-mm diameter mold, then (*b*) hot rolled to 12.5-mm plate (56×). (*c*) Photomicrograph and (*d*) microradiograph of banding in a 0.35%C–1.8%Mn steel, as-hot-rolled (200×). (Courtesy of R.L. Miller.)

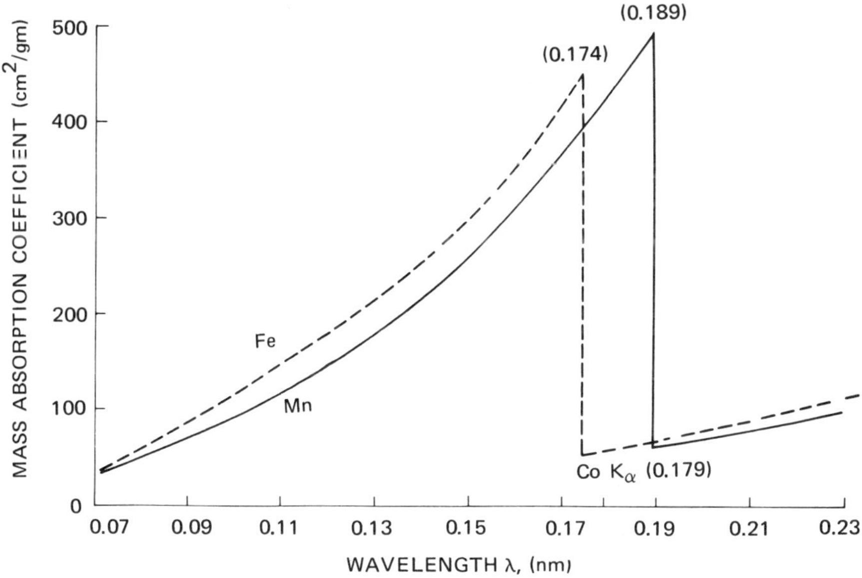

Figure V-20. The K absorption edges of iron and manganese, showing the wavelength of cobalt K_α radiation falling between them.

The wavelength of the bands is taken as $2l$. The maximum concentration in the bands would then be $C_0 + C_m$ and the minimum $C_0 - C_m$.

Given the opportunity, the Mn will flow from regions of high concentrations to regions of lower concentration. We will make the additional assumption, quite reasonable in this instance, that D, the diffusion coefficient, is not a function of concentration. Then after time t at constant temperature the concentration at any point X is given by

$$C_{(x,t)} = C_0 + \left[C_m \sin\left(\frac{\pi X}{l}\right)\right] \exp\left(-\pi^2 \frac{Dt}{l^2}\right) \tag{V-24}$$

The change of C at any point toward the final value C_0 is given by an equation of the form

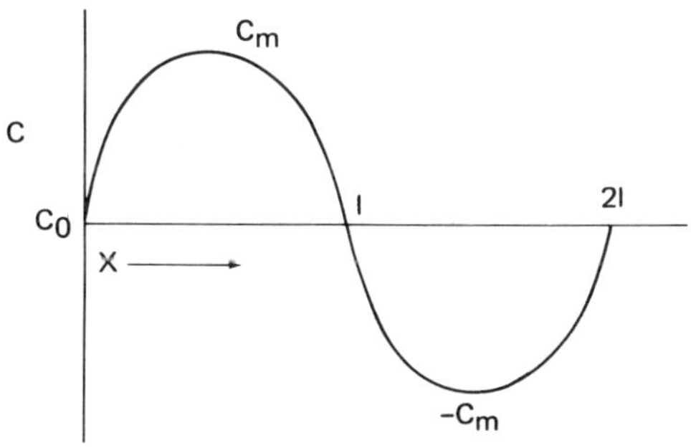

Figure V-21. Schematic sinusoidal variation of composition within a steel plate.

$$C - C_0 \cong \exp\left(-\frac{t}{\tau}\right) \quad \text{(V-25)}$$

where τ, called the relaxation time because it has the dimensions of time, is given by

$$\tau = \frac{l^2}{\pi^2 D} \quad \text{(V-26)}$$

Equation V-24 satisfies Fick's second law:

$$\frac{\delta c}{\delta t} = D\frac{\delta^2 C}{\delta X^2} \quad \text{(V-27)}$$

At $X = 0$ and $X = l$, $C - C_0$ is zero for all values of t and

$$C - C_0 = C_m \sin\left(\frac{\pi X}{l}\right) \qquad \text{when} \qquad t = 0 \quad \text{(V-28)}$$

The amplitude of the bands thus becomes a fraction $1/f$ of the original amplitude when

$$\exp\left(-\frac{\pi^2 Dt}{l^2}\right) = \frac{1}{f} \quad \text{(V-29)}$$

Taking logarithms,

$$0.4343\frac{\pi^2 Dt}{l^2} = \log f \quad \text{(V-30)}$$

We will assume that bands will become insignificant when reduced to 1/10 of their original amplitude; setting $f = 10$,

$$\frac{Dt^2}{l^2} = \frac{1}{0.4343\pi^2} = 0.233 \quad \text{(V-31)}$$

From this expression and a knowledge of the diffusivity of Mn in austenite, we can obtain the time required at a given temperature to relieve the segregation. From Wells and Mehl,[1]

$$D^{\gamma}_{\text{Mn}} = 0.65\exp\left(-\frac{276\ \text{kJ/mole}}{RT}\right)\ \text{cm}^2/\text{s} \quad \text{(V-32)}$$

If we are dealing with a steel slab, a value of l (the half-wavelength of the segregation) of 4 mm is reasonable.

To reach 90% of homogenization, the times given in Table V-2 are required. If the slab is reduced in thickness so that $l = 0.02$ mm, then the time required for 90% homogenization at 1200°C is only about 26 h. The lesson is obvious: whenever homogenization is to be attempted, first reduce the spacing between the segregations as much as possible.

[1] C. Wells and R.F. Mehl, *Met. Technol. (NY)*, 1941, Tech. Pub. 1282.

Table V-2 Time required for 90% homogenization of Mn in a steel slab

		Time	
Temperature (°C)	D_{Mn} (cm²/s)	Seconds	Days
1100	1.7×10^{-11}	22×10^{8}	25,400
1200	9.8×10^{-11}	3.8×10^{8}	4,400
1300	37×10^{-11}	1.0×10^{8}	1,150

Manganese is not the only element that segregates in bands in hot-rolled steels; Defilippi and Chao[1] used microradiography to reveal Mo and Cr segregation in Type 434 stainless steel (17%Cr, 1%Mo).

Directionality of Properties

Although banding in ferrite–pearlite steels would seem to be an obvious source of directionality of mechanical properties, nonmetallic inclusions have much greater effects,[2] so attention will be devoted principally to this phenomenon. Deformation and annealing textures can also lead to anisotropy of mechanical properties in wrought products.[3] Such textural effects are minor at present, but if trends continue toward the increasing use of controlled rolling, with finishing temperatures in the ferrite range, textures may be consciously applied to control fracture properties.[4]

In rolled products, variation of mechanical properties with testing direction is determined from test bars cut in the longitudinal (rolling) direction (*L* in Fig. V-22), the transverse direction *X*, and in the through-thickness, or short transverse, direction *Z*.[5] Often the product is not sufficiently thick to allow a full-length test piece in

[1] J.D. Defilippi and H.C. Chao, *Met. Trans.*, 2:3209 (1971).
[2] R.A. Grange, *Met. Trans.*, 2:417 (1971).
[3] D.V. Wilson, *Metals Tech.*, 2:8 (1975).
[4] D. Dulieu and I.G. Davies, *Metals Tech.*, 2:178 (1975).
[5] W.B. Morrison, *Metals Tech.*, 2:33 (1975).

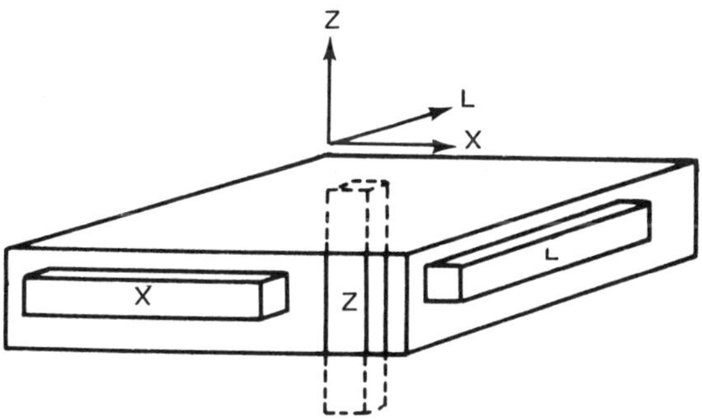

Figure V-22. Test bar orientations related to principal hot-working directions.

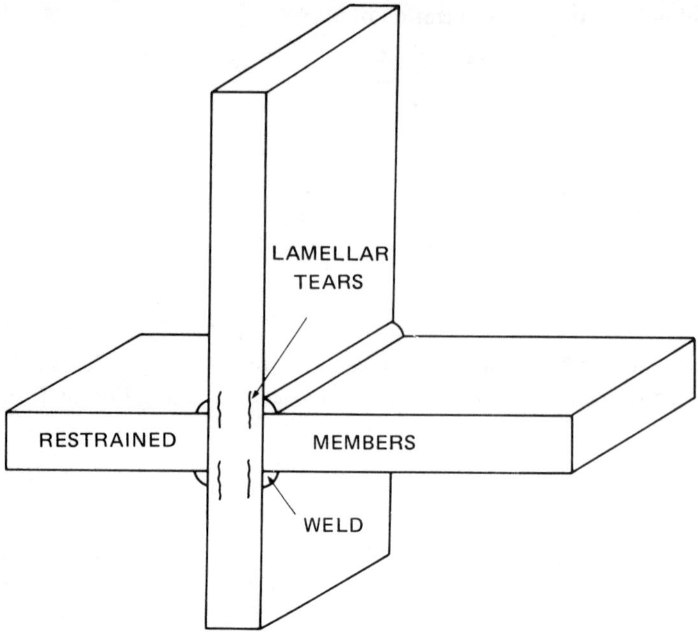

Figure V-23. Lamellar tears in heavy welded construction, occasioned by lack of through-thickness ductility in the plates.

the through-thickness direction, so extensions must be welded to the short specimen. Ductility in the through-thickness direction has become more important in recent years with the advent of heavy welded construction in which the members are constrained, as in very large ships and off-shore drilling platforms. In conditions shown in Fig. V-23 plates with inadequate through-thickness ductility are subject to lamellar tearing.

To return to the relative importance of banding and inclusions on directionality of properties, Grange[1] developed a short-time, high-temperature treatment that eliminated ferrite–pearlite banding without markedly changing the shape of inclusions, allowing him to compare the effects of inclusions in one steel in the banded and unbanded condition. When many elongated inclusions were present, removal of banding did little to improve transverse ductility. In the absence of inclusions, removal of banding improved transverse ductility (reduction of area in a tensile test). In improving transverse and through-thickness ductility of wrought products, elimination of inclusions must be the primary goal.

The effects of inclusions are exercised through their size, distribution, volume fraction and shape. The pronounced effect of volume fraction is indicated in Fig. V-24. Figure V-25 shows the pronounced effect of aspect ratio on toughness in the through-thickness direction. Some inclusions are elongated during hot deformation of

[1] R.A. Grange, *Met. Trans.*, 2:417 (1971).

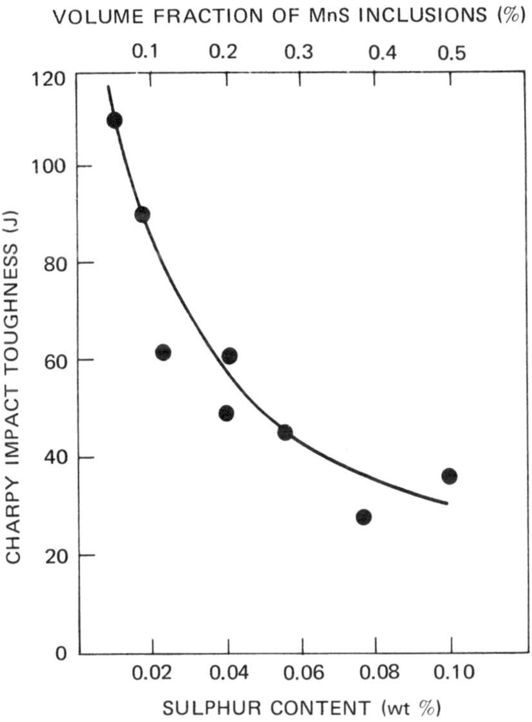

Figure V-24. Effect of volume fraction of inclusions on transverse impact toughness. (T.J. Baker, K.G. Gove, and J.A. Charles, *Metals Tech.*, 3:183 (1976), by permission.)

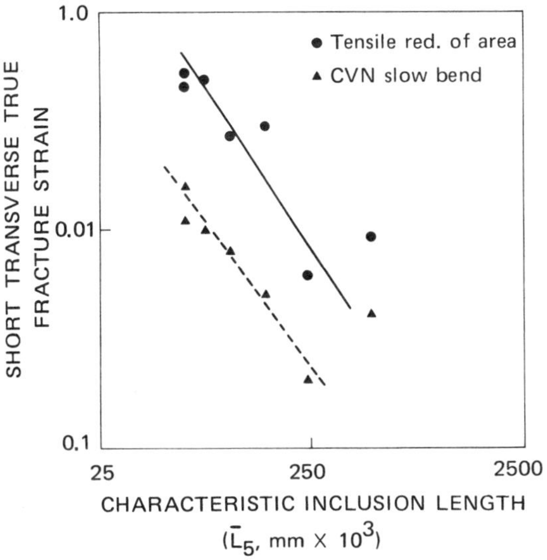

Figure V-25. Effect of inclusion length and strain rate on the short transverse fracture strain. (A.J. DeArdo and E.G. Hamburg, in *Sulfide Inclusions in Steel*, Am. Soc. Met., 1975, by permission.)

the steel, the most common of these being MnS. There are three types of MnS[1]: Type I, which forms as isolated globules with a high oxygen content and probably forms as a liquid; Type II, the dominant type, which takes the form of rods or chains precipitated in grain boundaries and also forms as a liquid; and Type III, which occurs as equiaxed particles with planar faces. It has the lowest oxygen content of the three and is usually found in killed steels.

If similar dispersions and volume fractions are obtained, silicate, sulfide, or alumina inclusions have similar effects on toughness. However, the sulfide inclusions are more likely to appear as long "stringers." The deformation of silicate inclusions during hot rolling depends upon the composition of the silicate glass, which in turn depends upon the deoxidation practice. Alumina particles are the least likely to be deformed during hot rolling, but these are precipitated as dendrites in the cast steel, and the reorientation of such configurations can cause effects as damaging to ductility as MnS stringers.

Inclusion Shape Control

It is obvious from the foregoing that to obtain reasonably isotropic mechanical properties in wrought steels it is necessary either to reduce S and O in the steels to very low levels, or to ensure that any nonmetallic inclusions are present as small equiaxed or globular particles. The desirability of either or both of these precautions increases as the strength of the steel increases. To obtain equiaxed particles in the finished steel it is necessary to precipitate in the cast steel equiaxed particles that resist deformation during subsequent hot and cold rolling. This can be done by adding to the steel elements that form refractory sulfides, such as Ca, Zr, or Ce (or mixed rare earths).[2,3] These must be added to thoroughly deoxidized steel if they are to be effective. The presence of such equiaxed sulfides can greatly improve the through-thickness ductility.

Effects of Sulfur on Machinability and Fatigue Life

Although sulfur and sulfides can have very detrimental effects on toughness of steels and can cause problems in hot working,[4] they are beneficial in at least two instances: improving machinability[5] and improving fatigue life of bearing steels.[6] To improve machinability sulfur must be present as MnS. MnS, MnSe, MnTe, and Pb, all of which improve machinability, form soft, easily deformable inclusions in steels. The MnS should be present as large globules, achieved by maintaining a high oxygen content in the steel.

[1] C.E. Sims, H.A. Saller, and F.W. Boulger, *Trans. Am. Foundrymens Soc.*, 57:233 (1949).

[2] T.M. Banks and T. Gladman, *Metals Tech.*, 6:81 (1979).

[3] P.E. Waudby, *Int. Metal. Rev.*, 23:74 (1978).

[4] W.J.McG. Tegart and A. Gittins, in *Sulfide Inclusions in Steel, Am. Soc. Met.*, Metals Park, Ohio, 1975, p. 198.

[5] C.W. Kovach, in *Sulfide Inclusions in Steel, Am. Soc. Met.*, Metals Park, Ohio, 1975, p. 459.

[6] C.M. Lyne and A. Kazak, *Trans. Am. Soc. Met.*, 61:10 (1968).

Sulfur improves fatigue life because MnS inclusions generate compressive stresses in the surrounding steel matrix. Their coefficient of thermal expansion is greater than that of the steel.[1] In contrast, silicates and alumina, having lower coefficients of thermal expansion than ferrite, generate tessellated tensile stresses in the matrix around them, thereby promoting the formation of fatigue cracks. By adding sulfur to the steel, silicate and alumina inclusions can be coated with MnS, thus reducing the tensile stresses.

Overheating and Burning

The reheating of steels for forging must be done with some care, lest either the ductility of the subsequent forging be slightly reduced or the forging break up under the hammer. The first phenomenon is termed *overheating.* It is caused by the solution of MnS in austenite at high temperatures and its subsequent precipitation during cooling as fine (~0.5 μm) particles of MnS in austenite grain boundaries.[2,3] If the steel is later broken at room temperature, the fracture follows the path of the former austenite grain boundaries. Although ductility is reduced, the fracture is ductile, resulting in only small reductions in fracture toughness. Overheating can be eliminated by increasing the Mn content, which in turn reduces the concentration of S in solid solution in austenite at any given temperature, or by adding Ca or Ce to the steel.

Burning is a distinctly different process which consists of the formation of a film of sulfur-rich liquid at austenite grain boundaries. If this exists during forging, the forging will break. Burning occurs at a higher temperature than overheating. It can be avoided in the same manner: by treating with Ca, Ce, or mixed rare earth addition to form refractory, less soluble sulfides.

Wire-Drawing, Patenting, and the Work Hardening of Pearlite

The strongest engineering material available in commercial quantities is drawn pearlitic steel wire, made to a carbon content between 0.80 and 1.00%. The product has been with us for 100 years. With due care in steel-making and wire-drawing, tensile strengths up to 4.8 GPa can be obtained, in company with a reduction in area of about 20%. The production process involves patenting and wire-drawing. Patenting is always done continuously. It consists of austenitizing, cooling rapidly, then holding in a temperature range in which fine pearlite, with perhaps some upper bainite, will form. This can be done in two baths of lead, one held at the austenitizing temperature, the other at the transformation temperature. In the newer, double-cascade-quench process, the wire from the lead austenitizing bath is quenched by liquid salt to a temperature below Ar_1, then passed into a second cascade quench at a temperature between 400 and 565°C and finally enters a holding furnace where the transformation is completed. Wire can be patented several times during processing, being drawn between each patenting, as in Table V-3.

[1] D. Brooksbank and K.W. Andrews, *J. Iron Steel Inst.*, 206:595 (1968).
[2] D.R. Glue, C.H. Jones, and H.K.M. Lloyd, *Metals Tech.*, 2:416 (1975).
[3] T.J. Baker, in *Sulfide Inclusions in Steel, Am. Soc. Met.*, Metals Park, Ohio, 1975, p. 135.

Table V-3 Typical sequence of patenting

Diameter (mm)	Austenitizing temperature (°C)	Prequench temperature (°C)	Transformation temperature (°C)
8.53	925	290	540
5.59	925	290	540
3.23	925	290	540
1.68	925	290	540
0.84	925	290	540

For eutectoid carbon steel, the strength after patenting can be in the range 1240–1450 MPa, with pearlite interlamellar spacings as small as 40 nm. It is at this strength that drawing of high-strength wire begins. In a 0.95%C steel as much as 50% of the structure can be upper bainite.

When pearlitic steel wire is drawn, the ferrite takes up a ⟨110⟩ fiber texture, just as in drawn iron wire. The presence of the cementite does not alter the texture of the ferrite. Transverse sections of the wire show the same "curly grain" structure as in drawn ferrite,[1,2] as indicated in Fig. V-26. The difference is in the much higher rate of work hardening. With pearlitic wire there is a linear relation between ln $(\sigma - \sigma_0)$ and ϵ,

[1] G. Langford, *Met. Trans. A,* 8A:861 (1977).
[2] N. Lambert and T. Greday, *Metall. Rep. CRM,* 40:43 (Sept. 1974).

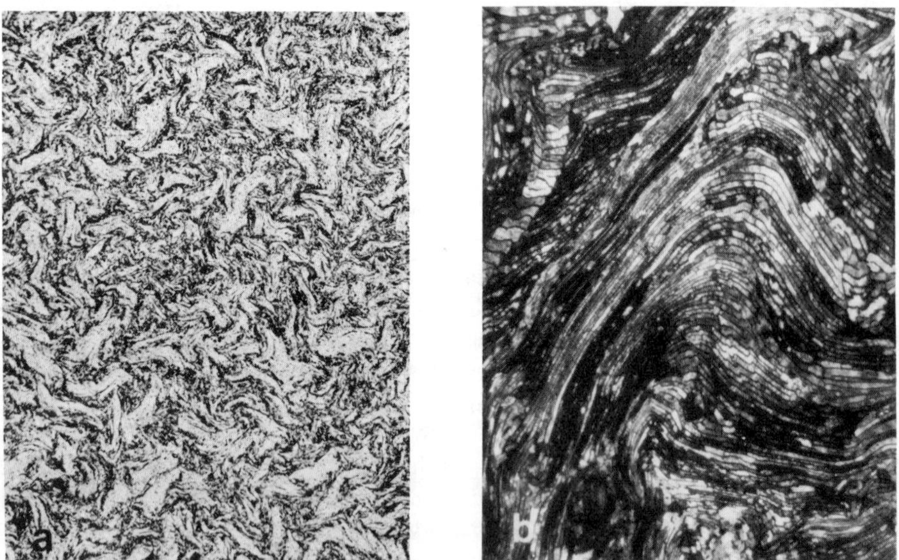

Figure V-26. Curly grain structure in transverse sections of cold-drawn wire. (*a*) Iron (light micrograph, true strain 2.7, 2% nital etch, 200X). (*b*) pearlitic steel (transmission electron micrograph, true strain 3.2, 20,000X). Longitudinal sections of both show fibrous structure. (G. Langford, *Met. Trans. A,* 8A:861 (1977), by permission.)

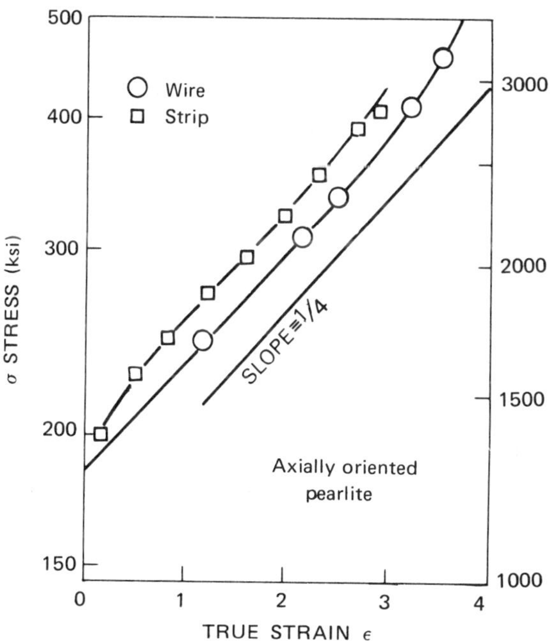

Figure V-27. Strain hardening of axially oriented pearlite in wire drawing and strip drawing. (G. Langford, *Met. Trans. A*, 8A:861 (1977), by permission.)

as shown by Fig. V-27. The true strain in wire-drawing of pearlite,

$$\epsilon = 2 \ln \frac{D_{\text{initial}}}{D_{\text{final}}}$$

is limited by fracture to about 4.

Starting with randomly oriented pearlite lamellae, all are oriented parallel to the wire axis after a true strain of about 1. After a true strain of about 4, the cementite lamellae can be reduced to a thickness of about 2 nm. A unit cell of cementite (orthorhombic) has lattice constants

$$a = 0.509 \text{ nm}$$

$$b = 0.647 \text{ nm}$$

$$c = 0.452 \text{ nm}$$

so there are only a few unit cells across the width of a lamella; the lamella is nearly "all surface."

During deformation, C atoms must leave the cementite for more favored sites at dislocations, aided by adiabatic heating during drawing. This dissolved C pins dislocations, which act as barriers to additional dislocation motion. It is well known that the wire must be kept cool during drawing to prevent fracture. Drawing at about -100°C produces lower strength. A high rate of dislocation pinning leads to a very high rate of work hardening.[1]

[1] V.K. Chandhok, A. Kasak, and J.P. Hirth, *Trans. Am. Soc. Met.*, 59:288 (1966).

Cold Forming

If a steel part such as a bolt can be formed cold rather than being forged hot or machined from a blank, there are obvious advantages to be gained. The surface finish, dimensional control, and strength are better than after hot forging and the waste of material inherent in machining is avoided. For these reasons, use of the cold-forming process is expanding rapidly.

Cooksey[1] conducted an extensive study of the compositional and microstructural variables affecting the extrusion pressure in backward extrusion of spheroidized steel slugs of 1-in diameter and length, with an extrusion ratio of 2:1. The extrusion ratio is the ratio of the initial cross-sectional area to the cross-sectional area after extrusion, $R = A_0/A_f$. The slugs were lubricated with zinc phosphate and soap. The results apply to steels containing a maximum of 0.4%C, 1.5%Mn, 0.35%Si, 0.6%Cr, 0.5%Mo, 0.6%(Ni + Cr). Extrapolation beyond these limits should be done cautiously. Cooksey derived a multiple regression equation for extrusion pressure:

$$\begin{aligned} P\,(\text{MPa}) = 15.44[&46.52 + 2.73\text{Mn} + 16.14\text{Si} + 0.95\text{Cr} + 6.0\text{Mo} \\ &+ 3.52(\text{Ni} + \text{Cu}) + 0.33\,(\%\,\text{pearlite}) + 0.26\,(\text{ferrite grain size})^{-1/2} \\ &+ 0.18\,(\text{mean free ferrite path})^{-1}] \end{aligned} \tag{V-33}$$

In this equation, grain size and mean free ferrite path are in millimeters. The 95% confidence limits were stated to be ±67.9 MPa.

It is obvious that to minimize extrusion pressure the carbon content must be kept to the minimum necessary to attain the desired strength in the part, whether it is to be used in the cold-formed state or subsequently heat-treated. It is also obvious that elements that segregate in ferrite in preference to the carbide, and strengthen the ferrite, should be avoided. Silicon is particularly harmful in this respect, so Al-killed steels are preferred over Si-killed. If it is necessary to increase hardenability without an undue penalty in extrusion pressure, manganese is the preferred element, on the bases of cost and effect on extrusion pressure. If still greater hardenability is needed, an addition of boron should be considered. Of the other common alloying elements, chromium is the least harmful, because of its very small solid solution strengthening effect in ferrite and its tendency to segregate in the carbide.

Cooksey[1] found that Brinell hardness was the laboratory test that gave the best correlation with extrusion pressure. He suggested the equation

$$P\,(\text{MPa}) = 15.44(20 + 0.36H_{B3000}) \tag{V-34}$$

Later work by Hurley et al.[2] confirmed the results of Cooksey's study with the exception of the effect of nickel, which Hurley et al. found to have little influence on extrusion pressure. In view of the strengthening of ferrite by nickel during backward extrusion (Fig. V-28), this claim must be regarded with some suspicion.

[1] R.J. Cooksey, *Met. Form.*, 35:98 (1968), No. 4.
[2] J.L. Hurley, A.L. Kitchin, and R.R. Crawford, *Met. Eng. Q.*, 11:36 (1971), No. 3.

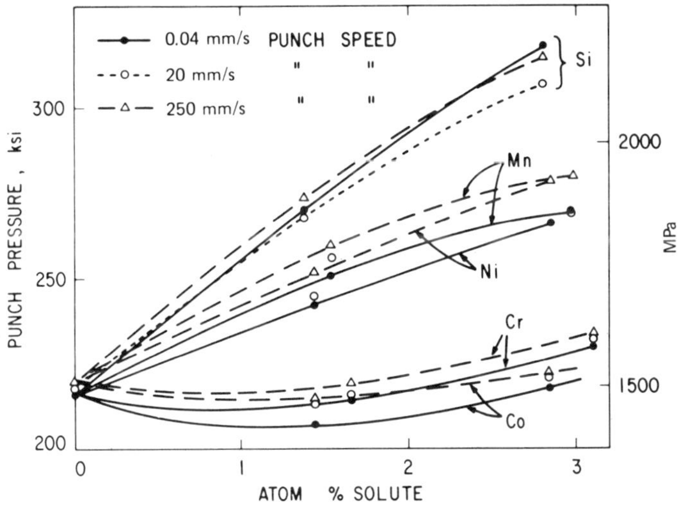

Figure V-28. The effects of elements in solid solution in iron on the punch pressure required for the backward extrusion of small cups. (L. Kashar, cited in *Met. Trans.*, 3:5 (1972).)

Superplasticity in Carbon Steels

Superplasticity is the name given to a behavior characterized by very large elongations of alloys in tension tests, which can exceed 1000%. To obtain such behavior it is necessary for the alloy to contain fine grains and to maintain them at the test temperatures, and for the alloy to have a large dependence of flow stress on the strain rate. The strain rate dependence of the flow stress is usually expressed as

$$\sigma = K\dot{\epsilon}^m \tag{V-35}$$

although this relation is often not closely followed. The temperature at which superplastic behavior occurs is usually somewhat above half the melting temperature of the alloy, in kelvins, and the strain rate required is between creep and the strain rate of conventional hot-working processes. Grain boundary sliding and grain rotation are involved.[1] The value of m, $d \ln \sigma / d \ln \dot{\epsilon}$, is generally in the neighborhood of 0.5.

Superplasticity in carbon steels can be achieved by[2]

1. Working in the $\alpha + \gamma$ range so that grain growth is minimized
2. Making additions of Al, P, or Si to expand the $\alpha + \gamma$ region to higher temperatures

When these things are done, total elongations exceeding 300% can be attained at strain rates of about $10^{-4}\ s^{-1}$ and at temperatures near 900°C. Because of the special compositions, high forming temperatures, and low strain rates involved, superplastic

[1] J.W. Edington, K.N. Melton, and C.P. Cutler, *Prog. Mater. Sci.*, 21:61 (1976).
[2] W.B. Morrison, *Trans. Am. Soc. Met.*, 61:423 (1968).

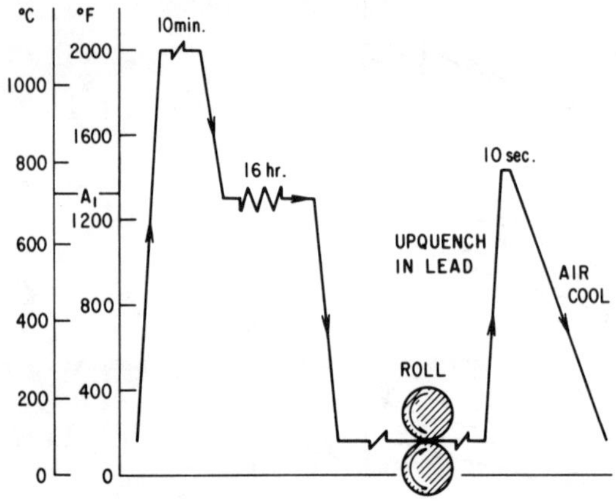

Figure V-29. Thermomechanical process for making oriented coarse pearlite. (R.A. Grange, *Proc. 2nd ICSMA*, Am. Soc. Met., 1970, by permission.)

forming is of little technological interest for a low-cost material such as carbon steels, but it can be of great interest in the production of hard-to-form aircraft parts of expensive titanium alloys.

Control of Microstructure in Carbon Steels

Some microstructures having very interesting mechanical properties can be developed in carbon steels by ingenious control of processing.[1] One such structure is oriented pearlite. In a eutectoid carbon steel the carbide lamellae occupy about 1/7 of the total volume. If these could be oriented, we would obtain a composite material. A thermomechanical process for orienting pearlite is shown in Fig. V-29. The process was begun with plates about 7.6 mm thick. These were austenitized, then transformed to coarse pearlite at 700°C. They were then air-cooled to ambient temperature and subsequently cold-rolled to 75% reduction in thickness, which oriented the pearlite lamellae. Rapid austenitizing, holding only about 10 s in the austenite range, dissolved the carbide lamellae, but did not allow the homogeneous redistribution of carbon. A pattern of carbon segregation remained, in the form of the earlier coarse pearlite. After air-cooling, oriented coarse pearlite appeared where fine pearlite would have

[1] R.A. Grange, *Proc. 2nd ICSMA, Am. Soc. Met.*, Cleveland, Ohio, 1970, p. 861.

Table V-4 Mechanical properties of oriented coarse pearlite

Structure	Yield strength (0.2% offset, MPa)	Ultimate tensile strength (MPa)	Elongation in 25mm (%)
Randomly oriented coarse pearlite	336	671	10.5
Oriented coarse pearlite	544	690	18.5

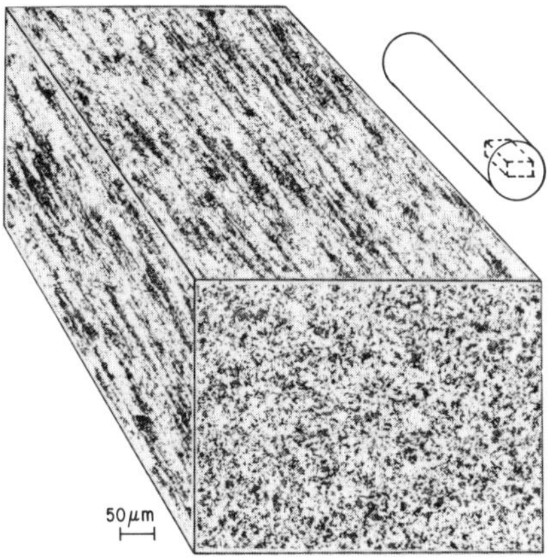

Figure V-30. Fibrous martensite in matrix of ferrite in a 1016 steel rod, swaged cold to 90% reduction of area, rapidly heated into the $\alpha + \gamma$ region, then quenched. (R.A. Grange, *Proc. 2nd ICSMA*, Am. Soc. Met., 1970, by permission.)

formed during conventional processing. The properties obtained, in the longitudinal direction, are listed in Table V-4.

Unfortunately the mechanical properties of coarse pearlite, even when oriented, are not particularly attractive and the process does not work for fine pearlite.

Other kinds of processing produce oriented martensite fibers. If the nonhomogeneous austenite produced by rapid austenitizing (Fig. V-29) is quenched, the resulting martensite will be inhomogeneous, with high-carbon martensite fibers in a low-carbon martensite matrix. This process gave the results shown in Table V-5.

Usually structures consisting of mixtures of ferrite and martensite have inferior mechanical properties, but old Japanese swords were made by forging and folding steel and iron into a composite. After hardening, they were composed of alternate layers of martensite and soft iron.[1] This structure can be produced in any hypoeutectoid steel by two different kinds of treatment. In one, a mixture of about 50% ferrite, 50% martensite can be developed by austenitizing, cooling into the $\alpha + \gamma$ region, rolling to about 70% reduction, then quenching. The product has about the same strength as the same steel produced by austenitizing and quenching, but has improved toughness.

[1] E.C. Bain, *J. Iron Steel Inst.*, 200:265 (1962).

Table V-5 Mechanical properties of martensite–ferrite composite

Structure	Yield strength (0.2% offset, MPa)	Ultimate tensile strength (MPa)	Elongation in 25mm (%)
Conventional (homogeneous) martensite	1270	1296	<1
Fiber martensite	1455	1517	5

A more interesting process uses steels containing about 0.2%C. The hot-rolled steel is cold-rolled to about 70% reduction of area. It is then rapidly heated to a low temperature in the $\alpha + \gamma$ region and quenched. Austenite forms in the flattened, extended regions occupied by the deformed pearlite, and quenching transforms the austenite pools to fibrous martensite. In this manner, low-carbon steels can be strengthened in much larger sections than can be quenched to 100% martensite. This process could be applied to the treatment of cold-drawn steel bars. The outstanding feature of the structure (Fig. V-30) is that even at −196°C, the temperature of liquid nitrogen, it still produces a ductile fracture in a notch impact test.

CHAPTER

SIX

HIGH-STRENGTH, LOW-ALLOY STEELS

VI-1. DEFINITION AND GRADES

The American Iron and Steel Institute[1] states that high-strength, low-alloy steels

> . . . comprise a specific group of steels with chemical composition specially developed to impart higher mechanical properties values, and, in certain of these steels, materially greater resistance to atmospheric corrosion resistance than is obtainable from conventional carbon steels. HSLA steel is generally produced with emphasis on mechanical properties requirements rather than to chemical composition limits. It is not considered to be alloy steel, even though the utilization of any intentionally added alloy content would technically qualify it as such.

HSLA steels can be produced in the as-hot-rolled condition with yield strengths in the range 290-550 MPa and ultimate tensile strengths in the range 415-700 MPa. Because of their low carbon content, they are readily weldable. Their development was spurred by the demand for strong, tough, weldable steels for natural gas transmission lines, ships, and off-shore drilling platforms. More recently, they are being used to reduce the weight, and hence the fuel consumption, of automotive equipment. Table VI-1 lists the grades available for automotive use. Note that these grades are specified by yield strength rather than by chemical composition.

The strengthening mechanisms applied in HSLA steels include:

1. grain refinement
2. precipitation
3. dislocation substructure
4. solid solution strengthening
5. strain aging

[1] *High-Strength Low-Alloy Steel and High-Strength Intermediate Manganese Steel,* AISI Steel Products Manual 17, Oct. 1967.

Table VI-1 High-strength, low-alloy steels for automotive applications[1]

Grade	Chemical composition (ladle analysis) (% max)[2]			Form	Minimum mechanical properties				
					Yield strength		Tensile strength		Elongation
	C	Mn	P		(ksi)	(MPa)	(ksi)	(MPa)	(% in 50 mm)
942X	0.21	1.35	0.04	Plates, shapes, and bars to 100 mm incl.	42	290	60	414	24
945A	0.15	1.00	0.04	Sheet and strip	45	310	60	414	22
945C	0.23	1.40	0.04	Plates, shapes, and bars to 12.4 mm incl.	45	310	65	448	22
				12.5–37.5 mm incl.	42	290	62	427	24
				37.5–75 mm incl.	40	276	62	427	24
945X	0.22	1.35	0.04	Sheet and strip	45	310	60	414	25
				Plates, shapes, and bars to 37.5 mm incl.	45	310	60	414	22
950A	0.15	1.30	0.04	Sheet and strip	50	345	70	483	22
950B	0.22	1.30	0.04	Plates, shapes, and bars to 12.5 mm incl.	50	345	70	483	22
950C	0.25	1.60	0.04	12.5–37.5 mm incl.	45	310	67	462	24
950D	0.15	1.00	0.15	37.5–75 mm incl.	42	290	63	434	24
950X	0.23	1.35	0.04	Sheet and strip	50	345	65	448	22
				Plates, shapes, and bars to 37.5 mm incl.	50	345	65	448	–
955X	0.25	1.35	0.04	Sheet and strip	55	380	70	483	20
				Plates, shapes, and bars to 37.5 mm incl.	57	393	70	483	–
960X	0.26	1.45	0.04	Sheet and strip	60	414	75	517	18
				Plates, shapes, and bars to 37.5 mm incl.	60	414	75	517	–
965X	0.26	1.45	0.04	Sheet and strip	65	448	80	552	16
				Plates, shapes, and bars to 19 mm incl.	65	448	80	552	–
970X	0.25	1.65	0.04	Sheet and strip	70	483	85	586	14
				Plates, shapes, and bars to 19 mm incl.	70	483	85	586	–
980X	0.26	1.65	0.04	Sheet and strip	80	552	95	655	12
				Plates to 9.5 mm incl.	80	552	95	655	–

[1] Adapted from *Metal Progress Databook*, Vol. 116, June 1979, p. 34.
[2] 0.05 S max, 0.90 Si max.

Their production can entail:

1. alloying with small amounts of strong carbide and nitride-forming elements (microalloying)
2. controlled rolling
3. controlled cooling
4. inclusion shape control

These factors can be employed singly or in combination to produce the desired properties.

The development of HSLA steels is an interesting example of the interplay of technology and economics. The use of small amounts of Nb to strengthen ferrite-pearlite steels was patented by the Union Carbide Corporation in 1936, but at that time the cost of Nb and the lack of demand for steels of this type made the process little more than a curiosity. In the late 1950s, however, a drop in the price of Nb and a simultaneous demand for improved strength, toughness, and weldability in steels for pipelines led to a reawakening of interest.[1,2]

VI-2. GRAIN REFINEMENT IN HSLA STEELS

Grain refinement is accomplished by the addition of Nb, V, Ti, or Al.[3] Nb or V can be added to semikilled steels having a higher ingot-to-slab yield than the killed steels, thus reducing the cost. Of the two, Nb is favored. Because the solubility of Nb (C, N) in austenite is less than the solubility of V carbide and nitride (Table IV-1), grain refinement can be achieved with a smaller addition. Ferrite grain refinement in ferrite-pearlite steels is accomplished through restricting the growth of austenite grains during hot rolling and/or by inhibiting the recrystallization of austenite during hot rolling so that the $\gamma \rightarrow \alpha$ transformation occurs in unrecrystallized austenite.

In the hot rolling of semikilled carbon steels, recrystallization of austenite occurs at temperatures down to about 760°C. A 30% reduction at 815°C is sufficient for about 10% recrystallization. In a similar steel containing 0.03%Nb, 10% recrystallization occurred after the steel was reduced 50% at 925°C. This explains why it is so difficult to refine grains in carbon steels until the rolling temperature drops below 815°C. It is usually less expensive to normalize a carbon steel than to obtain grain refinement by controlled rolling. In contrast, grain refinement in niobium steels can be obtained at finishing temperatures as high as 925°C.

In most instances, all Nb, C, and N are in solution at the start of the hot rolling of austenite, but precipitation occurs during the rolling as the temperature of the steel drops. The precipitate particles hinder growth of austenite grains, and at still lower

[1] C.A. Beiser, ASM preprint No. 138, 1959.
[2] C.L. Altenburger, *Am. Iron Steel Inst., Regional Tech. Meetings*, vol. 59, 1960.
[3] F.B. Pickering, in *Microalloying 75*, Union Carbide Corp., New York, 1976, p. 9.

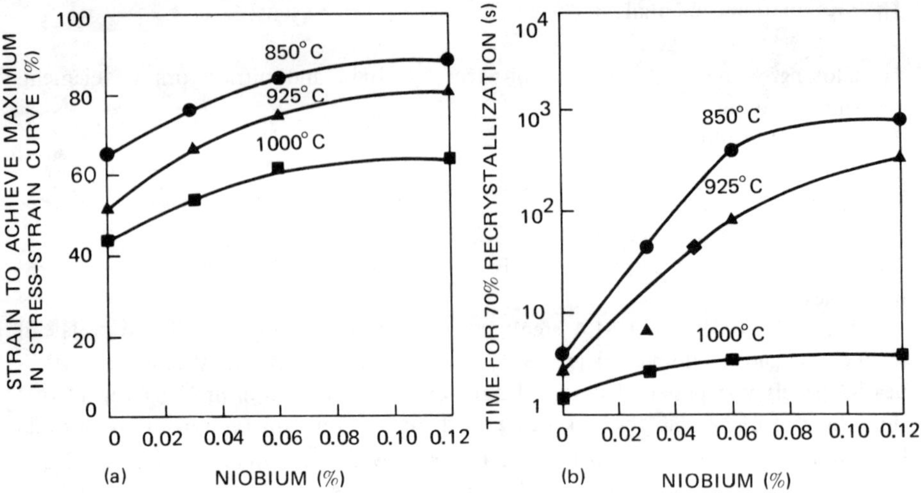

Figure VI-1. Effect of niobium on the recrystallization of a 0.05%C–1.8%Mn steel. (*a*) Dynamic recrystallization. (*b*) Static recrystallization. (L. Meyer, F. Heisterkamp, and W. Mueschenborn, in *Microalloying 75*, Union Carbide Corp., New York, 1976, p. 153, by permission.)

temperatures the particles (or preprecipitation clusters) inhibit recrystallization of the deformed austenite grains. These effects of Nb on recrystallization are shown in Fig. VI-1. As indicated in Fig. VI-2, the effectiveness of microalloying elements in refining ferrite grains is in the same order as the solubility of their carbides in austenite (Fig. VI-3).

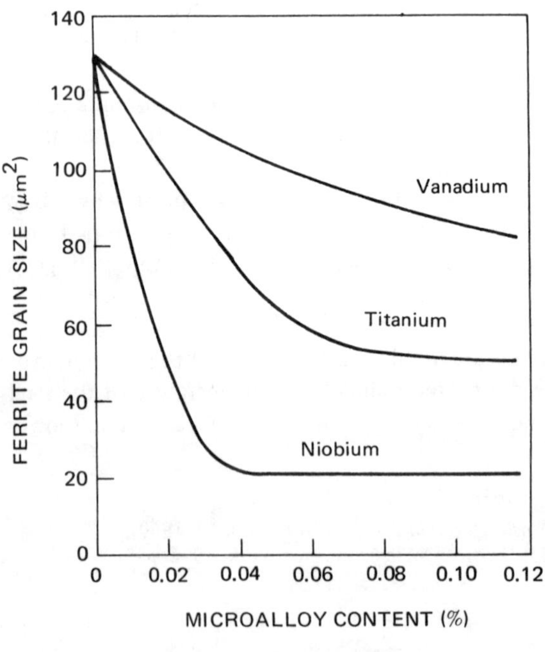

Figure VI-2. Effect of microalloy content (niobium, vanadium, or titanium) on the ferrite grain size of as-hot-rolled low-carbon steel strip. (L. Meyer, F. Heisterkamp, and W. Mueschenborn, in *Microalloying 75*, Union Carbide Corp., New York, 1976, p. 153, by permission.)

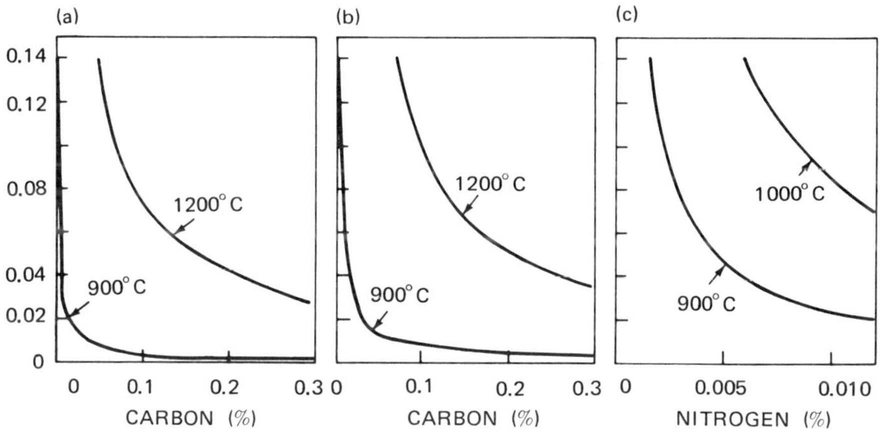

Figure VI-3. Solubility isotherms of (*a*) niobium carbide, (*b*) titanium carbide, and (*c*) vanadium nitride in austenite. (L. Meyer, F. Heisterkamp, and W. Mueschenborn, in *Microalloying 75*, Union Carbide Corp., New York, 1976, p. 153, by permission.)

VI-3. PRECIPITATION STRENGTHENING IN HSLA STEELS

Grain refinement is the preferred mode of strengthening in HSLA steels because it also improves toughness, but precipitation strengthening is the next most preferred mechanism. The particles that form at high temperatures in austenite, although they are effective in controlling grain growth, do not cause strengthening because they are too large and widely spaced. The strengthening particles are those that form at low temperatures in austenite, at the α-γ interface during transformation, and in ferrite during cooling. Because of its higher solubility in austenite, VN tends to precipitate in ferrite where it is an effective strengthener.

For a sparingly soluble precipitate such as NbC, the greatest amount of precipitate is available at the stoichiometric composition, but in most HSLA steels the Nb/C ratio is much less than stoichiometry.

Studies of precipitation in these steels led to the discovery of interphase precipitation,[1] which has since been shown to occur in many systems. At the phase transformation $\gamma \rightarrow \alpha$, there is a discontinuous decrease in solubility of the carbide or nitride. Precipitate nucleation and growth occurs at the $\gamma \rightarrow \alpha$ interface, which normally moves discontinuously.[2] The result is the configuration shown in Fig. VI-4, in which the particles form on the interface planes and in lines on those planes. The lines mark former positions of ledges on the interface.

The spacing of the precipitate planes decreases as cooling rate increases. The growth of particles occurs by diffusion of Nb in ferrite, which is assumed to be 100 times more rapid in ferrite than in austenite. In the latter, diffusivity is given by

[1] J.M. Gray and R.B.G. Yeo, *Trans. Am. Soc. Met.*, 61:255 (1968).

[2] E. Eichen and J.W. Spretnak, *Trans. Am. Soc. Met.*, 61:454 (1959).

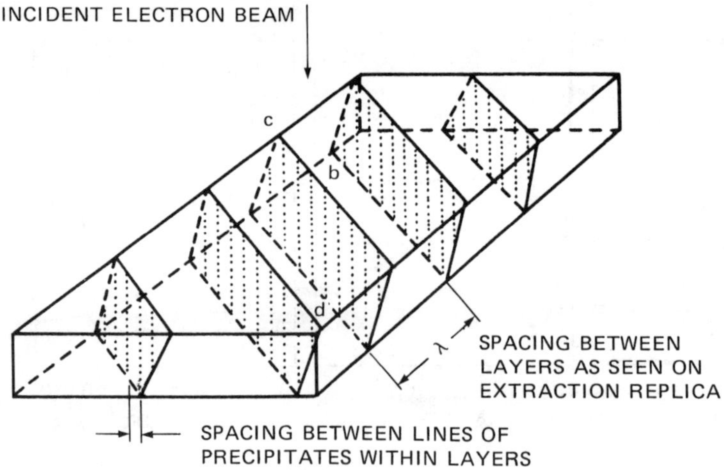

Figure VI-4. Schematic diagram of three-dimensional precipitate distribution. (J.M. Gray and R.B.G. Yeo, *Trans. Am. Soc. Met.*, 61:255 (1968), by permission.)

$$D = 400 \exp\left(-\frac{293 \text{ kJ/mol}}{RT}\right) \tag{VI-1}$$

In Fig. VI-5 the Nb atoms are assumed to move from the line XY, midway between the last row of precipitates to form and the new position of the α–γ interface. The distances that the Nb atom can travel in ferrite during the normal rates of air cooling are consistent with the spacing of the precipitate planes, 80 to 400 nm.

Interphase precipitation also has been seen in steels containing V, W, Mo, Ti, and Cr.[1] Davenport and Honeycombe[2] have modified the model of Gray and Yeo to include a buildup of carbon in the austenite ahead of the moving boundary, which aids nucleation of carbides at the boundary. Such nucleation also can be assisted by drag of substitutional solutes by the advancing boundary. When carbides precipitate, the carbon content of the austenite adjacent to the boundary is depleted, which increases the driving force for transformation to ferrite. These steps are illustrated in Fig. VI-6.

Honeycombe[3] has reviewed the direct evidence for the nucleation of alloy carbide particles at γ–α interfaces during transformation of austenite. Such interfaces within a single austenite grain are seldom curved; rather, they are composed of planar facets joined by ledges. The orientation relationship between ferrite and austenite is the Kurdjumov–Sachs

$$\{111\}_\gamma \parallel \{110\}_\alpha$$

$$\langle 110\rangle_\gamma \parallel \langle 111\rangle_\alpha$$

[1] R.W.K. Honeycombe and F.B. Pickering *Met. Trans.*, 3:1099 (1972).
[2] A.T. Davenport and R.W.K. Honeycombe, *Proc. Roy. Soc. (London)*, A322:191 (1971).
[3] R.W.K. Honeycombe, *Met. Trans. A*, 7A:915 (1976).

which matches the closest packed planes in both lattices. The regularly spaced sheets of precipitate particles are nucleated at the interface and grow in the ferrite, as shown in Fig. VI-7. The ledges are probably noncoherent and move too rapidly for nucleation of carbides to occur on them. Nucleation occurs on the slower moving, low-energy planar facets. The ledge height is often uniform, as shown in Fig. VI-7*a*, and between 5 and 50 nm, but sometimes it may vary, as shown in Fig. VI-7*b*. This nucleation process, occurring across an austenite grain, produces the astonishingly uniform sheets of particles that are frequently observed.

The particles formed during interphase precipitation, or in ferrite after the phase transformation, can be very small, on the order of 5 nm, and are therefore very effective as strengthening agents. Figure VI-8 illustrates the strengthening that can be attained by additions of Nb or Ti to hot-rolled strip. Niobium is a more effective strengthener at low concentrations than is Ti. Cold rolling and box annealing lower the strength of the Nb steel by agglomeration of carbide particles. The Ti steel suffers less from this process.

Because of the distinct contribution of precipitation to the strength of many HSLA steels, it is proper to consider the theory of precipitation strengthening and

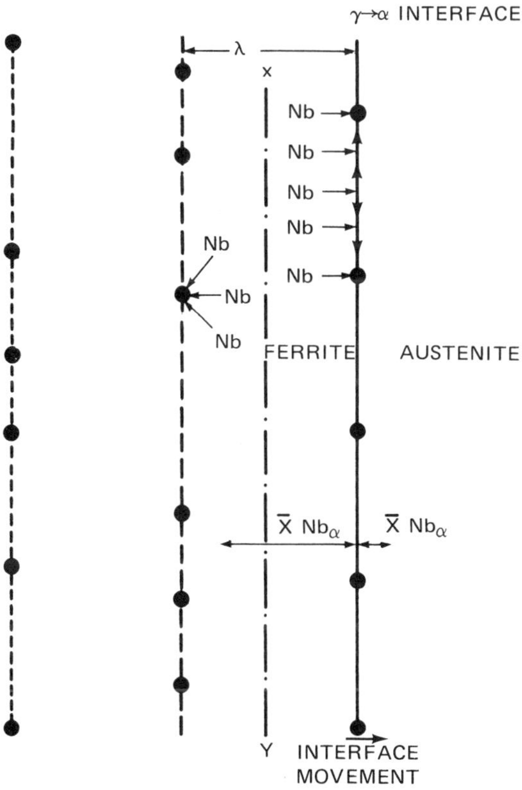

Figure VI-5. Diffusional model for the interphase precipitation of niobium carbide. (J.M. Gray and R.B.G. Yeo, *Trans. Am. Soc. Met.*, 61:255 (1968), by permission.)

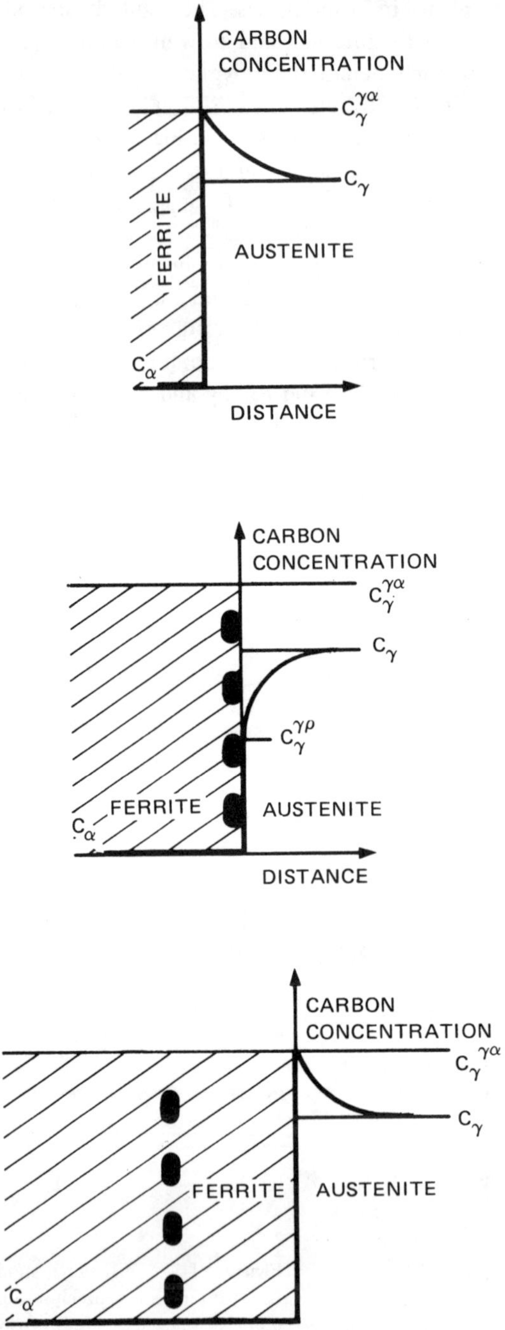

(a)

(b)

(c)

Figure VI-6. Schematic model of interphase precipitation (according to A.T. Davenport and R.W.K. Honeycombe, *Proc. Roy. Soc. (London)*, A322:191 (1971), by permission). (*a*) The carbon concentration in austenite increases at the ferrite-austenite boundary. (*b*) The carbon concentration promotes carbide nucleation on the ferrite side of the boundary, thereby pinning the boundary and depleting the carbon concentration in the austenite. The depletion of carbon increases the driving force for the transformation of austenite, moving the boundary away from the precipitate particles. (*c*) The process begins again.

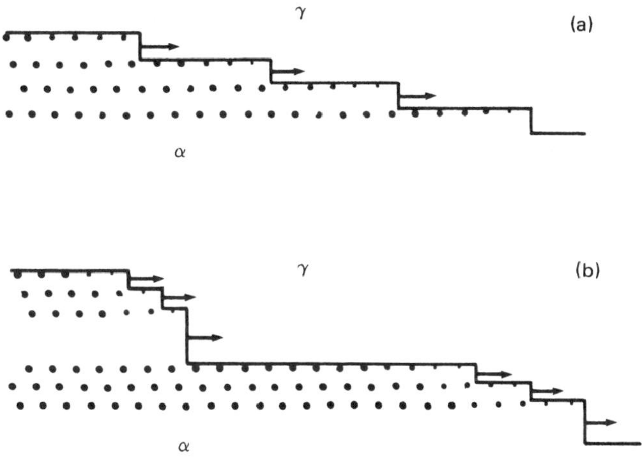

Figure VI-7. The mechanism of nucleation and growth of carbides on the γ–α interface. (R.W.K. Honeycombe, *Met. Trans. A*, 7A:915 (1976), by permission.)

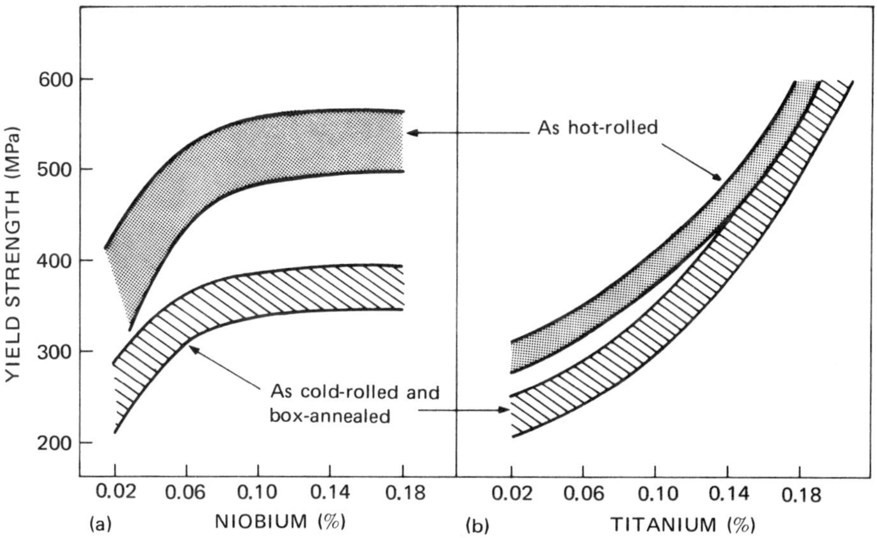

Figure VI-8. Strengthening effect of (*a*) niobium and (*b*) titanium in hot strip and cold-rolled sheet after temper rolling. (L. Meyer, F. Heisterkamp, and W. Mueschenborn, in *Microalloying 75*, Union Carbide Corp., New York, 1976, p. 153, by permission.)

its agreement with observed strengthening. Following Gladman et al.,[1] we adopt Ashby's[2] modification of Orowan's model, in which a disolocation line bows out between two neighboring particles on a slip plane. The model is also based on a random distribution of precipitate particles, which is clearly not the case when interphase precipitation occurs. In this model, the shear stress required for a dislocation to move through a field of hard, nondeformable particles is given by

$$\tau = \frac{1}{1.18}\left(\frac{1.2Gb}{2\pi L}\right)\ln\left(\frac{\bar{X}}{2b}\right) \tag{VI-2}$$

where τ = resolved shear stress
$\bar{X}$ = mean linear intercept diameter of a precipitate particle
L = distance from surface to surface of particles, equal to $(1/\sqrt{n_s}) - \bar{X}$ where n_s is the number of particles per unit area of slip plane
G = shear modulus, 8.065 GPa for iron
b = magnitude of Burgers vector in the slip direction, 0.248 nm for iron

With L and $\bar{X}$ in m, Eq. VI-3 becomes

$$\tau\ (\text{MPa}) = \frac{2.6}{L}\ln\left(\frac{\bar{X}}{2.5 \times 10^{-4}}\right) \tag{VI-3}$$

Assuming in the usual manner that $\sigma = 2\tau$,

$$\sigma\ (\text{MPa}) = \frac{5.2}{L}\ln\left(\frac{\bar{X}}{2.5 \times 10^{-4}}\right) \tag{VI-4}$$

Also assuming that particles are spherical and that their volume fraction is given by

$$f = n_s\left(\frac{\pi\bar{X}^2}{4}\right) \tag{VI-5}$$

then $$L = \bar{X}\left(\sqrt{\frac{\pi}{4f}} - 1\right) \tag{VI-6}$$

In microalloyed steels, $L \gg \bar{X}$, so Eq. VI-6 can be rewritten as

$$L = \bar{X}\sqrt{\frac{\pi}{4f}} \tag{VI-7}$$

Substituting in Eq. VI-4 gives

$$\sigma\ (\text{MPa}) = \frac{5.9\sqrt{f}}{\bar{X}}\ln\left(\frac{\bar{X}}{2.5 \times 10^{-4}}\right) \tag{VI-8}$$

[1] T. Gladman, D. Dulieu, and I.D. McIvor, in *Microalloying 75*, Union Carbide Corp., New York, 1976, p. 32.

[2] M.F. Ashby, in *Oxide Dispersion Strengthening*, AIME, New York, 1966, p. 61.

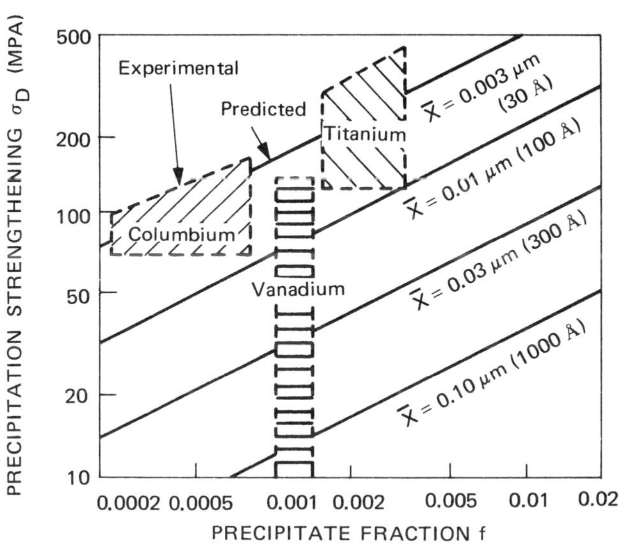

Figure VI-9. The dependence of precipitation strengthening on precipitate size ($\bar{X}$) and fraction, from the Ashby–Orowan model, compared with observations on microalloyed steels. (T. Gladman, D. Dulieu, and I.D. McIvor, in *Microalloying 75*, Union Carbide Corp., New York, 1976, p. 32, by permission.)

From this it can be seen that alloy strengthening from particles increases with reduction in particle size and with increase in the volume fraction of particles. Figure VI-9 shows quite good agreement between calculated and observed strengthening by precipitates.

The precipitates that strengthen HSLA steels need not be confined to carbides or nitrides. Steels can be made with Cu contents between 1.0 and 1.5% so that hardening occurs by the clustering of Cu atoms in ferrite, as described in Chap. III-8.

VI-4. CONTROLLED ROLLING AND CONTROLLED COOLING

The purpose of controlled hot rolling of ferrite-pearlite HSLA steels is to develop a fine, uniform ferrite grain size coupled with precipitation hardening. Having fine austenite grains, or very thin unrecrystallized austenite grains, is a prerequisite for fine ferrite grains in steel microalloyed with Nb. The situation demands heavy deformation and low finishing temperatures in the austenite region, below about 950°C. To finish at such temperatures may require interrupted rolling, which decreases the productivity of a rolling mill. The low finishing temperature and heavy deformation also increase the load on the finishing stands of a mill, sometimes exceeding the design limits of older installations. These factors explain the desire to obtain HSLA steels that can be produced economically on existing mills.

During most controlled rolling, all the deformation occurs in austenite. If rolling

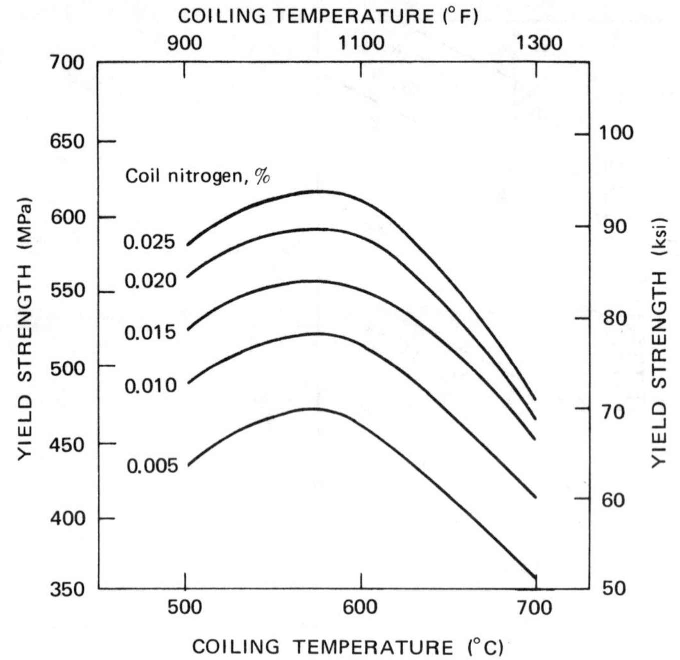

Figure VI-10. Effect of coiling temperature and total nitrogen content on the yield strength of 6.3-mm-thick sheet, finished at 900°C then cooled at 17°C per second. The base composition was 0.13%C–1.4%Mn–0.5%Si–0.12%V. (J.D. Grozier, in *Microalloying 75*, Union Carbide Corp., New York, 1976, p. 241, by permission.)

is done in the austenite-ferrite region, certain consequences follow, not all of them desirable.[1] Hot-working HSLA steels in the two-phase region can produce sharp increases in strength, but these are accompanied by increases in notch-impact transition temperature because of the dislocation substructure retained in the ferrite. In contrast, as the finishing temperature is lowered in the austenite range, the subsequent yield strength increases and the transition temperature decreases because of grain refinement.

Steels rolled in the two-phase region are also subject to splitting fracture because of enhanced banding and development of a (100)[100] texture in ferrite.

Particularly when working with V-treated steels, it is important to cool rapidly after the finishing pass, to prevent overaging of VN particles. This rapid cooling generally necessitates the installation of an enlarged water quenching system on new or existing mills.[2] Figure VI-10 shows the relation between yield strength, coiling temperature, and nitrogen content in a vanadium-treated steel.

[1] G.R. Speich and D.S. Dabkowski, in *Hot Deformation of Austenite,* AIME, New York, 1977, p. 577.

[2] J.D. Grozier, in *Microalloying 75,* Union Carbide Corp., New York, 1976, p. 241.

VI-5. ACICULAR FERRITE STEELS

Another approach to the development of HSLA steels is to obtain the required strength through an acicular ferrite structure. This structure, in which the ferrite grains are highly irregular in shape with a high dislocation density, about 10^{10} cm^{-2}, forms by either a bainitic or massive transformation of austenite to ferrite. This transformation can be obtained by quenching, or by alloying so that transformation during air cooling is retarded. The latter seems the more practical. Manganese is the least expensive addition for retarding transformation, and it can be supplemened with Mo.[1] To obtain high toughness, the austenite grains are kept small by an addition of Nb and the C content is kept at about 0.08% or less. A typical composition[2] in wt % is:

C	Mn	Si	Mo	Nb	Al	N
0.08	2.18	0.10	0.32	0.069	0.020	0.007

When hot-rolled to 13-mm-thick plates, the structure of this steel was acicular ferrite containing some islands of martensite. If the C and Mn contents are reduced to 0.05 and 1.85% respectively, the structure becomes predominantly polygonal ferrite, and at still lower concentrations of these elements it becomes entirely polygonal ferrite.

When steels with acicular ferrite structures are strained in tension (Fig. VI-11), their flow curves are smooth, in contrast to those of low-carbon (pearlite-reduced) steels with polygonal ferrite structures. The acicular ferrite provides a large supply of mobile dislocations and the martensite islands provide residual stresses; the combination gives continuous yielding and a high rate of work hardening. If plastic deformation is followed by aging, the yield strength is greatly increased and yielding becomes discontinuous. The yield strength (0.2% offset) of 0.08%C steel in the as-hot-rolled condition was 487 MPa. After straining 2.5% and aging at 620°C the lower yield stress was 590 MPa. The yield characteristics of acicular ferrite steels are desirable for pipe forming.

Because the austenite transforms at a low temperature, the precipitation of Nb (C, N) in these acicular ferrite steels is not complete, so reheating can lead to further precipitation strengthening.

VI-6. BAINITIC STEELS

By increasing the C content to about 0.15%, retarding the transformation of austenite to ferrite by adding elements such as Mo and B, and controlling grain size by Nb, hot-rolled plates can be made with a bainitic structure and with yield strengths in the range 450 to 900 MPa. The object is to avoid transformation to either ferrite or martensite without hindering the transformation to bainite, allowing the

[1] R.L. Cryderman, A.P. Coldren, Y.E. Smith, and J.L. Mihelich, *Mechanical Working and Steel Processing X,* AIME, New York, 1972, p. 114.

[2] G. Tither and M. Lavite, *J. Met.,* 27:15 (Sept. 1975).

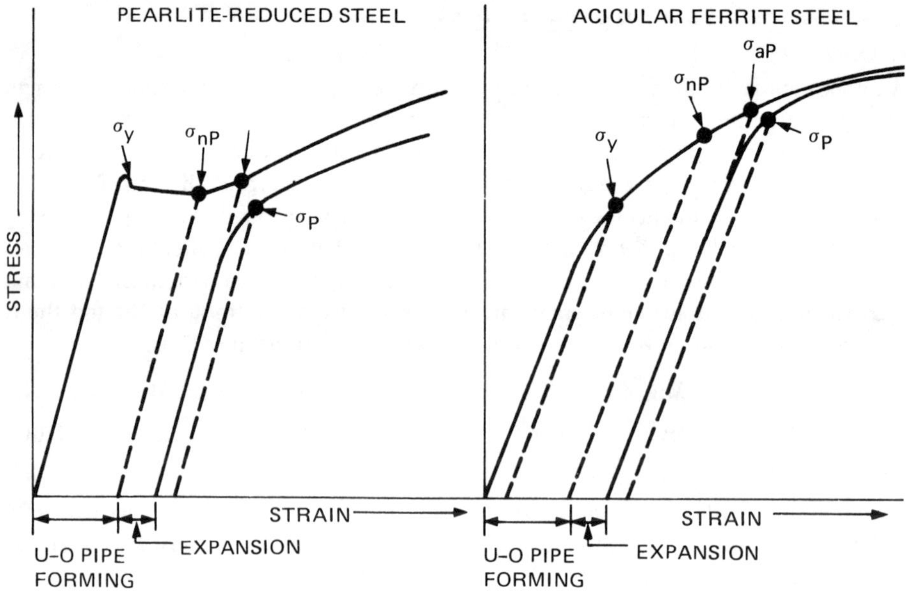

Figure VI-11. Schematic representation of the effect of yielding behavior of pearlite-reduced and acicular ferrite steels on the yield strength of pipe: σ_y = yield strength of plate, σ_{nP} = "actual" yield strength of nonexpanded pipe, σ_{aP} = "actual" yield strength of expanded pipe, σ_P = yield strenth of expanded pipe (flattened tensile specimen). The loss in strength due to the Bauschinger effect is $\sigma_{aP} - \sigma_P$. (G. Tither and M. Lavite, *J. Met.*, 27:15 (Sept. 1975), by permission.)

transformation to bainite to occur over a wide range of cooling rates, as indicated in Fig. VI-12. The strength obtained can be controlled by reducing the banite start temperature, B_s, which varies with composition according to the equation[1]:

$$B_s\,(^\circ\mathrm{C}) = 830 - 270(\%\mathrm{C}) - 90(\%\mathrm{Mn}) - 37(\%\mathrm{Ni}) - 70(\%\mathrm{Cr}) - 83(\%\mathrm{Mo}) \qquad \text{(VI-9)}$$

The temperatures for 50% transformation, B_{50}, and for complete transformation to bainite, B_f, are related to the bainite start temperature by

$$B_{50}\,(^\circ\mathrm{C}) = B_s - 60$$

$$B_f\,(^\circ\mathrm{C}) = B_s - 120$$

The strengths of bainitic structures are linearly related to the temperature for 50% transformation, regardless of composition (Fig. VI-13). The structure changes from upper to lower bainite at a transformation temperature of about 550°C, with no discontinuity in the strength.

The choice of alloying elements to obtain any desired strength and transformation

[1] W. Steven and A.G. Haynes, *J. Iron Steel Inst.*, 183:349 (1956).

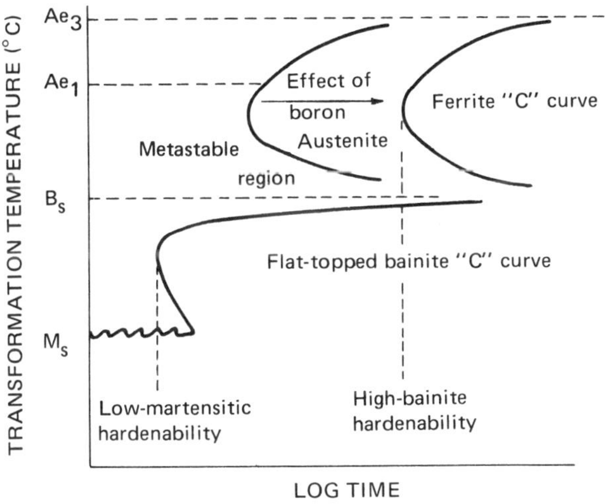

Figure VI-12. Schematic diagram showing transformation requirements for bainitic steel. (F.B. Pickering, in *Transformation and Hardenability in Steels*, Climax Molybdenum Co., Ann Arbor, Michigan, 1967, p. 109, by permission.)

temperature can be made on the basis of cost, toughness, weldability, and formability. The last three factors demand low carbon content. The lower bainite structures have much better toughness than upper bainite, as shown in Fig. VI-14.

VI-7. DUAL–PHASE HSLA STEELS

Dual-phase carbon steels were mentioned in Chap. V, and such steels, with structures of islands of martensite or lower bainite in ferrite, are deserving of more attention.

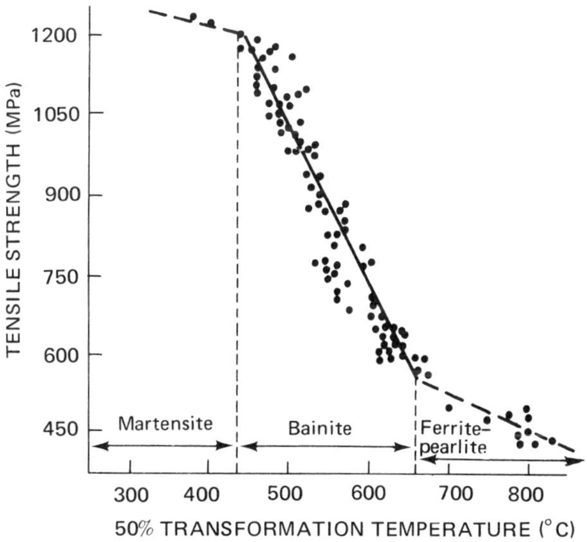

Figure VI-13. The effect of 50% transformation temperature on tensile strength of bainitic steels. (F.B. Pickering, in *Transformation and Hardenability in Steels*, Climax Molybdenum Co., Ann Arbor, Michigan, 1967, p. 109, by permission.)

They are of interest primarily to potential users of high-strength steel sheet, either hot-rolled or cold-rolled. A typical dual-phase steel could be an HSLA steel containing about 1.5%Mn and 0.5%Si, strengthened by an addition of about 0.05%V. The sheet is hot-rolled, cold-rolled, then continuously annealed at a temperature in the $\alpha + \gamma$ region (in the neighborhood of 760°C), then cooled so that the islands of austenite transform to martensite or lower banite.[1-4] The transformation produces a dispersion of hard particles in a matrix of ferrite. In addition, it creates a pattern of tesselated residual stresses in the ferrite surrounding each particle of martensite. The resulting tensile strength generally follows the rule of mixtures, as shown in Fig. VI-15. This linear relation between tensile strength and the volume fraction of martensite may not always be valid, as there is an indication[4] that the tensile strength also depends upon the carbon content of the martensite.

Other features are of greater interest than the ultimate tensile strength, among these being a low yield stress, continuous yielding, a high rate of work hardening, and a large uniform elongation. A typical steel can have a yield strength of 370 MPa, a tensile strength of 665 MPa, and a uniform elongation of 22%. These are desirable characteristics for cold forming. In these steels, the relation

$$\sigma = K\epsilon^n$$

is not followed; that is, a plot of log σ vs. log K is a curve rather than a straight line. The value of n, the strain-hardening exponent, decreases with increasing strain.[4] For cold forming, the adverse factor in these steels is an $\bar{r}$ of 1.00 or less.

[1] M.S. Rashid, *SAE Trans.*, 85:938, Sec. 2 (1976).

[2] S. Hayami and T. Furukawa, in *Microalloying 75*, Union Carbide Corp., New York, 1976, p. 311.

[3] J.H. Bucher and E.G. Hamburg, *SAE Trans.*, 86:730, Sec. 1 (1977).

[4] G.R. Speich and R.L. Miller, in *Structure and Properties of Dual-Phase Steels*, AIME, Warrendale, Pennsylvania, 1979, p. 145.

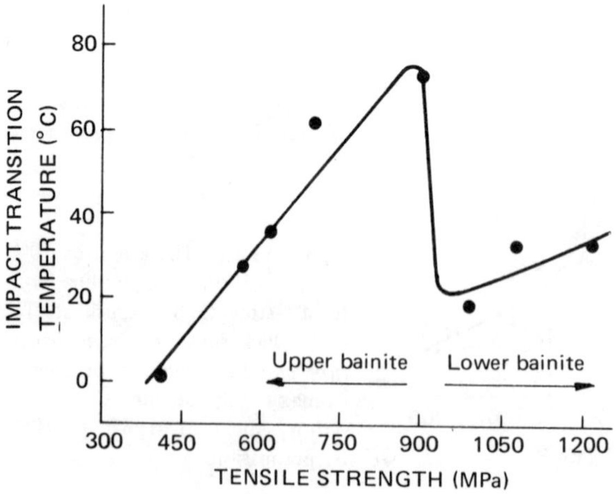

Figure VI-14. Effect of tensile strength on impact-transition temperature of low-carbon bainitic steels. (K.J. Irvine and F.B. Pickering, *J. Iron Steel Inst.*, 201:518 (1963), by permission.)

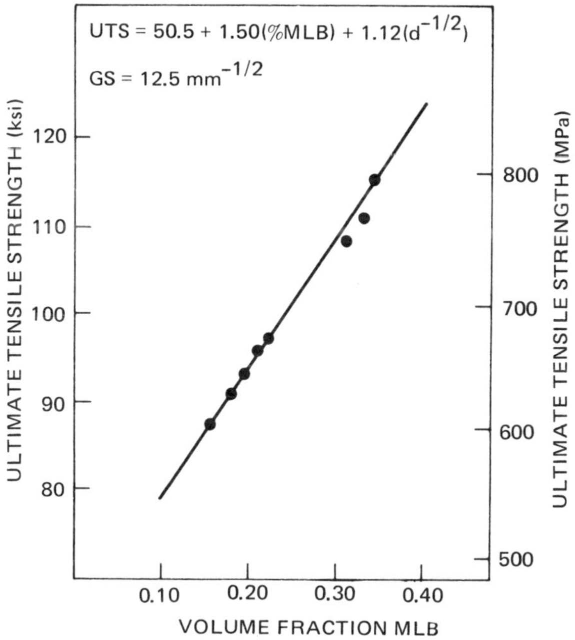

Figure VI-15. The effect of the volume fraction of martensite plus lower bainite on tensile strength of dual-phase steels. (J.H. Bucher and E.G. Hamburg, *SAE Trans.*, 86:730, Sec. 1 (1977), by permission.)

The dual-phase steels strengthen rapidly during deformation, because of their high work-hardening rate. In addition, they can be markedly strengthened by static or dynamic strain aging.[1] The high work-hardening rate and strain-aging response are attributed to the dispersion of a hard phase, the presence of tesselated residual stresses before straining, and the ready availability of carbon to pin discloations. After forming, the yield strength of dual-phase steels can be increased sharply by tempering at 300 to 400°C. The magnitude of this strengthening in an undeformed steel is shown in Fig. VI-16.

VI-8. DESULFURIZATION, DEOXIDATION, AND INCLUSION SHAPE CONTROL

Desulfurization and deoxidation operations are becoming increasingly common in the production of HSLA and heat-treatable alloy steels because of the large gains that can be made in formability and toughness. This is illustrated in Fig. VI-17, which shows the improvement in reduction of area in the through-thickness direction as the total oxygen plus sulfur content is reduced.

The most effective desulfurization and deoxidation treatments are done by injecting lime, calcium carbide, powdered Ca–Si alloy, iron-coated Ca wire, or Mg

[1] J.A. Yaker, C.-C. Li, and W.C. Leslie, *SAE Trans.*, 88, paper 79009, 1979.

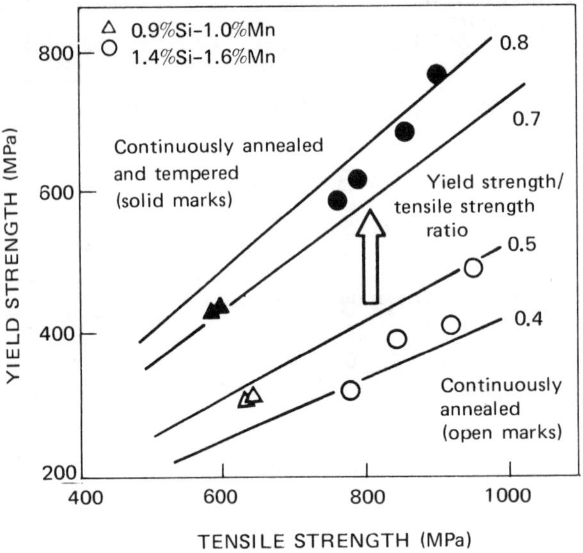

Figure VI-16. The relation between yield strength and tensile strength of dual-phase steels before and after tempering. The steels were continuous-annealed at 800°C for 1 to 5 min, then air cooled. Tempering was done at 400°C for 5 min. (S. Hayami and T. Furukawa, in *Microalloying 75*, Union Carbide Corp., New York, 1976, p. 311, by permission.)

into the liquid steel in the ladle or transfer car before the pouring of ingots.[1] This is done after Al deoxidation. Stable solid oxides and sulfides of Ca and Mg tend to float out into the slag; in this manner heats can be made with less than 0.005%S and 0.002%O in solution in the steel. Consideration must be given to the ladle lining so that reoxidation of the steel does not occur.

[1] H. Pircher and W. Klapdar, in *Microalloying 75*, Union Carbide Corp., New York, 1976, p. 232.

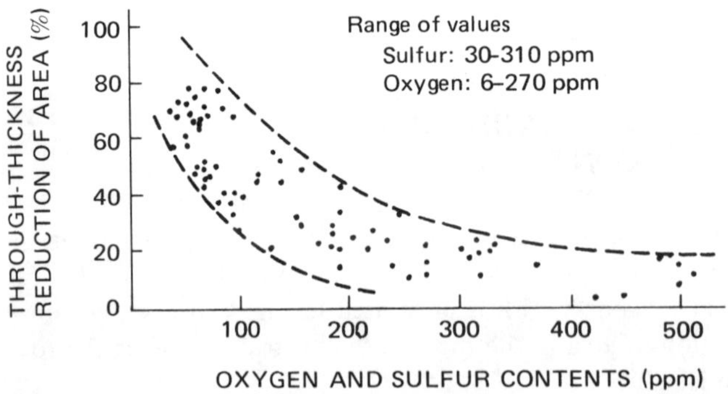

Figure VI-17. Effect of oxygen and sulfur contents on reduction of area in the through-thickness direction. (K. Norring, H. Harwig, and B. Lindvall, in *Microalloying 75*, Union Carbide Corp., New York, 1976, p. 684, by permission.)

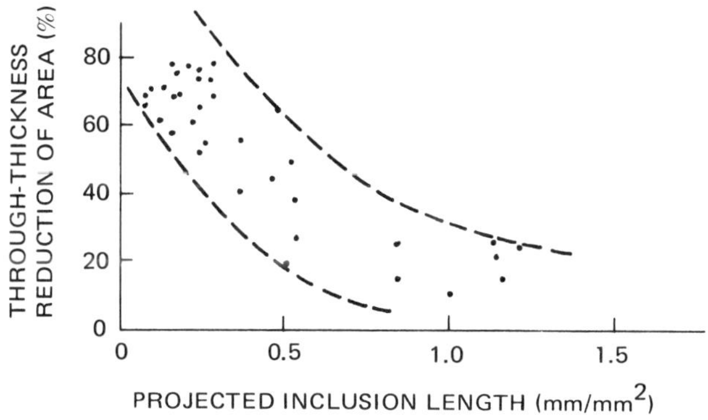

Figure VI-18. The relation between reduction of area in the through-thickness direction and projected inclusion length. (K. Norring, H. Harvig, and B. Lindvall, in *Microalloying 75*, Union Carbide Corp., New York, 1976, p. 684, by permission.)

Such treatments greatly reduce the number of inclusions in the solid steel, and those remaining are globular, often containing more than one phase. As shown in Fig. VI-18, the correlation of through-thickness toughness is better with the inclusion length per unit area than with total O and S. With few inclusions, and these present as equiaxed particles, the tensile properties in the through-thickness direction become indistinguishable from those in the longitudinal direction. The steel can have isotropic mechanical properties even when S and O are not reduced to very low levels.

The shelf energy—that is, the energy absorbed in a ductile fracture in a Charpy V-notch impact test—usually decreases as the strength of the steel increases. Therefore, inclusion shape control is more urgent in high-strength steels than in low-carbon ferrite-pearlite steels (Fig. VI-19). The shape of inclusions can be modified by additions of Ca (as already mentioned), Ce, mixed rare-earth metals, Zr, or Ti. The shape change is accomplished by altering the composition of the oxides or sulfides so that they form as globules with lower plasticity at the hot-rolling temperature. Cerium and rare-earth sulfides form almost undeformable particles, with the result that when added in the proper concentration, they produce a steel with isotropic properties (Fig. VI-20).

When a steel is to be strengthened by precipitation of VN, inclusion shape control with Zr must be avoided because Zr forms a highly stable nitride,[1] thereby reducing the concentration of N available for hardening. When Ti is added in comparatively large amounts (~0.20% or more) it serves both to control inclusion shape and as a strengthening agent.

[1] A. McLean and D.A.R. Kay, in *Microalloying 75*, Union Carbide Corp., New York, 1976, p. 215.

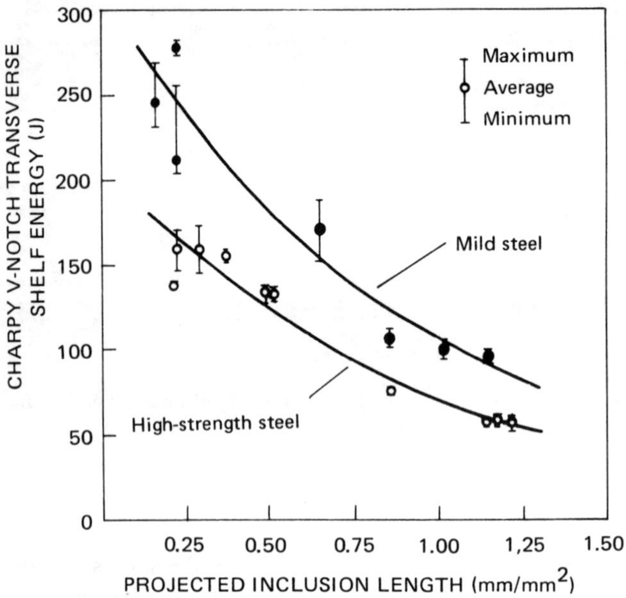

Figure VI-19. The effect of projected inclusion length on transverse shelf energy. (K. Norring, H. Harvig, and B. Lindvall, in *Microalloying 75*, Union Carbide Corp., New York, 1976, p. 684, by permission.)

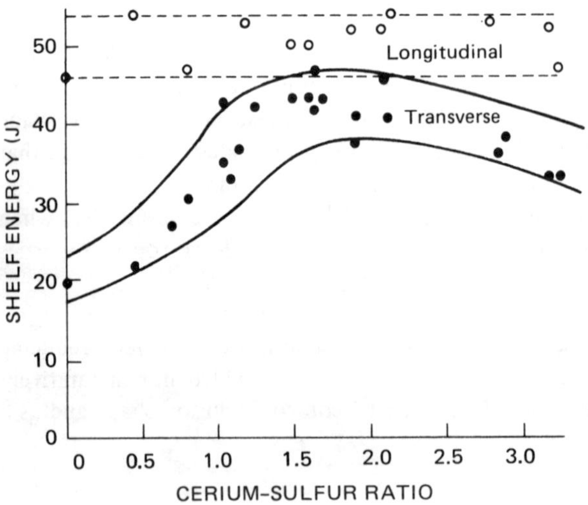

Figure VI-20. Effect of cerium additions on longitudinal and transverse Charpy shelf energy. (F.B. Pickering, in *Microalloying 75*, Union Carbide Corp., New York, 1976, p. 9, by permission.)

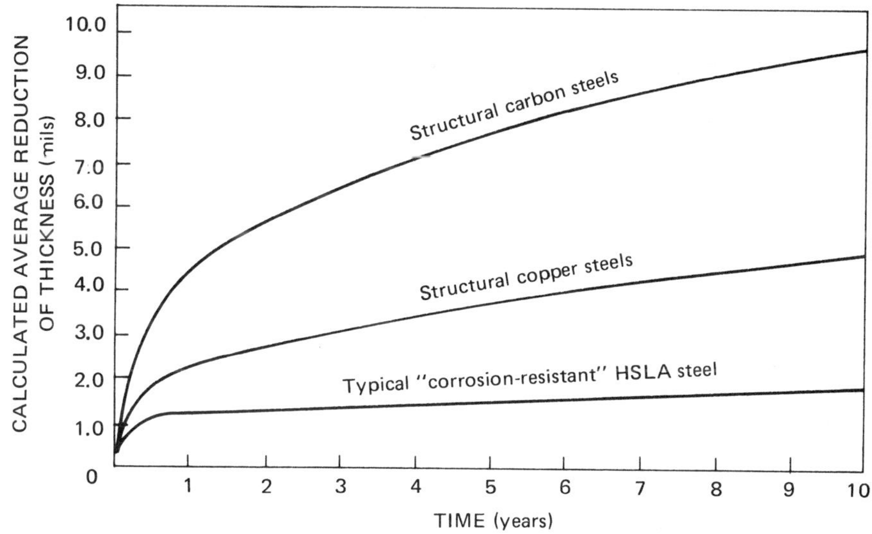

Figure VI-21. Comparative corrosion of steel specimens in an industrial atmosphere. (*Making, Shaping, and Treating of Steel*, 9th ed., p. 1152, copyright 1971, U.S. Steel Corp., by permission.)

VI-9. WEATHERING STEELS

Enthusiasm for the newer kinds of HSLA steel should not lead us to overlook the older types that have been available for nearly 50 years. These have ferrite-pearlite microstructures and gain their increased strength principally through elements in solid solution. Yield strengths are usually slightly in excess of 345 MPa, as compared to 248 for carbon steels. Because of this greater hardness, abrasion resistance is improved; also, if Cu and P are added to the steel, the resistance to atmospheric corrosion is improved to such a degree that the steel need not be painted. It can develop an attractive "patina" after a period of exposure, but it should not be used above light-colored masonry.

A typical composition (wt %) for such weathering steel is

C	Mn	Si	Cu	P	Ni	Cr
0.10	0.35	0.50	0.40	0.10	0.40	0.80

The combination of Cu and P provides greater corrosion resistance than does either element alone, as indicated by Fig. VI-21. Misawa et al.[1] proposed a mechanism by which the improved corrosion resistance of weathering steels is accomplished. As these steels rust, Cu^{2+} and PO_4^{3-} ions are slowly concentrated at the rust-metal inter-

[1] T. Misawa, T. Kyuno, W. Suetaka, and S. Shimodaira, *Corros. Sci.*, 11:35 (1971).

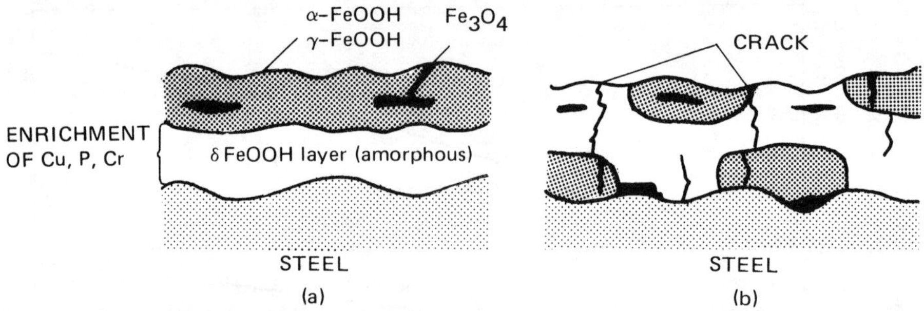

Figure VI-22. Schematic representation of the rust layers on steels after exposure to the atmosphere for long periods: (*a*) weathering steel, (*b*) low-carbon steel. (Reprinted with permission from T. Misawa, T. Kyuno, W. Suetaka, and S. Shimodaira, *Corros. Sci.*, 11:35 (1971). Copyright Pergamon Press, Ltd.)

face. These ions catalyze the oxidation of Fe(II)–hydroxo complexes to amorphous δ FeOOH, which forms as a cohesive, crack-free layer at the metal surface, thereby limiting the access of O_2 and H_2O to the metal surface and migration of Fe^{2+} ions away from the metal. The difference between the rusting of carbon steels and weathering steels is shown schematically in Fig. VI-22.

CHAPTER

SEVEN

HEAT-TREATED STEELS

VII-1. INTRODUCTION

To this point we have been concerned with the properties of steels that are used in the cold-rolled and annealed or hot-rolled condition; they are not heat-treated by either the steel producer or by the user. As control of steel properties by the producer improves, and as a larger range of strength becomes available, steel users will increasingly use steels in the as-received condition, to eliminate the cost and uncertainties of heat treatment. Nevertheless, steels hardened by quenching and tempering will continue to be essential for highly stressed parts. Alloying elements are added to steels principally to improve mechanical properties after quenching and tempering; the current standard alloy steels are listed in Table VII-1. The object of quenching a heat-treatable steel from the austenite range is to obtain a structure of martensite, or largely martensite, throughout the stressed portion of the part, which can then be tempered to achieve the optimum combination of strength and ductility. Therefore, it is appropriate to begin our study of the properties of heat-treated steels by examining the structure and strength of martensite.

VII-2. THE STRUCTURE OF MARTENSITE

In the general sense, martensite is the product of a phase transformation that occurs by shear. In steels, the parent phase is nearly always austenite, but the daughter phase can be body-centered cubic (bcc), body-centered tetragonal (bct), or hexagonal close packed (hcp). In low- or medium-carbon steels, the common structure is lath martensite. The units of the structure are in the shape of laths, grouped into packets or

Table VII-1 AISI-SAE standard alloy steels[1]

AISI–SAE No.	UNS No.	Composition (%)[2,3]							
		C	Mn	P (max)	S (max)	Si	Ni	Cr	Mo
1300	G13300	0.28–0.33	1.60–1.90	0.035	0.040	0.15–0.30	–	–	–
1335	G13350	0.33–0.38	1.60–1.90	0.035	0.040	0.15–0.30	–	–	–
1340	G13400	0.38–0.43	1.60–1.90	0.035	0.040	0.15–0.30	–	–	–
1345	G13450	0.43–0.48	1.60–1.90	0.035	0.040	0.15–0.30	–	–	–
4023	G40230	0.20–0.25	0.70–0.90	0.035	0.040	0.15–0.30	–	–	0.20–0.30
4024	G40240	0.20–0.25	0.70–0.90	0.035	0.035–0.050	0.15–0.30	–	–	0.20–0.30
4027	G40270	0.25–0.30	0.70–0.90	0.035	0.040	0.15–0.30	–	–	0.20–0.30
4028	G40280	0.25–0.30	0.70–0.90	0.035	0.035–0.050	0.15–0.30	–	–	0.20–0.30
4037	G40370	0.35–0.40	0.70–0.90	0.035	0.040	0.15–0.30	–	–	0.20–0.30
4047	G40470	0.45–0.50	0.70–0.90	0.035	0.040	0.15–0.30	–	–	0.20–0.30
4118	G41180	0.18–0.23	0.70–0.90	0.035	0.040	0.15–0.30	–	0.40–0.60	0.08–0.15
4130	G41300	0.28–0.33	0.40–0.60	0.035	0.040	0.15–0.30	–	0.80–1.10	0.15–0.25
4137	G41370	0.35–0.40	0.70–0.90	0.035	0.040	0.15–0.30	–	0.80–1.10	0.15–0.25
4140	G41400	0.38–0.43	0.75–1.00	0.035	0.040	0.15–0.30	–	0.80–1.10	0.15–0.25
4142	G41420	0.40–0.45	0.75–1.00	0.035	0.040	0.15–0.30	–	0.80–1.10	0.15–0.25
4145	G41450	0.43–0.48	0.75–1.00	0.035	0.040	0.15–0.30	–	0.80–1.10	0.15–0.25
4147	G41470	0.45–0.50	0.75–1.00	0.035	0.040	0.15–0.30	–	0.80–1.10	0.15–0.25
4150	G41500	0.48–0.53	0.75–1.00	0.035	0.040	0.15–0.30	–	0.80–1.10	0.15–0.25
4161	G41610	0.56–0.64	0.75–1.00	0.035	0.040	0.15–0.30	–	0.70–0.90	0.25–0.35
4320	G43200	0.17–0.22	0.45–0.65	0.035	0.040	0.15–0.30	1.65–2.00	0.40–0.60	0.20–0.30
4340	G43400	0.38–0.43	0.60–0.80	0.035	0.040	0.15–0.30	1.65–2.00	0.70–0.90	0.20–0.30
E4340[4]	G43406	0.38–0.43	0.65–0.85	0.025	0.025	0.15–0.30	1.65–2.00	0.70–0.90	0.20–0.30
4615	G46150	0.13–0.18	0.45–0.65	0.035	0.040	0.15–0.30	1.65–2.00	–	0.20–0.30
4620	G46200	0.17–0.22	0.45–0.65	0.035	0.040	0.15–0.30	1.65–2.00	–	0.20–0.30
4626	G46260	0.24–0.29	0.45–0.65	0.035	0.040	0.15–0.30	0.70–1.00	–	0.15–0.25
4720	G47200	0.17–0.22	0.50–0.70	0.035	0.040	0.15–0.30	0.90–1.20	0.35–0.55	0.15–0.25
4815	G48150	0.13–0.18	0.40–0.60	0.035	0.040	0.15–0.30	3.25–3.75	–	0.20–0.30

4817	G48170	0.15–0.20	0.40–0.60	0.035	0.040	0.15–0.30	3.25–3.75	–	0.20–0.30
4820	G48200	0.18–0.23	0.50–0.70	0.035	0.040	0.15–0.30	3.25–3.75	–	0.20–0.30
5117	G51170	0.15–0.20	0.70–0.90	0.035	0.040	0.15–0.30	–	0.70–0.90	–
5120	G51200	0.17–0.22	0.70–0.90	0.035	0.040	0.15–0.30	–	0.70–0.90	–
5130	G51300	0.28–0.33	0.70–0.90	0.035	0.040	0.15–0.30	–	0.80–1.10	–
5132	G51320	0.30–0.35	0.60–0.80	0.035	0.040	0.15–0.30	–	0.75–1.00	–
5135	G51350	0.33–0.38	0.60–0.80	0.035	0.040	0.15–0.30	–	0.80–1.05	–
5140	G51400	0.38–0.43	0.70–0.90	0.035	0.040	0.15–0.30	–	0.70–0.90	–
5150	G51500	0.48–0.53	0.70–0.90	0.035	0.040	0.15–0.30	–	0.70–0.90	–
5155	G51550	0.51–0.59	0.70–0.90	0.035	0.040	0.15–0.30	–	0.70–0.90	–
5160	G51600	0.56–0.64	0.75–1.00	0.035	0.040	0.15–0.30	–	0.70–0.90	–
E51100[4]	G51986	0.98–1.10	0.25–0.45	0.025	0.025	0.15–0.30	–	0.90–1.15	–
E52100[4]	G52986	0.98–1.10	0.25–0.45	0.025	0.025	0.15–0.30	–	1.30–1.60	–
6118	G61180	0.16–0.21	0.50–0.70	0.035	0.040	0.15–0.30	–	0.50–0.70	0.10–0.15V
6150	G61500	0.48–0.53	0.70–0.90	0.035	0.040	0.15–0.30	–	0.80–1.10	0.15 V min
8615	G86150	0.13–0.18	0.70–0.90	0.035	0.040	0.15–0.30	0.40–0.70	0.40–0.60	0.15–0.25
8617	G86170	0.15–0.20	0.70–0.90	0.035	0.040	0.15–0.30	0.40–0.70	0.40–0.60	0.15–0.25
8620	G86200	0.18–0.23	0.70–0.90	0.035	0.040	0.15–0.30	0.40–0.70	0.40–0.60	0.15–0.25
8622	G86220	0.20–0.25	0.70–0.90	0.035	0.040	0.15–0.30	0.40–0.70	0.40–0.60	0.15–0.25
8625	G86250	0.23–0.28	0.70–0.90	0.035	0.040	0.15–0.30	0.40–0.70	0.40–0.60	0.15–0.25
8627	G86270	0.25–0.30	0.70–0.90	0.035	0.040	0.15–0.30	0.40–0.70	0.40–0.60	0.15–0.25
8630	G86300	0.28–0.33	0.70–0.90	0.035	0.040	0.15–0.30	0.40–0.70	0.40–0.60	0.15–0.25
8637	G86370	0.35–0.40	0.75–1.00	0.035	0.040	0.15–0.30	0.40–0.70	0.40–0.60	0.15–0.25
8640	G86400	0.38–0.43	0.75–1.00	0.035	0.040	0.15–0.30	0.40–0.70	0.40–0.60	0.15–0.25
8642	G86420	0.40–0.45	0.75–1.00	0.035	0.040	0.15–0.30	0.40–0.70	0.40–0.60	0.15–0.25

[1] *Steel Products Manual*, American Iron and Steel Institute, revised January 1979; *ASM Metal Progress Databook*, Vol. 116, June 1979, p. 18.

[2] Small quantities of certain elements are present which are not specified or required. These incidental elements may be present to the following maximum amounts: Cu, 0.35%; Ni, 0.25%; Cr, 0.20%; and Mo, 0.06%.

[3] Standard alloy steels can also be produced with a lead range of 0.15–0.35%. Such steels are identified by inserting the letter "L" between the second and third numerals of the AISI or SAE number, e.g., 41L40.

[4] Electric furnace steel.

Table VII-1 AISI–SAE standard steels *(Continued)*

AISI-SAE No.	UNS No.	Composition (%) C	Mn	P (max)	S (max)	Si	Ni	Cr	Mo
8645	G86450	0.43–0.48	0.75–1.00	0.035	0.040	0.15–0.30	0.40–0.70	0.40–0.60	0.15–0.25
8655	G86550	0.51–0.59	0.75–1.00	0.035	0.040	0.15–0.30	0.40–0.70	0.40–0.60	0.15–0.25
8720	G87200	0.18–0.23	0.70–0.90	0.035	0.040	0.15–0.30	0.40–0.70	0.40–0.60	0.20–0.30
8740	G87400	0.38–0.43	0.75–1.00	0.035	0.040	0.15–0.30	0.40–0.70	0.40–0.60	0.20–0.30
8822	G88220	0.20–0.25	0.75–1.00	0.035	0.040	0.15–0.30	0.40–0.70	0.40–0.60	0.30–0.40
9260	G92600	0.56–0.64	0.75–1.00	0.035	0.040	1.80–2.20	–	–	–
Standard boron grades[5]									
50B44	G50441	0.43–0.48	0.75–1.00	0.035	0.040	0.15–0.30	–	0.40–0.60	–
50B46	G50461	0.44–0.49	0.75–1.00	0.035	0.040	0.15–0.30	–	0.20–0.35	–
50B50	G50501	0.48–0.53	0.75–1.00	0.035	0.040	0.15–0.30	–	0.40–0.60	–
50B60	G50601	0.56–0.64	0.75–1.00	0.035	0.040	0.15–0.30	–	0.40–0.60	–
51B60	G51601	0.56–0.64	0.75–1.00	0.035	0.040	0.15–0.30	–	0.70–0.90	–
81B45	G81451	0.43–0.48	0.75–1.00	0.035	0.040	0.15–0.30	0.20–0.40	0.35–0.55	0.08–0.15
94B17	G94171	0.15–0.20	0.75–1.00	0.035	0.040	0.15–0.30	0.30–0.60	0.30–0.50	0.08–0.15
94B30	G94301	0.28–0.33	0.75–1.00	0.035	0.040	0.15–0.30	0.30–0.60	0.30–0.50	0.08–0.15

[5] 0.0005–0.003%B.

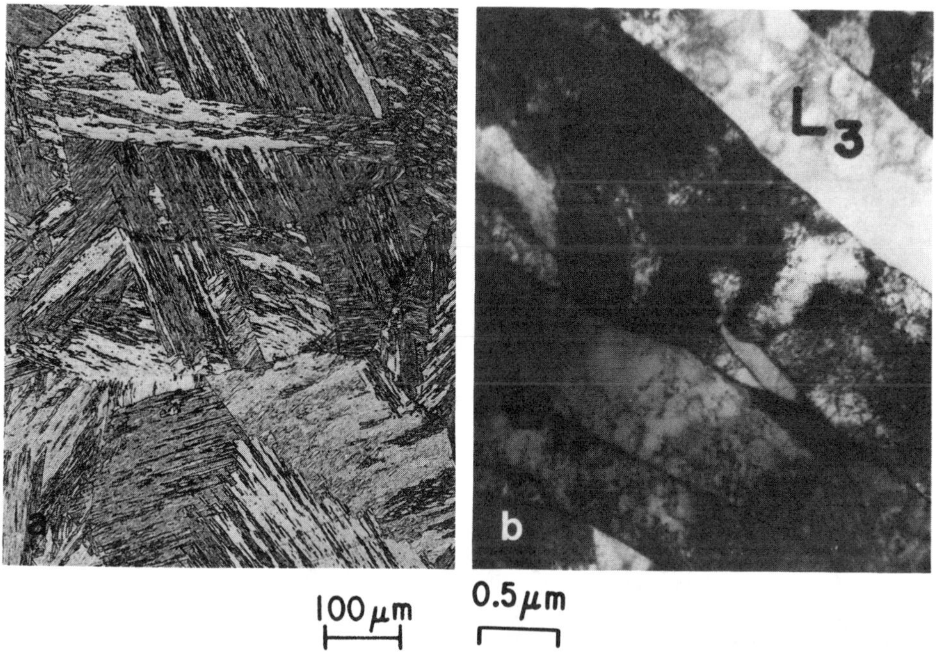

Figure VII-1. Structure of lath martensite in a 0.02%C–2.0%Mn steel. L_1, L_2, and L_3 distinguish separate laths. The laths are separated by high-angle boundaries. Each lath contains a high density of dislocations. (*a*) Light micrograph. (*b*) Transmission electron micrograph. (G.R. Speich and W.C. Leslie, *Met. Trans.*, 3:1043 (1972).)

sheaves. The substructure consists of a very high density of dislocations (Fig. VII-1). There has been no precise measurement of the density, but it is usually considered to be in the neighborhood of 10^{11} to 10^{12} cm^{-2}. Each lath is composed of many dislocation cells with an average width of about 250 nm. The crystal structure of lath martensite site is usually bcc, but at C contents in excess of 0.20% the structure becomes tetragonal. The reason for this change will be discussed in the section dealing with tempering.

The martensite start temperature can be decreased by higher carbon or alloy content, accordingly to the equation[1]

$$\text{Ms}\ (^\circ\text{C}) = 561 - 474(\%\text{C}) - 33(\%\text{Mn}) - 17(\%\text{Ni}) - 17(\%\text{Cr}) - 21(\%\text{Mo}) \qquad \text{(VII-1)}$$

When the Ms temperature is reduced to about 200°C or below, the martensite changes to the plate form. The martensite units form as individual lenticular plates, with a substructure of very fine twins spaced about 5 nm apart. These twins generally do not

[1] W. Steven and A.G. Haynes, *J. Iron Steel Inst.*, 183:349 (1956).

Figure VII-2. Structure of plate martensite in Fe–32%Ni alloy, cooled to –195°C; 12% retained austenite. Each plate contains closely spaced twins. (*a*) Light micrograph. (*b*) Transmission electron micrograph. (W.C. Leslie, D.W. Stevens, and M. Cohen, in *High Strength Materials*, V.F. Zackay, ed., John Wiley and Sons, New York, 1965, p. 382.)

extend to the sides of the plate, but gradually merge into an array of dislocations near the periphery of the plate (Fig. VII-2). The crystal structure of plate martensite can be either body-centered tetragonal or body-centered cubic, depending upon the carbon content.

The transition from lath to plate martensite as composition changes is not abrupt. Figure VII-3 shows the change as the C content is increased. Of course, the cooling rates employed in determining this plot were those obtained in normal brine quenching, except for the very low carbon contents. It is obvious that steels containing carbon contents in the range 0.5 to 0.8% have complicated microstructures as quenched, with lath martensite, plate martensite, and retained austenite all present.

VII-3. STRENGTH OF MARTENSITE

The strength of martensite in the as-quenched state depends primarily upon the carbon content. Alloying elements, within the limits of commercial steels, have little or no effect on the maximum hardness attainable. The relation between carbon content and hardness of martensite is shown in Fig. VII-4. Note that the maximum diamond

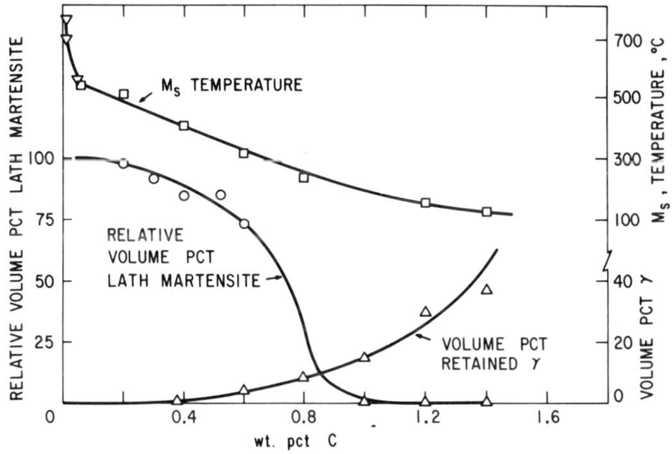

Figure VII-3. Effect of carbon content on relative volume pct of lath and plate martensite, Ms temperature, and volume % of retained austenite. (G.R. Speich and W.C. Leslie, *Met. Trans.*, 3:1043 (1972).)

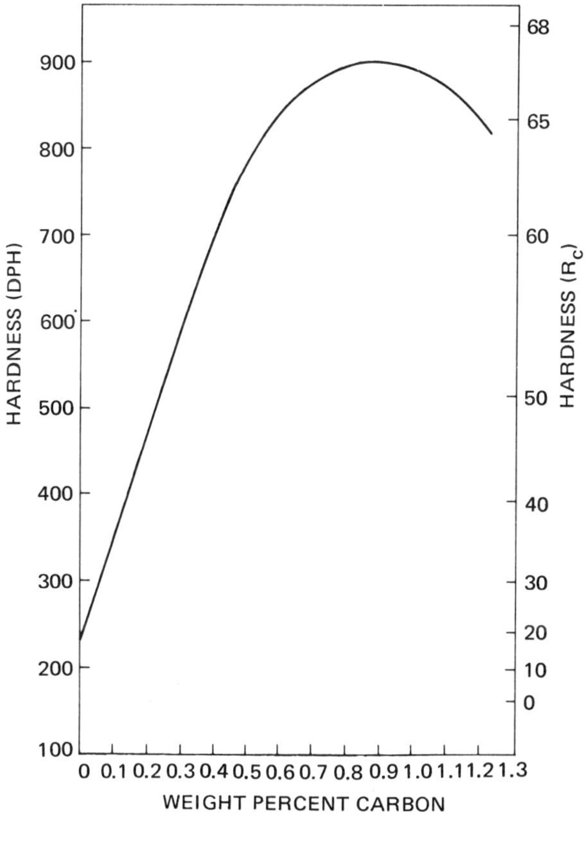

Figure VII-4. Hardness of martensite as a function of carbon content in Fe–C alloys and steels. (G. Krauss, in *Hardenability Concepts with Applications to Steel*, AIME, Warrendale, Pennsylvania, 1978, p. 235; by permission.)

pyramid hardness (DPH) attainable increases linearly with C content, up to a C content of about 0.40%, at which point retained austenite appears in the microstructure (Fig. VII-3) and the rate of hardening begins to decrease. The hardness of the duplex martensite-retained austenite structure reaches a maximum of about 900 DPH at about 0.9%C and decreases thereafter as the amount of retained austenite increases. The course of the C content–hardness curve above 0.4%C depends upon the amount of retained austenite present, which in turn depends upon the grain size, alloy content, and use of subzero treatments. The curve of Fig. VII-4 represents the hardness of martensite when retained austenite is minimized. An extrapolation of the line to zero C content gives a hardness of 240 DPH, in excellent agreement with the hardness of C-free martensite in shock-transformed iron.

If the diamond pyramid hardness of as-quenched martensite increases linearly with C content, it is reasonable to expect that the yield strength will also increase linearly. This expectation is borne out by Fig. VII-5. Data from a variety of steels fall within the dash lines.

Doane[1] has tabulated hardness values obtained from microstructures of 99.9% martensite, at 1.6 mm from the quenched ends of Jominy bars, with results presented in Table VII-2. The hardness in this table for any given C content is slightly below the maximum shown in Fig. VII-4.

This relation between C content and hardness of as-quenched martensite is well known and generally accepted, but because of their complex nature the mechanisms by which the hardness is attained are not as familiar. The strength of as-quenched

[1] D.V. Doane, *Hardenability Concepts with Applications to Steel,* AIME, Warrendale, Philadelphia, 1978, p. 369.

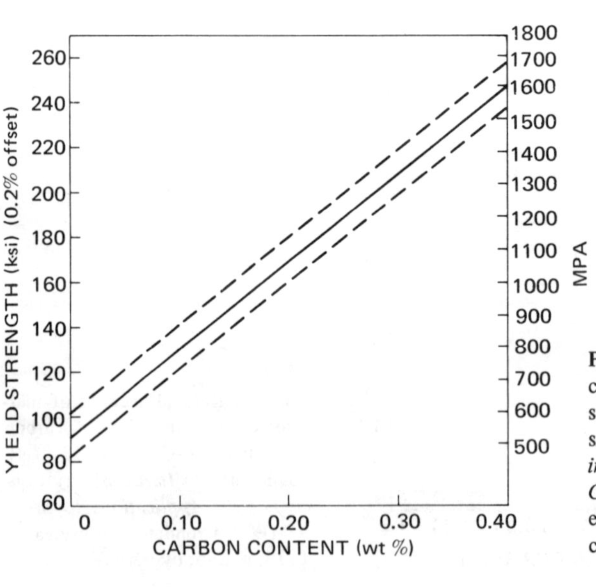

Figure VII-5. The effect of carbon content on the yield strength of untempered martensites. (W.C. Leslie, in *Strengthening Mechanisms, Metals and Ceramics*, Burke, Reed, and Weiss, eds., Syracuse Univ. Press, Syracuse, New York, 1966, p. 46.)

Table VII-2 Carbon hardness factors

Carbon (%)	Maximum hardness (HRC)[1]	Carbon (%)	Maximum hardness (HRC)
0.10	38	0.36	54
0.11	39	0.37	55
0.12	40	0.38	55
0.13	40	0.39	56
0.14	41	0.40	56
0.15	41	0.41	57
0.16	42	0.42	57
0.17	42	0.43	58
0.18	43	0.44	58
0.19	44	0.45	58
0.20	44	0.46	59
0.21	45	0.47	59
0.22	45	0.48	59
0.23	46	0.49	60
0.24	46	0.50	60
0.25	47	0.51	60
0.26	48	0.52	61
0.27	49	0.53	61
0.28	49	0.54	61
0.29	50	0.55	61
0.30	50	0.56	61
0.31	51	0.57	62
0.32	51	0.58	62
0.33	52	0.59	62
0.34	53	0.60	62
0.35	53		

[1] Rockwell C scale.

carbon martensite is derived from several sources[1,2]: the defect structure, dynamic strain aging during testing at ambient temperature, work hardening during testing, solid solution strengthening by C, and rearrangement of C atoms and dislocations during the quench. The problem becomes one of sorting out these variables and assigning relative magnitudes to them as the carbon content is changed. To do this it is appropriate to begin with the strength of high-purity bcc Fe and its variation with temperature and strain rate, to examine the strengthening when C is present in solid solution in "defect free" alpha iron, to determine the effect of introducing a martensitic structure into high-purity α Fe, and to examine the effects on strength of adding C and/or substitutional solutes to iron with a martensitic defect structure.

[1] W.C. Leslie and R.J. Sober, *Trans. Am. Soc. Met.*, 60:459 (1967).

[2] F.B. Pickering, *Hardenability Concepts with Applications to Steel,* AIME, Warrendale, Pennsylvania, 1978, p. 179.

The first two of the foregoing steps have been discussed (Figs. I-13, I-14, and II-26). The structure of iron that has undergone two pressure-induced shear transformations, $\alpha \rightarrow \epsilon \rightarrow \alpha$, is shown in Fig. I-36*a*. The temperature and strain-rate dependence of yield and flow stress in iron with this lath martensite structure is illustrated in Fig. VII-6. A comparison of the data in this figure with those of Fig. I-13 indicates, as would be expected, that the strengthening imparted by the martensitic structure of shock-transformed iron is independent of temperature. This athermal component of strength is about equal to 540 MPa. The 0.2% offset yield strength of the shock-transformed iron at ambient temperature, 655 MPa, can be taken as an upper limit for the strength of martensite with zero carbon content.

When a high concentration of substitutional solutes (~29 at. %) is contained in iron with a martensitic structure, in the absence of C in solution, as in a maraging steel in the as-quenched condition, then the yield strength is increased only slightly (about 150 MPa) over that of unalloyed martensite. Solid solution strengthening can be reduced by a high concentration of lattice defects; this is to say, their effects are not additive.

When carbon content is combined with martensitic structure, as in AISI 4330 steel, some new effects appear, as seen in Fig. VII-7:

1. At room temperature and at low strain rates the strain rate dependence of the flow stress is negative, denoting dynamic strain aging during testing.
2. The temperature dependence of the flow stress is slightly negative, with the flow stress at −120°C being slightly less than at room temperature.
3. The rate of work hardening is independent of both temperature and strain rate; but, although not shown in Fig. VII-7, it increases greatly with increasing carbon content.

Although it is doubtful that the components of the strength of martensite are precisely additive, there appears to be no other feasible approach to estimating the relative importance of each of these components. The information available indicates that

1. Strengthening from fine structure is independent of C content or strain rate. The magnitude of 480 to 625 MPa can be varied by controlling austenite grain size.
2. A thermal component, arising from the temperature dependence of the yield stress of iron, is important only at low temperatures.
3. Dynamic strengthening during testing is nearly independent of C content but strongly dependent upon test temperature and strain rate. It vanishes at temperatures below −60°C, when C becomes immobile. At room temperature and customary strain rates its magnitude is 205 to 275 MPa.
4. Work hardening during the test depends strongly on C content and only slightly upon temperature and strain rate. At 0.40%C, work hardening between strains of 0.001 and 0.002 adds about 200 MPa to the strength.
5. Solid solution strengthening by immobile carbon atoms is strongly dependent on

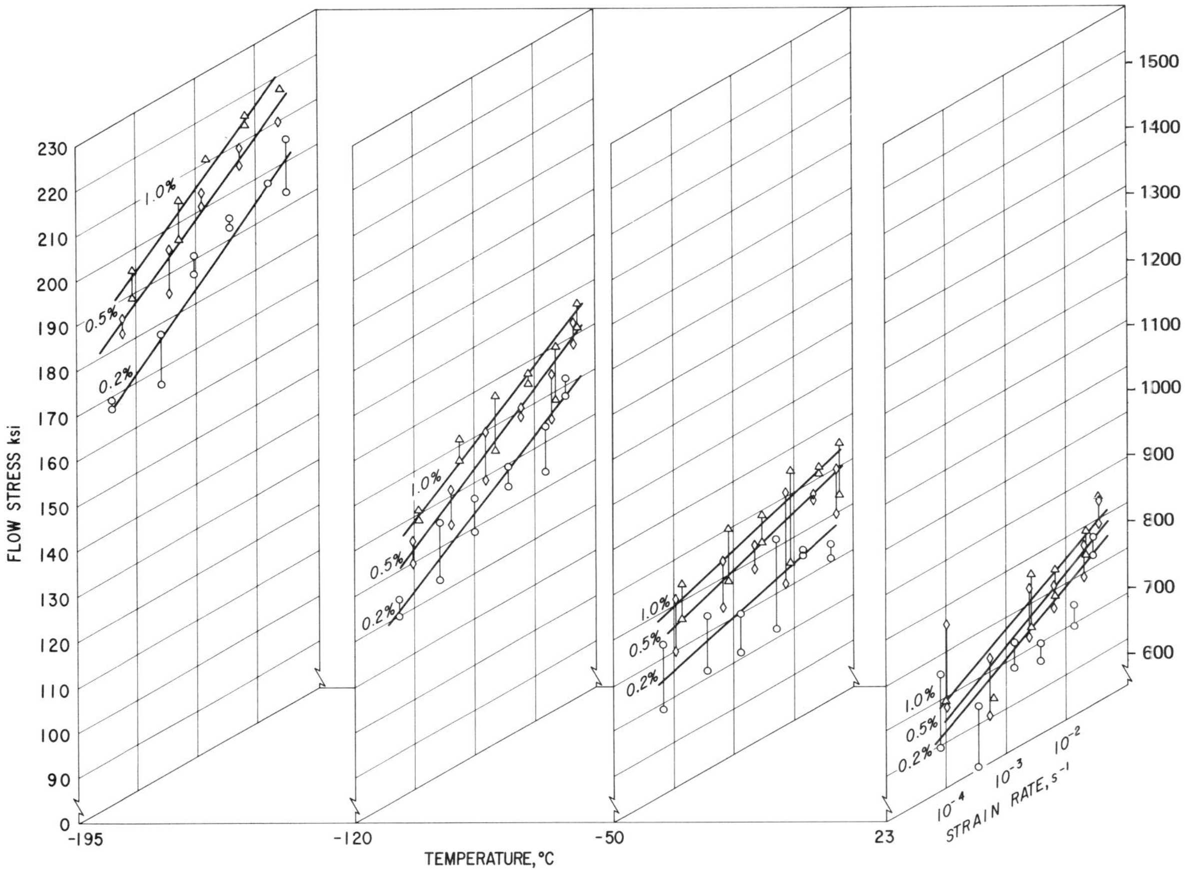

Figure VII-6. Temperature and strain-rate dependence of flow stress of Fe–0.15%Ti, shock loaded to 24–26 GPa. (W.C. Leslie and R.J. Sober, *Trans. Am. Soc. Met.*, 60:459(1967).)

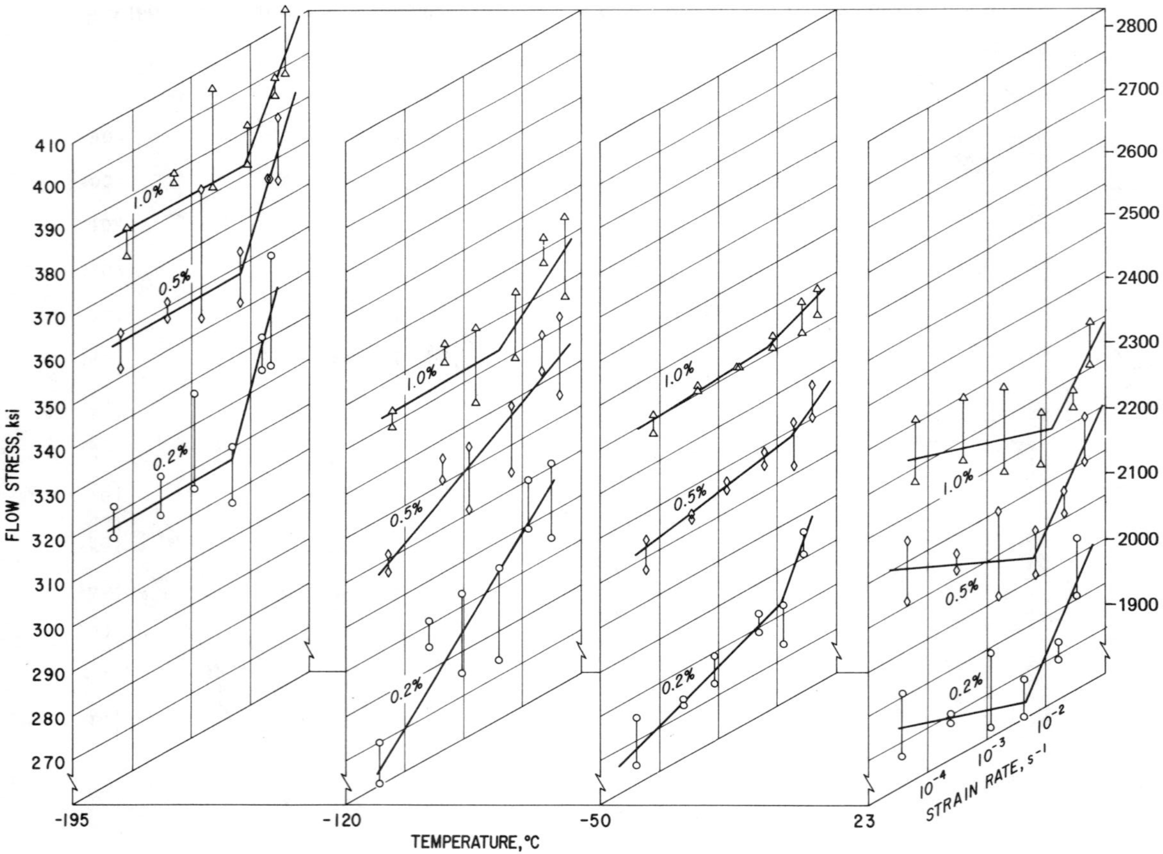

Figure VII-7. Temperature and strain-rate dependence of flow stress of AISI 4330, quenched from 900°C. (W.C. Leslie and R.J. Sober, *Trans. Am. Soc. Met.*, 60:459 (1967).

Table VII-3 Components of the strength of as-quenched martensite

AISI 4310	Component	AISI 4340
620 MPa	Fine structure	620 MPa
205 MPa	Dynamic strengthening during the test	205 MPa
	Work hardening	240 MPa
345 MPa	Rearrangement of C atoms during quench	760 MPa
	Solid solution strengthening by C	415 MPa
1170 MPa	0.2% offset yield strength	2240 MPa

C content. Its magnitude can be estimated only at −60°C or below. When this is done, the strengthening is small.

6. Rearrangement of C atoms occurs during quenching, after the transformation to martensite, when steels have an Ms temperature well above ambient. The magnitude of this effect depends strongly upon carbon content and upon the Ms temperature. At intermediate and high carbon contents it is the dominant strengthening effect.

Having said that the components of the strength of martensite are probably not independent additive entities, we will now treat them as such. Table VII-3 lists these components and their relative magnitudes for a low-carbon and intermediate-carbon steel.

VII-4. STRENGTH DIFFERENTIAL EFFECT

To compound an already complicated situation, it was discovered during a study of the strength of martensite[1] that carbon martensite had no unique strength; the yield and flow stresses were lower when measured in tension than when measured in compression (Fig. VII-8). Subsequent studies[2,3] verified the observation and established the magnitude of the fractional strength differential, $2[(|\sigma_c| - |\sigma_T|)/(|\sigma_c| + |\sigma_T|)]$, at 0.10 to 0.15. As the yield strength is decreased by tempering the martensite or by reducing the carbon content, the absolute magnitude of the difference decreases, as shown in Fig. VII-9. The differential persists to at least a true strain of 0.7[3] and is substantially constant over this range. The phenomenon occurs not only in martensite but also in lower bainite, upper bainite, Widmanstätten ferrite, and ferrite-martensite mixtures. The magnitude decreases in the order martensite, lower bainite, upper bainite, and Widmanstätten ferrite. The common feature of all these structures is a high density of dislocations. No strength differential (SD) effect appears in

[1] W.C. Leslie and R.J. Sober, *Trans. Am. Soc. Met.*, 60:459 (1967).
[2] G.C. Rauch and W.C. Leslie, *Met. Trans.*, 3:373 (1972).
[3] R. Chait, *Met. Trans.*, 3:365 (1972).

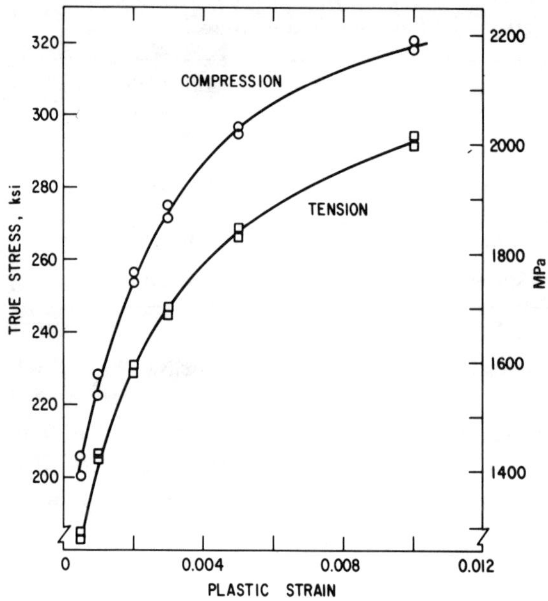

Figure VII-8. True stress–plastic strain curves for as-quenched martensite in AISI 4330 steel, obtained from tension-compression specimens. (G.C. Rauch and W.C. Leslie, *Met. Trans.*, 3:373 (1972).)

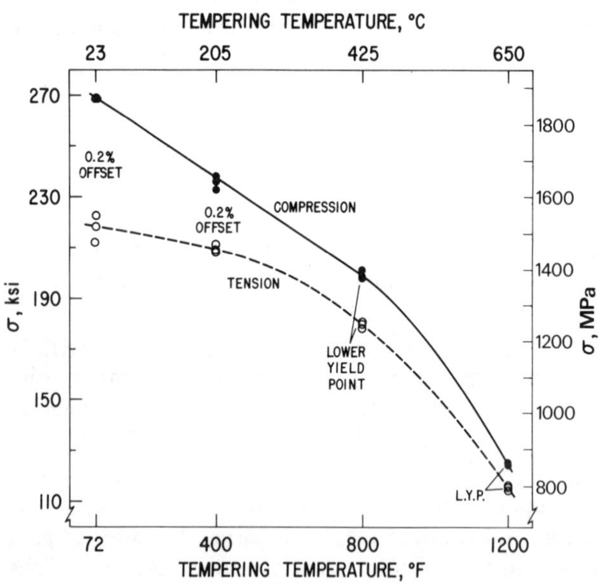

Figure VII-9. The effect of tempering on the yield strength of AISI 4330, quenched from 900°C. Time at 205, 425, or 650°C, 1h; time at room temperature, 4 days. (W.C. Leslie and R.J. Sober, *Trans. Am. Soc. Met.*, 60:459 (1967).)

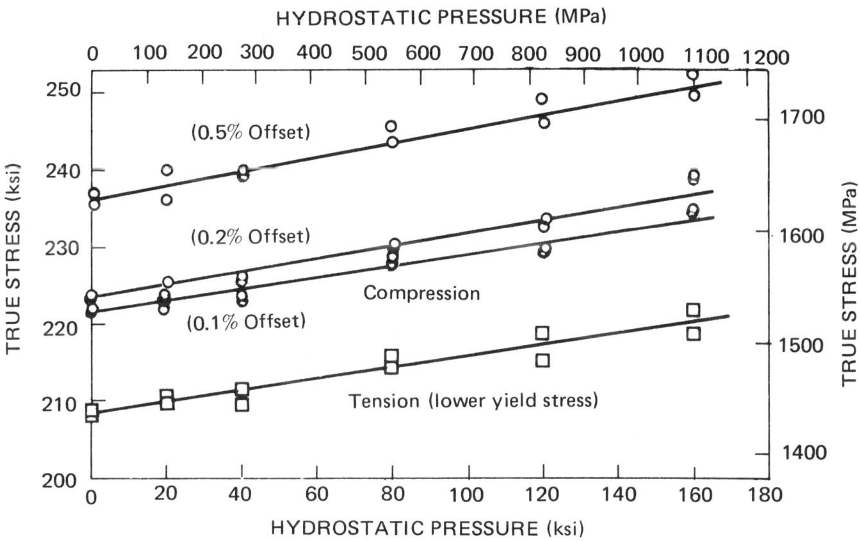

Figure VII-10. Effect of hydrostatic pressure on the yield stress of 4330 steel, quenched from 900°C and then tempered 1 h at 260°C. (Reprinted with permission from W.A. Spitzig, R.J. Sober, and O. Richmond, *Acta Met.*, 23:885 (1975), copyright Pergamon Press, Ltd.)

annealed ferrite or ferrite-pearlite mixtures. The SD effect increases with decreasing test temperature.

Because of its small magnitude in quenched and tempered steels and because specifications based on tensile properties are conservative with respect to yielding in compression, the SD effect is of little practical moment; but it is disturbing because it indicates that we do not understand fully the yielding and plastic flow of metals. Attempts to explain the SD effect have followed two avenues, one of an atomic mechanism[1] and the other of continuum mechanics.[2–5] The former is based on nonlinear elastic strains around carbon atoms in interstitial positions, but the authors also consider that the motion of dislocations can be affected by stress normal to the glide plane. The nonlinear elastic contribution to the conventional linear elastic interaction energy between a carbon atom and a dislocation is about 4%, and it was concluded that such nonlinear elastic interactions could explain a strength differential effect of 3 to 6%.

[1] J.P. Hirth and M. Cohen, *Met. Trans.*, 1:3 (1970).
[2] D.C. Drucker, *Met. Trans.*, 4:667 (1973).
[3] G.C. Rauch, R.L. Daga, S.V. Radcliffe, R.J. Sober, and W.C. Leslie, *Met. Trans. A*, 6A:2279 (1975).
[4] W.A. Spitzig, R.J. Sober, and O. Richmond, *Acta Metall.*, 23:885 (1975).
[5] W.A. Spitzig, R.J. Sober, and O. Richmond, *Met. Trans. A*, 7A:1703 (1976).

Measurements of the pressure dependence of yielding and flow, and of the volume changes occurring during plastic deformation, have been made on as-quenched martensites,[1] quenched and tempered martensites,[2] and maraging steels.[3] In every instance the pressure dependence of the yield and flow stress (Fig. VII-10) was much greater than expected from the small volume change during plastic flow. As shown in Fig. VII-10, the magnitude of the SD effect is independent of hydrostatic pressure or strain. Any model of plasticity that would fit this situation must be one in which the octahedral shear yield stress is proportional to the mean pressure, but the volume change is negligible. Yielding and flow under hydrostatic pressure can be represented by the relation

$$I_2 = C - aI_1 \tag{VII-2}$$

where $I_1 = (\sigma_1 + \sigma_2 + \sigma_3)$
$I_2 = \frac{3}{2}[(\sigma_1')^2 + (\sigma_2')^2 + (\sigma_3')^2]$
a = pressure coefficient
C = strength coefficient
$\sigma_1, \sigma_2, \sigma_3$ = principal stresses
$\sigma_1', \sigma_2', \sigma_3'$ = principal deviator stresses

but there is still no dislocation model that explains these macroscopic aspects of the flow of metals.

VII-5. THE TEMPERING OF MARTENSITE

Mechanisms

Tempering is the process of heating martensitic steels so that they become more ductile. The several processes involved are segregation of C to lattice defects, precipitation of carbides, transformation of retained austenite, and recovery and recrystallization of the martensitic structure. These processes are often concurrent and they occur on a very fine scale, so their study is not easy. They will be discussed here as they occur in carbon steel with increasing temperature; then the effects of alloying elements will be considered.[4]

Most steels have Ms temperatures above ambient, so that during quenching there is a brief period during which C atoms can redistribute themselves to low-energy sites associated with dislocations or vacancies, a process akin to strain aging. This is called self-tempering or autotempering. This redistribution of carbon atoms can continue

[1] G.C. Rauch, R.L. Daga, S.V. Radcliffe, R.J. Sober, and W.C. Leslie, *Met. Trans. A,* 6A:2279 (1975).

[2] W.A. Spitzig, R.J. Sober, and O. Richmond, *Acta Metall.,* 23:885 (1975).

[3] W.A. Spitzig, R.J. Sober, and O. Richmond, *Met. Trans. A,* 7A:1703 (1976).

[4] G.R. Speich and W.C. Leslie, *Met. Trans.,* 3:1043 (1972).

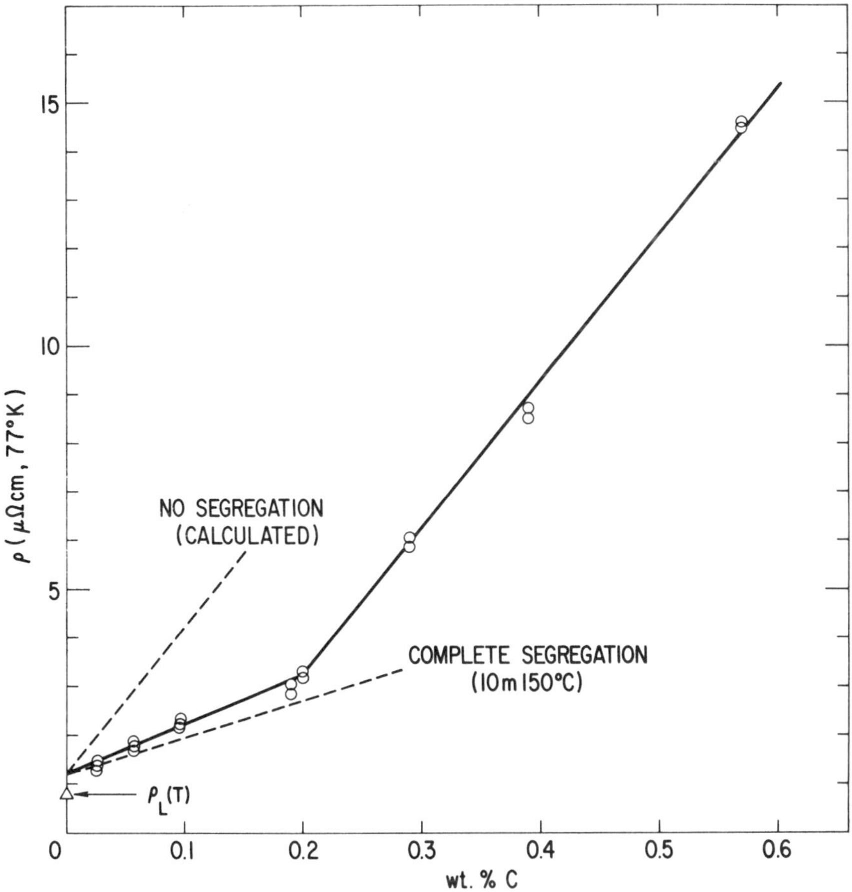

Figure VII-11. Carbon segregation during quenching of iron–carbon martensites, as detected by electrical resistivity measurements. (G.R. Speich, *Trans. AIME*, 245:2553 (1969), by permission.)

during tempering at low temperatures. Such segregation cannot be observed by light or electron metallography, but it can be detected by changes in electrical resistivity or internal friction when C atoms move from "normal" lattice positions to positions at dislocations or vacancies, with results shown in Fig. VII-11. When Fe–C alloys are quenched to martensite, then rapidly chilled to the temperature of liquid nitrogen, their resistivity at that temperature can be divided into two segments. In the first, in alloys containing from 0 to 0.20%C, the resistivity increases with carbon content at the rate of 10 μΩ-cm/wt % C. At a carbon content of 0.20%, the rate of change suddenly increases to 30 μΩ-cm/wt % C. The lower resistivity arises from the segregation of carbon to dislocations during the quench. At concentrations of C below 0.2%, about 90% is segregated to dislocations during quenching. Above 0.20%C these defect

Table VII-4 Carbides claimed to form during tempering of Fe-C martensite

Designation	Name and structure		Composition
η, eta		orthorhombic	Fe_2C[1,2,3]
χ, chi	Hägg carbide	monoclinc	$Fe_{2.2}C$[4]
ϵ, epsilon		hexagonal	$Fe_{2.4}C$[5]
θ, theta	Cementite	orthorhombic	Fe_3C

[1] Y. Hirotsu, S. Nagakura, and S. Oketani, *Proc. Int. Conf. Tech. Iron Steel,* Tokyo, 1971, p. 1140.

[2] Y. Hirotsu and S. Nagakura, *Acta Metall.,* 20:645 (1972).

[3] D.L. Williamson, K. Nakazawa, and G. Krauss, *Met. Trans. A,* 10A:1351 (1979).

[4] J. Chipman, *Met. Trans.,* 3:55 (1972).

[5] E.W. Langer, *Met. Sci. J.,* 2:59 (1968).

sites became almost saturated and carbon atoms remain in "normal" interstitial sites. Unless the rate of cooling during the quench is increased greatly, 0.20%C is also the point at which tetragonality of martensite can be detected. This conclusion is supported by studies by Kurdjumov[1] on Snoek peaks in internal friction of martensite.

When C concentration is increased to about 0.8% and above, the fine structure of martensite is changed to twins, as described previously, and the density of dislocations is reduced. In these circumstances, C atoms segregate as clusters and resistivity increases because of the lattice strains associated with the clusters.[2]

When a carbon martensite is warmed to about 100°C, C atoms cluster along {100} prior to precipitation of ϵ carbide.[3,4]

Our understanding of the precipitation of carbides from martensite, and especially from high-carbon martensite, is in an unsatisfactory state. Table VII-4 lists the various carbides that have been identified by several authors.

Hirotsu et al.[5,6] claim that in a 1.1%C steel, tempered at 120°C, the first carbide to precipitate is not ϵ, but η. The difficulties in differentiating the two are formidable because the particles are only 3 to 5 nm thick and are often irregular in shape, making habit plane and growth direction difficult to determine. Later work by Williamson et al.[7] confirms formation of η. It was identified in a high-purity Fe–1.2%C alloy,

[1] G.V. Kurdjumov, *Fiz. Met. Metalloved,* 24:909 (1967).

[2] J.M.R. Genin and P.A. Flinn, *Trans. AIME,* 242:1419 (1968).

[3] V.I. Izotov and L.M. Utevski, *Fiz. Met. Metalloved,* 25:98 (1968).

[4] D.W. Hoffman, *Acta Metall.,* 13:819 (1970).

[5] Y. Hirotsu, S. Nakagura, and S. Oketani, *Proc. Intern. Conf. Tech. Iron Steel,* Tokyo, 1971, p. 1140.

[6] Y. Hirotsu and S. Nakagura, *Acta Metall.,* 20:645 (1972).

[7] D.L. Williamson, K. Nakazawa, and G. Krauss, *Met. Trans. A,* 10A:1351 (1979).

austenitized, quenched to a structure of martensite plus retained austenite, then tempered 16 h at 150°C. The orientation relation between η carbide and martensite was found to be

$$(100)_\eta \parallel (010)_m \qquad [001]_\eta \parallel [100]_m$$

The χ, or Hägg, carbide is similarly difficult to distinguish from cementite during tempering at higher temperature. The Hägg carbide is apparently formed in some high-carbon steels tempered at 200 to 300°C. In low-carbon steels the place of the Hägg carbide in the tempering sequence is uncertain because the diffraction patterns from cementite and Hägg carbide are similar, and no data from Mössbauer spectroscopy seem to be available.

When steels containing more than about 0.2%C are tempered at temperatures between 100 and 200°C, ϵ carbide is the first precipitate. In plain carbon or low-alloy steels containing less than 0.2%C, no ϵ carbide is formed during tempering between these temperatures. As mentioned earlier, most of the C in such steels is at dislocation sites which provide a greater binding energy for C atoms than does ϵ carbide, so there is no driving force for precipitation. Before precipitation can occur some dislocations must be removed by recovery, but the C atoms inhibit recovery.

Figure VII-12 shows the progress of tempering of low-C Fe–C alloys at 150°C, as measured by electrical resistivity.[1] The region marked *A* is associated with the continuation of the segregation of C atoms to dislocations that began during quenching. Region *C* is associated with carbide precipitation, and *B* is a transition region. When the C content is increased to 0.4% the resistivity decreases continuously with time at all tempering temperatures, with no indication of three distinct regions. At these higher carbon contents not all of the C can be associated with dislocations and carbide precipitation occurs quickly even at 150°C.

Cementite forms when most steels are tempered at 250–700°C. The initial morphology of cementite in martensite is needle-like when formed either during deliberate tempering or when formed during slower cooling in the quenching of large sections. The common nucleation sites are martensite lath boundaries at low temperatures and ferrite grain boundaries at higher temperatures.

If retained austenite is present, it decomposes to bainite at tempering temperatures of 200 to 300°C. Retained austenite is present in appreciable quantities in low-alloy steels only when the C content exceeds 0.4% (Fig. VII-3). Therefore, this, reaction is important only in medium- or high-carbon steels.

It is difficult to determine when recovery of the defect structure begins during the tempering of martensite, but it becomes certainly an important feature of the tempering process above 400°C. At these temperatures the most striking feature of the early stages of tempering of steels containing 0.20%C or less is the sudden decrease in hardness and in total internal boundary area. These changes begin in a fraction of a second and quickly reach comparatively stable values that thereafter decrease slowly

[1] G.R. Speich, *Trans. AIME,* 245:2553 (1969).

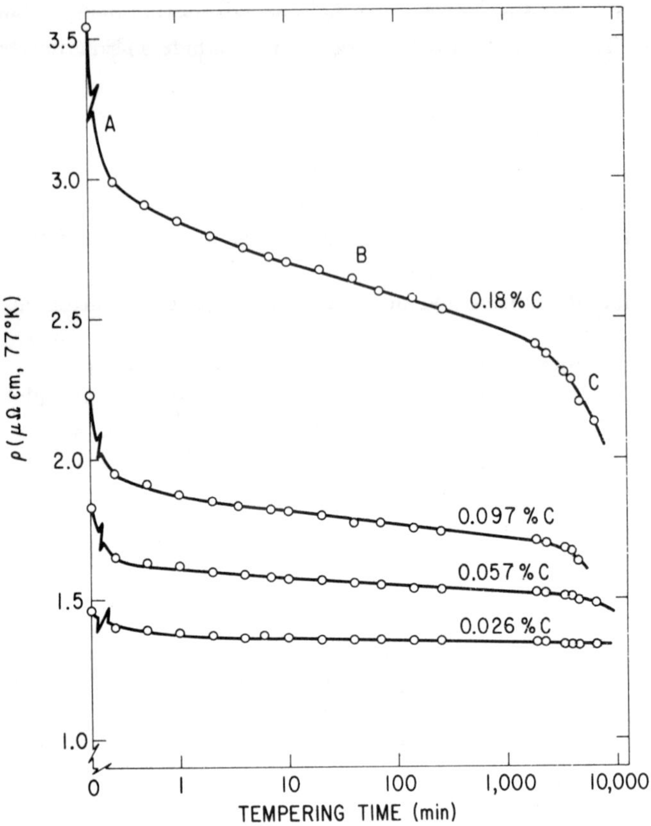

Figure VII-12. Tempering of low-carbon iron–carbon martensites at 150°C, as indicated by changes in electrical resistivity. (G.R. Speich, *Trans. AIME*, 245:2553 (1969), by permission.)

with time. During this rapid recovery, dislocation cell walls within martensite laths are annihilated and a fine, acicular grain structure is developed. The lath structure coarsens, but it retains the elongated packet morphology until recrystallization begins. Recrystallization occurs more rapidly in low- than in high-carbon steels because the recrystallization process is inhibited by the pinning action of carbides in boundaries.

In low-carbon steels tempered at high temperature, the stages are

1. Initial rapid softening (recovery)
2. An intermediate stage where recrystallization is suppressed because of the pinning of grain boundaries by carbide particles
3. A final stage in which grain growth occurs because of the coarsening of the carbide particles

The structure obtained after long tempering at high temperature consists of ferrite grains with carbides scattered throughout. After recrystallization is complete, growth of carbide particles and ferrite grains are the only kinetic processes that continue.

Figure VII-13 shows the changes in hardness occurring in Fe–C alloys initially quenched to martensite, then tempered 1 h at temperatures from 100 to 700°C. At 0.10%C at 100°C, no precipitation occurs and the hardness is constant. After long periods at 200°C, some precipitation occurs. At higher temperatures, precipitation occurs very quickly. In an alloy containing 0.18%C there is a definite increase in hardness during tempering at 100°C, presumably because of completion of segregation of C to dislocations. The various stages shown in Fig. VII-13 are the formation of rod-shaped Widmanstätten Fe_3C particles, recovery of the defect structure, spheroidization of carbide particles, recrystallization, and growth of carbide particles.

In plain carbon steels, as in the Fe–C alloys from which Fig. VII-13 was obtained, quenched martensite softens progressively as the temperature is increased in the range

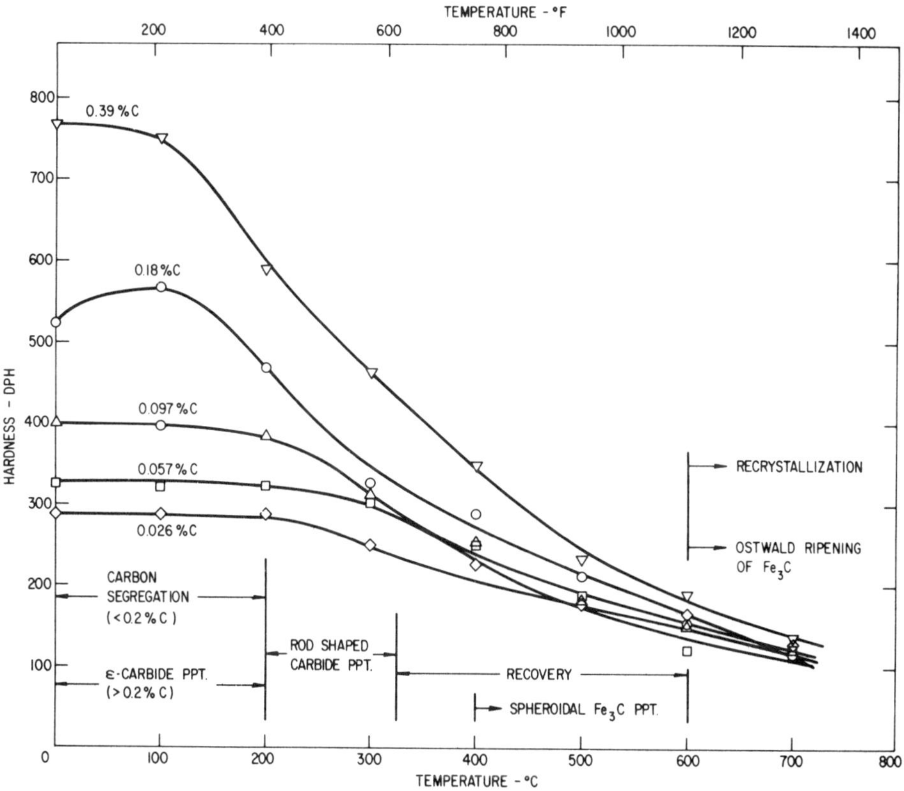

Figure VII-13. Hardness of iron-carbon martensites tempered 1 h at 100°C to 700°C. (G.R. Speich, *Trans. AIME*, 245:2553 (1969), by permission.)

100–700°C, as a result of the growth of cementite from a fine Widmanstätten array of rods to coarse spheroidal particles. However, if the carbide-forming elements Cr, Mo, V, Ti, or W are added to the steel, an additional and important strengthening mechanism called secondary hardening occurs in the temperature range from 500 to 600°C. As a result, hardness approaching that of the as-quenched alloy can be achieved, but with much improved toughness. This strengthening occurs because of the re-solution of Fe_3C and the precipitation of alloy carbides such as V_4C_3, Mo_2C, and W_2C. A large number of different alloy carbides are possible. These form initially as very fine coherent particles, primarily on dislocations inherited from the as-quenched martensite. At the peak of secondary hardening, Mo_2C needles are about 10 nm long and 1.5 nm in diameter. V_4C_3 forms as plates about 10 nm in diameter and 1 nm thick.[1] In these secondary hardening steels, an additional increment of strength can be obtained by adding less than 0.1%Nb,[2] because of formation of a fine dispersion of NbC which seems to precipitate in the matrix without regard to dislocations.

The effects of C and some alloying elements on the tempering of a 10%Ni steel are illustrated in Fig. VII-14.[3] Beginning with an alloy of Fe–10%Ni, successive additions of C, Co, Cr, and Mo were made. Co, although it does not enter the carbide, delays tempering. The addition of Cr provides an increase in strength during tempering, and Mo gives a sharp secondary hardening peak.

Estimation of Hardness after Tempering

In the manufacture of heat-treated steel articles, consistency of mechanical properties is highly desirable, and this consistency must be maintained despite the fluctuations of composition that are unavoidable in commercial steels. Therefore, some method of estimating the hardness of quenched and tempered steels from the chemical composition of incoming material is needed, to allow the tempering temperature to be varied to produce parts of constant hardness. The hardness of tempered martensite can be used to estimate the tensile strength.

Hollomon and Jaffe[4] developed the so-called "tempering parameter" to relate hardness, tempering temperature, and tempering time. Time and temperature are related by the rate equation

$$\frac{1}{t} = A \exp\left(-\frac{Q_1}{RT}\right) \tag{VII-3}$$

where t is the time required to reach a given hardness and Q_1 is an empirical "activation

[1] R.W.K. Honeycombe, H.J. Harding, and J.T. Irani, in *High Strength Materials,* Wiley, New York, 1965, p. 213.

[2] R.W.K. Honeycombe, in *Metallography 1963,* Spec. Report 80, Iron and Steel Inst., London, 1964, p. 245.

[3] G.R. Speich, D.S. Dabkowski, and L.F. Porter, *Met. Trans.,* 4:303 (1973).

[4] J.H. Hollomon and L.D. Jaffe, *Trans. AIME,* 162:223 (1945).

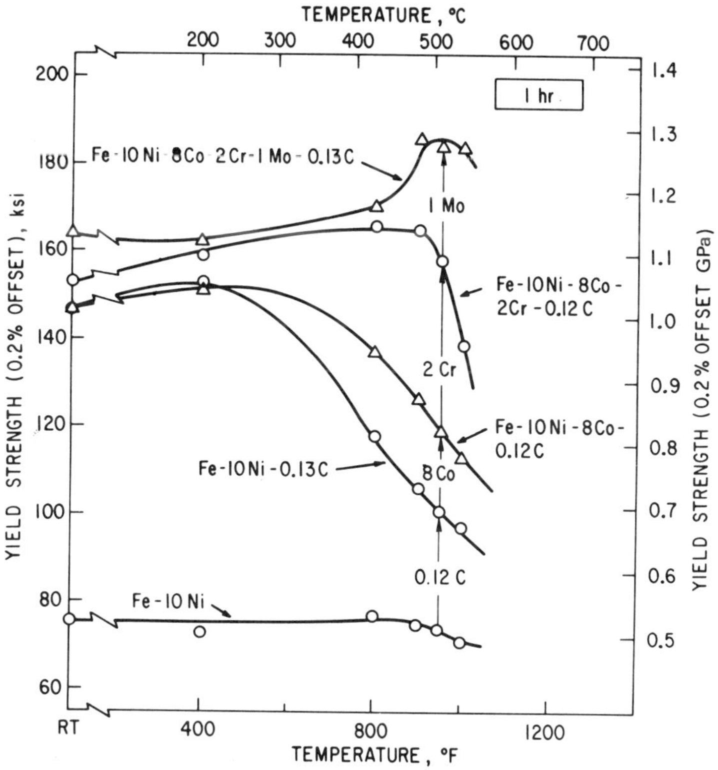

Figure VII-14. Effect of carbon, cobalt, chromium, and molybdenum on tempering of 10%Ni steels. (G.R. Speich, D.S. Dabkowski, and L.F. Porter, *Met. Trans.*, 4:303 (1973), by permission.)

energy" for the process. As might be expected in a process as complicated as tempering, Q is not constant, but changes with hardness. It was assumed that hardness H would be a function of time and temperature:

$$H = f\left[t \exp\left(-\frac{Q_1}{RT}\right)\right] \tag{VII-4}$$

It was found that $[t \exp(-Q_1/RT)]$ was constant, t_0.

Equating the "activation energies" of Eqs VII-3 and VII-4, and converting to the logarithmic form,

$$Q = T(\ln t - \ln t_0) = f(H) \tag{VII-5}$$

or

$$H = f\left(T \log \frac{t}{t_0}\right) \tag{VII-6}$$

and because t_0 is constant,

$$H = f[T(C + \log t)] \tag{VII-7}$$

This last quantity is the tempering parameter. Hardness should be measured as DPH because the plots then more closely approach straight lines. The exact value of C is not critical, but with temperature in °F, a value of 18 gives the least scatter of data.

Figure VII-15 can be used to convert from one tempering time and temperature to any other, on the basis that combinations of tempering temperature and time having the same value of the tempering parameter will produce the same hardness. Thus, about 50 h at 500°F (260°C) or 2 min at 700°F (371°C) will produce the same hardness as 1 h at 600°F (316°C).

The hardness of tempered martensite over the normal range of AISI carbon and alloy steel compositions can be calculated by the procedure developed by Grange,

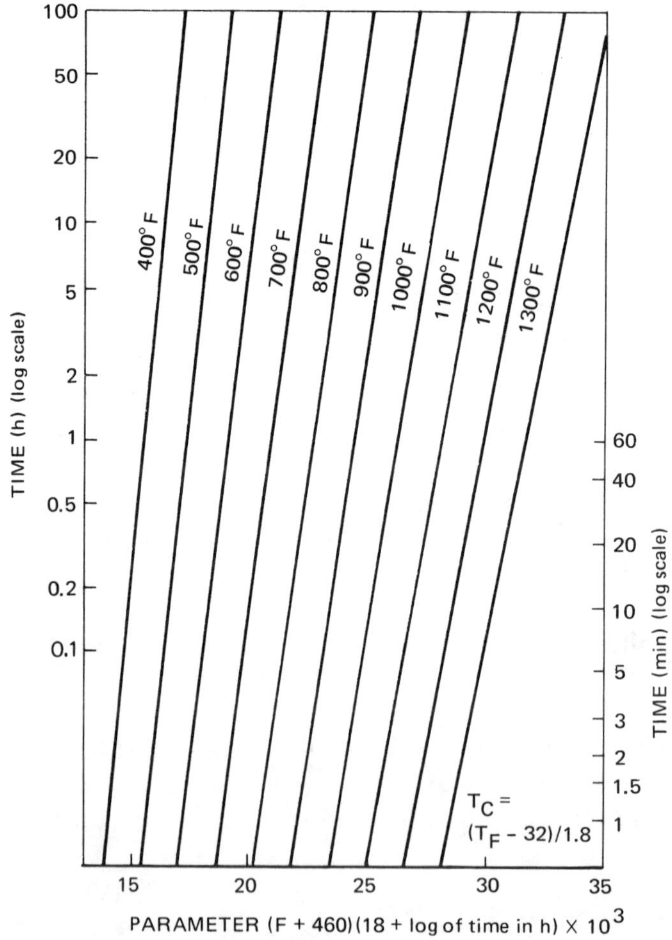

Figure VII-15. Chart showing relationship of tempering parameter to temperature and time. (R.A. Grange, C.R. Hribal, and L.F. Porter, *Met. Trans. A*, 8A:1775 (1977), by permission.)

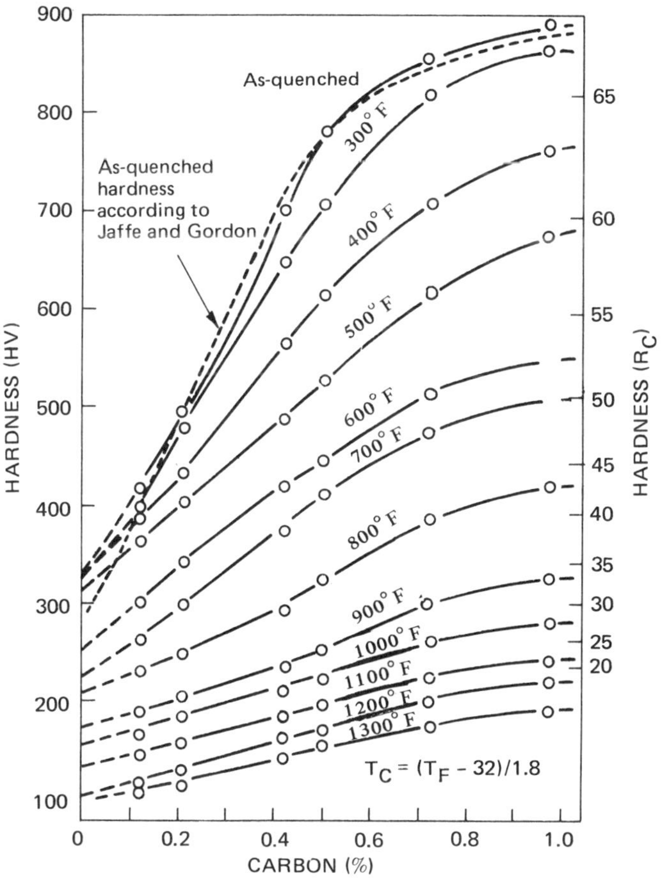

Figure VII-16. Hardness of tempered martensite in iron–carbon alloys. (R.A. Grange, C.R. Hribal, and L.F. Porter, *Met. Trans. A*, 8A:1775 (1977), by permission.)

Hribal, and Porter.[1] Fe–C alloys were austenitized at 927°C, brine quenched to 100% martensite, and tempered for 1 h at temperatures in the range 149 to 704°C; then hardness (DPH) was measured. The results are shown in Fig. VII-16. Some small amounts of retained austenite were present in alloys containing more than 0.50%C. The hardness of as-quenched martensite was not altered when alloying elements were added to the Fe–C base, but such additions had marked effects on the hardness after tempering. The alloying elements used were Mn, P, Si, Ni, Cr, Mo, and V. It was found that 0.5%Mn had about the same hardening effect at all C concentrations. In subsequent tests, the C content was held constant at 0.2% and the Mn at either 0.3 or 0.5%. To this base the other alloying elements were added in various concentrations and their hardening effects measured at a series of tempering temperatures. The assumption was

[1] R.A. Grange, C.R. Hribal, and L.F. Porter, *Met. Trans. A*, 8A:1775 (1977).

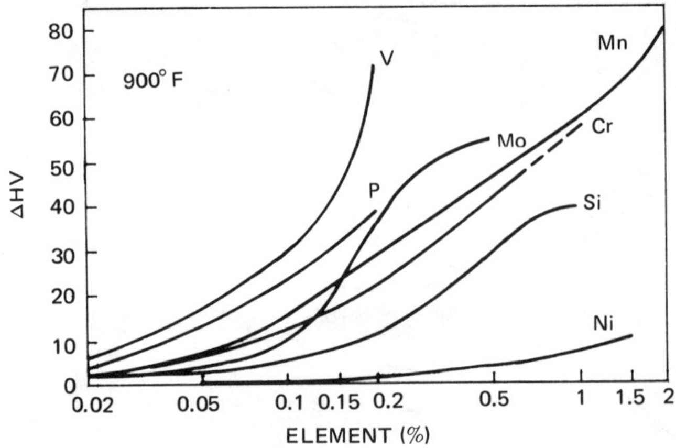

Figure VII-17. Effect of alloying elements on the hardness of martensite tempered at 482°C for 1h. (R.A. Grange, C.R. Hribal, and L.F. Porter, *Met. Trans. A.*, 8A:1775 (1977), by permission.)

made that the hardening effect of each alloying element was the same at all C contents. This assumption was later tested by comparing predicted and observed hardness for commercial steels and found to be reasonably valid.

The increase in hardness, ΔHV, produced by alloying elements was plotted for each tempering temperature in the manner shown in Fig. VII-17. A logarithmic scale is used for composition to allow accurate readings for small concentrations. To estimate the hardness of any given steel after tempering at any selected temperature, it is only necessary to read off the hardness of the Fe-C alloy of the same C content (Fig. VII-16), then to add to this the ΔHV values for each of the alloying elements in the steel, at the selected tempering temperature:

$$\text{Estimated HV} = HV_c + \Delta HV_{Mn} + \Delta HV_p + \Delta HV_{Si} + \Delta HV_{Ni} + \Delta HV_{Cr} + \Delta HV_{Mo} + \Delta HV_V \tag{VII-8}$$

Grange et al.[1] provide a set of curves similar to Fig. VII-17, obtained at intervals of 56°C over the range 205 to 705°C.

The effect of silicon on tempering has been mentioned previously; it had its greatest hardening effect during the tempering of a 0.19%Cr-0.5%Mn steel at 316°C. This can be attributed to the effect of Si in inhibiting the formation of Fe_3C (Table IV-2). V and Mo, because of their secondary hardening effects through precipitation of alloy carbides, show pronounced hardening at tempering temperatures above 500°C.

VII-6. EMBRITTLEMENT DURING TEMPERING

Two varieties of embrittlement serve to reduce ductility during tempering. The first has been called 500°F embrittlement, 350°C embrittlement, or one-step temper

[1] R.A. Grange, C.R. Hribal, and L.F. Porter, *Met. Trans. A*, 8A:1775 (1977).

embrittlement. It is a malady afflicting low-alloy steels that have been quenched to martensite, then tempered in the range 250–350°C. The hardness of the steel decreases continuously as the tempering temperature is raised, but the toughness passes through a minimum, as indicated in Fig. VII-18. When the steels are embrittled, the fracture becomes predominantly intergranular, along the prior austenite grain boundaries. The embrittlement is associated with the change in structure of carbides from ϵ to cementite. This embrittlement poses an awkward problem for the development of optimum strength and ductility, for often the strength desired can be obtained best by tempering near 250–300°C.

Although the embrittlement is concurrent with the precipitation of cementite, such precipitation is not in itself the cause of the loss of toughness. When P, Sb, Sn, As, and N are absent from a steel, the embrittlement does not occur. Also, the problem can be avoided by adding sufficient Si to the steel to inhibit formation of cementite in the critical temperature range. The process may begin with segregation of P or N to grain boundaries in austenite, perhaps promoted by Mn. The segregation that occurs is slight.[1] Subsequent precipitation of cementite at the former austenite grain boundaries may increase the impurity segregation by rejecting solutes from the growing carbide particles. The likely combination is that impurity segregation weakens adhesion at the prior austenite grain boundaries and cracking of carbides initiates intergranular fracture.

The second variety of embrittlement is commonly called temper embrittlement or, less often, two-step temper embrittlement. It is also a sickness of alloy steels.[2,3] The distinction between the two varieties is shown in Fig. VII-19. The one-step temper embrittlement occurs in high-strength steels, such as AISI 4130, quenched, then

[1] C.L. Briant and S.K. Banerji, *Met. Trans. A,* 10A:123 (1979).
[2] I. Olefjord, *Int. Metall. Rev.,* 23:149 (1978).
[3] C.J. McMahon Jr. and L. Marchut, *J. Vac. Sci. Technol.,* 15:450 (1978).

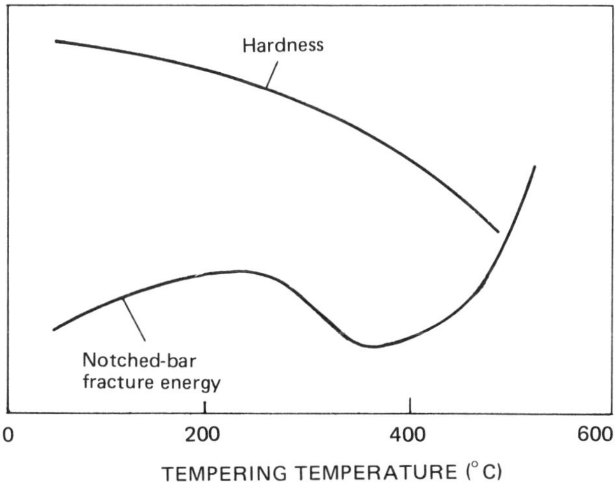

Figure VII-18. Schematic of change in hardness and room-temperature notched-bar fracture energy during tempering of a quenched steel. (C.L. Briant and S.K. Banerji, *Int. Metall. Rev.*, 23:164 (1978), by permission.)

tempered at a low temperature. The two-step temper embrittlement occurs in alloy steels having much lower yield strengths because of tempering at high temperatures, 600-700°C, followed by slow cooling through the range 600 to 350°C, or having a service temperature in that range. The latter poses a difficult problem, for embrittlement can occur slowly in items such as steam turbine rotors whose service temperatures lie in the critical range. The steady reduction of ductility can lead to existing flaws becoming larger than the critical flaw size for the conditions of stress, causing catastrophic fracture.

Temper embrittlement has certain characteristics:

1. It does not occur in carbon steels.
2. It is caused primarily by Sb and P, secondarily by Sn or As.
3. Additions of Ni, Mn, Cr, and Si enhance embrittlement.
4. It does not occur in alloy steels of high purity.
5. Mo, Ti, and Zr delay the onset of embrittlement.
6. The embrittlement is not removed by overaging up to 30,000 h.
7. It is reversible. It can be removed by heating to temperatures above 600°C followed by rapid cooling and restored by holding in the range 350-550°C.
8. There is no precipitation of the offending elements.

The embrittlement is manifested in three ways:

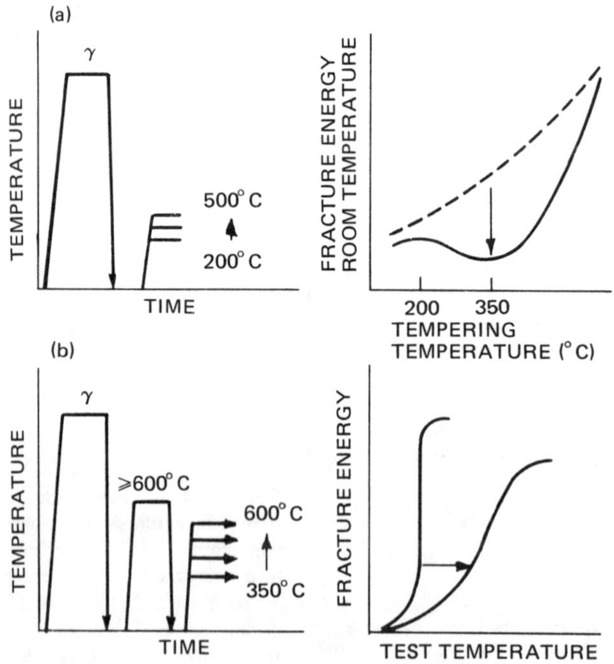

Figure VII-19. Heat treatments involved in (*a*) one-step and (*b*) two-step temper embrittlement. (C.L. Briant and S.K. Banerji, *Int. Metall. Rev.*, 23:164 (1978), by permission.)

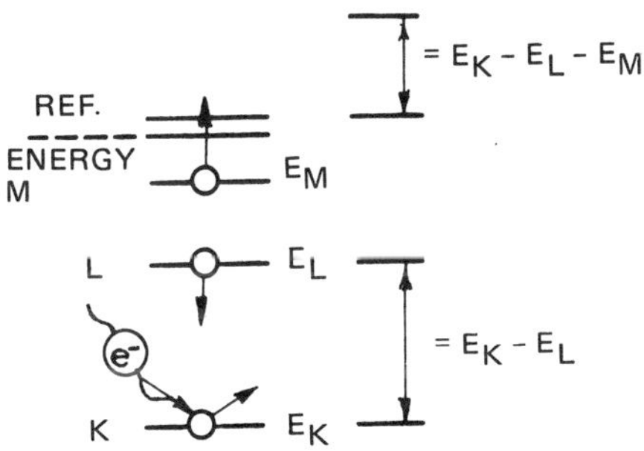

Figure VII-20. The Auger process and electron energy relations. (I. Olefjord, *Int. Metall. Rev.*, 23:149 (1978), by permission.)

1. The notch-impact transition temperature is raised very markedly. An Sb content of 0.04% (much more than is ever present in commercial steels) can raise the transition temperature of a 0.3%C, Cr-Ni steel from −40°C in the unembrittled condition to 600°C in the embrittled condition.
2. Fracture occurs in prior austenite grain boundaries.
3. A grain boundary etching effect appears after embrittlement. This can be put to good use in determining austenite grain size in alloy steels.

There are two landmarks in the long, frustrating study of temper embrittlement. The first was the work of Balajiva and Steven,[1] which proved conclusively that Sb, P, Sn, and As were potent embrittling agents and that high-purity steels were not embrittled. The second was the use of Auger electron spectroscopy to determine quantitatively the extent of segregation of the offending elements.[2,3] This technique is deserving of a brief explanation. A metal surface is bombarded with high-energy electrons (150–450 aJ). An inner K shell electron is rejected, leaving a hole that can be filled by an electron from an outer L shell. The energy difference between the K and L levels can be compensated by the emission of another electron, a so-called Auger electron from an outer electron level, such as an M shell (Fig. VII-20). The Auger electrons have discrete energy levels characteristic of the emitting element. An energy distribution curve of these electrons can be differentiated to obtain a curve of dN/dE vs. E, where N is the number of electrons emitted at the energy level E. The height of the dN/dE peak for a given element is a measure of the concentration of that element.

In Auger spectroscopy the beam of electrons can be focused to a diameter of 1 μm. The mean free path of electrons in the solid is only about 0.5 to 1.5 nm, so the analysis is confined to a layer of that thickness. Successive layers can be removed by sputtering and analyzed in turn to obtain a profile of the segregation. It is essential that the surface to be analyzed be clean, so it is customary to break a specimen in ultrahigh vacuum and to examine the fracture surface in that vacuum. Of course,

[1] W. Steven and K. Balajiva, *J. Iron Steel Inst.*, 193:141 (1959).
[2] P.W. Palmberg and H.L. Marcus, *Trans. Am. Soc. Met.*, 62:1016 (1969).
[3] R.A. Mulford, C.J. McMahon Jr., D.P. Pope, and H.C. Feng, *Met. Trans. A*, 7A:1183 (1976).

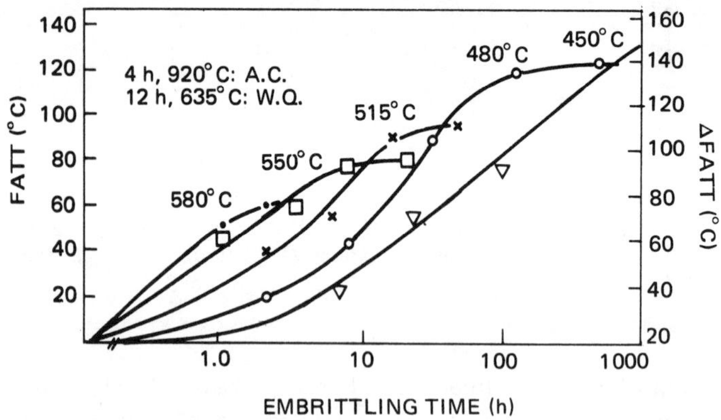

Figure VII-21. Effect of temperature on rate of isothermal embrittlement of a 0.34%C–3.1%Cr–0.59%Mo steel. (B.L. King and G. Wigmore, *Met. Trans. A*, 7A:1761 (1976), by permission.)

segregation to a free surface in vacuum can be determined without the necessity of fracture.

In temper embrittlement developed at constant temperature, the embrittlement increases with time; that is, the fracture appearance transition temperature (FATT) increases with time, in the manner shown in Fig. VII-21 for a steel of the composition (wt %):

C	Cr	Mo	Mn	Si	Ni	S	P	Sn	As	Sb
0.34	3.10	0.59	0.89	0.33	0.18	0.036	0.030	0.015	0.029	0.008

The maximum embrittlement increases as the aging temperature is reduced.[1]

To minimize embrittlement after tempering above 600°C, the cooling rate required through the embrittlement range, 600 to 400°C, is often so high (on the order of 1000 to 2000°C per hour) that residual stresses and distortion in large components become unacceptably high.

Although agreement is not complete, most workers in the field seem to agree that the characteristic features of temper embrittlement are best explained by a double equilibrium segregation (cosegregation) model. Impurity atoms (P, Sb) and alloying elements (Mn, Ni) segregate *in ferrite* to prior austenite grain boundaries. Our understanding of this process has been advanced through the development by Guttmann[2,3] of the ternary regular solution model of grain boundary segregation. There is a high degree of segregation of metalloid impurities from Groups IVA and VA of the periodic table because of strong attractive interactions between these elements and Ni and Mn, resulting in enhanced cosegregregation. The grain boundary (or free surface) compositions in a ternary regular solution are given by Guttmann[2] as:

[1] B.L. King and G. Wigmore, *Met. Trans. A*, 7A:1761 (1976).
[2] M. Guttmann, *Surf. Sci.*, 53:213 (1975).
[3] M. Guttmann, *Met. Sci.*, 10:337 (1976).

$$X_1^{\phi} = \frac{X_1^{\rm B} \exp(\Delta G_1/RT)}{1 - X_1^{\rm B} + X_1^{\rm B} \exp(\Delta G_1/RT) - X_2^{\rm B} + X_2^{\rm B} \exp(\Delta G_2/RT)} \qquad \text{(VII-9)}$$

$$X_2^{\phi} = \frac{X_2^{\rm B} \exp(\Delta G_2/RT)}{1 - X_1^{\rm B} + X_1^{\rm B} \exp(\Delta G_1/RT) - X_2^{\rm B} + X_2^{\rm B} \exp(\Delta G_2/RT)} \qquad \text{(VII-10)}$$

where X_1 is atom fraction of element 1 (solute)
X_2 is atom fraction of element 2 (solute)
ϕ refers to surface
B refers to bulk

The driving forces for segregation of elements 1 and 2 (ΔG_1 and ΔG_2) are

$$\Delta G_1 = \Delta G_1^0 + 2\alpha_{13}(X_1^{\rm B} - X_1^{\phi}) + (\alpha_{12} - \alpha_{23} - \alpha_{13})(X_2^{\phi} - X_2^{\rm B}) \qquad \text{(VII-11)}$$

$$\Delta G_2 = \Delta G_2^0 + 2\alpha_{23}(X_2^{\rm B} - X_2^{\phi}) + (\alpha_{12} - \alpha_{23} - \alpha_{13})(X_1^{\phi} - X_1^{\rm B}) \qquad \text{(VII-12)}$$

ΔG_1^0 is the driving force for segregation of solute 1 in solvent 3. It accounts for changes in nonfigurational enthalpy (elastic, vibrational, etc.) when a solute atom replaces a solvent atom at a boundary.

The $\alpha_{\rm ij}$ terms are defined by:

$$\alpha_{\rm ij} = N_0 Z\left(\epsilon_{\rm ij} - \frac{\epsilon_{\rm ii} + \epsilon_{\rm ij}}{2}\right)$$

where N_0 = Avagadro's number
Z = coordination number in the crystal lattice
$\epsilon_{\rm ij}$ = energy of an *ij* bond, with the assumption that $\epsilon_{\rm ij}^{\phi} = \epsilon_{\rm ij}^{\rm B}$

McMahon and Marchut[1] give a lengthy derivation of these equations.

Segregation at prior austenite grain boundaries rather than at various interfaces in martensite or bainite occurs presumably because the former are high-angle boundaries and the latter are principally low-angle boundaries. The atoms with large size misfit will tend to segregate at high-angle boundaries. The beneficial effects of Mo, Ti, and Zr on inhibiting embrittlement are due to their scavenging action on impurity atoms. These actions only slow the rate of embrittlement, however, because the Mo, Ti, and Zr slowly react with C to form stable carbides, releasing the impurity atoms to segregate to boundaries.

The rate of embrittlement is consistent with the rate of diffusion of P in alpha iron. McLean[2] developed the equation

$$t_{1/2} = \frac{9\alpha^2 d^2}{64D} \qquad \text{(VII-13)}$$

for the time needed to reach a boundary concentration equal to half the equilibrium value. In Eq. VII-13,

[1] C.J. McMahon Jr. and L. Marchut, *J. Vac. Sci, Technol.*, 15:450 (1978).
[2] D. McLean, *Grain Boundaries in Metals*, Oxford Univ. Press, 1957, p. 133.

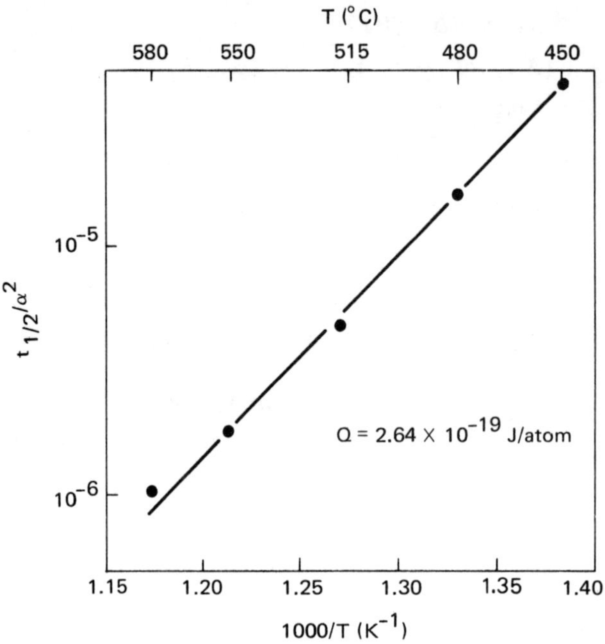

Figure VII-22. Activation energy for embrittlement of a 0.34%C–3.1%Cr–0.59%Mo steel. (B.L. King and G. Wigmore, *Met. Trans. A*, 7A:1761 (1976), by permission.)

α = ratio of the equilibrium grain boundary concentration at the tempering temperature to the bulk concentration
d = thickness of the grain boundary, taken as three lattice spacings
D = diffusivity of the embrittling species.

The diffusivity of P is represented by[1]

$$D = 7.1 \times 10^{-3} \exp\left(-\frac{20{,}000}{T}\right) \text{ cm}^2/\text{s} \qquad \text{(VII-14)}$$

With values of α ranging from 600 to 1160, excellent agreement was obtained between the calculated $t_{1/2}$ and the time required for ΔFATT to reach half the limiting value.[2] The activation energy for embrittlement can be obtained by plotting log $(t_{1/2}/\alpha^2)$ vs. $1/T$, as shown in Fig. VII-22. It agrees with the activation energy for volume diffusion of P in alpha Fe.

VII-7. HARDENABILITY

The hardenability of steel can be defined as that property which determines the depth and distribution of hardness induced by quenching.[3] It can also be defined on the basis

[1] P.L. Gruzin and V.V. Minal, *Fiz. Met. Metalloved*, 16:551 (1963).
[2] B.L. King and G. Wigmore, *Met. Trans. A*, 7A:1761 (1976).
[3] *Metals Handbook*, 8th ed., vol. 1, Am. Soc. Met., Metals Park, Ohio, 1961, p. 20.

of the microstructure developed by quenching, such as by the depth to which austenite transforms to 50% martensite + 50% other transformation products, or to 90% martensite. The 50% martensite criterion is the most commonly used because it is the easiest to determine by metallography and it is the point at which hardness changes most rapidly. The determination of hardenability is a determination of how rapidly the steel must be cooled to avoid formation of pearlite or upper bainite.

During cooling from a temperature above the Ac_3 of a steel, austenite undergoes transformation to various products: ferrite, pearlite, upper bainite, lower bainite, or martensite, in the order of increasing cooling rate and increasing hardenability. The cooling rate depends upon the ability of the cooling medium to extract heat from the steel and upon the thermal conductivity, thermal diffusivity, and volume specific heat of the steel.

Newton's law of heating or cooling states that the rate of heat transfer is proportional to the difference in temperature between the surface of a body and its surroundings, or

$$q = S(T_1 - T_2) \tag{VII-15}$$

where q = rate of heat transfer per unit surface area
T_1 = temperature of the surroundings
T_2 = temperature of the surface of the body
S = proportionality constant

A factor is required that includes properties of the solid body as well as the rate of heat transfer; this is H, the coefficient of surface transfer, defined by

$$H = S/K \tag{VII-16}$$

where K is the thermal conductivity in $\mathrm{J/m \cdot s \cdot {}^\circ C}$ and
S is the proportionality constant in Newton's Law

The severity of a quench can be expressed in terms of H.

As a steel cools and transforms during quenching, its thermal conductivity, thermal diffusivity, and volume specific heat all change with temperature, as indicated in Figs VII-23 and VII-24; therefore, H cannot be constant. The thermal diffusivity a, the diffusion coefficient of temperature, is defined as

$$a = \frac{\text{thermal conductivity}}{\text{volume specific heat}} \tag{VII-17}$$

This, also, is not a constant.

Solutions to problems involving unsteady state heat transfer can be obtained from Russell's tables,[1] based on the assumption of constant values of a and H; but in quenching steel, H is a function of both the size of the part and the quenching medium (Fig. VII-25).

[1] J.B. Austin, *Flow of Heat in Metals,* Am. Soc. Met., Metals Park, Ohio, 1942.

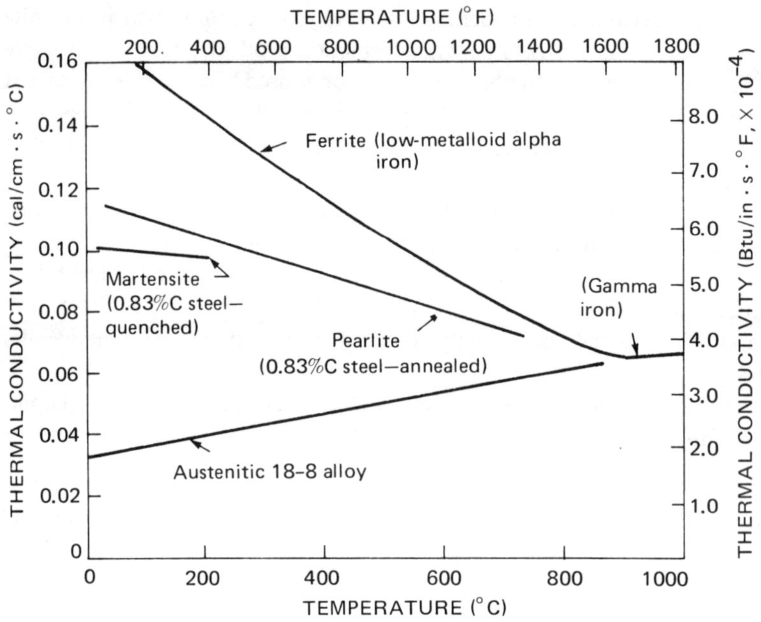

Figure VII-23. The thermal conductivity of ferrite, pearlite, martensite, and austenite. (J.B. Austin, *Flow of Heat in Metals*, Am. Soc. Met., Metals Park, Ohio, 1942, by permission.)

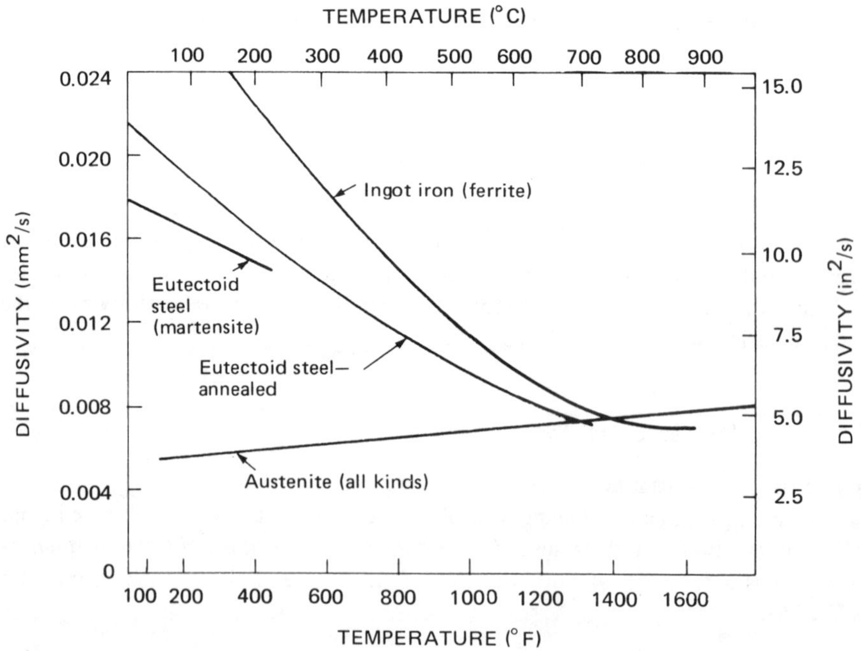

Figure VII-24. Thermal diffusivity of various structures. (J.B. Austin, *Flow of Heat in Metals*, Am. Soc. Met., Metals Park, Ohio, 1942, by permission.)

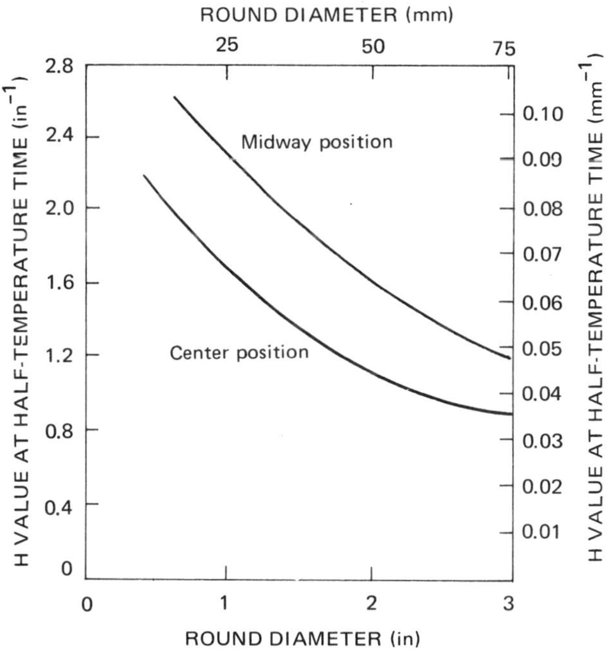

Figure VII-25. Variation of *H* value at half-temperature time in round bars water-quenched from 845°C. (D.J. Carney, *Trans. Am. Soc. Met.*, 46:882 (1954), by permission.)

The severity of quench, *H*, was employed by Grossmann et al.[1] in developing their concepts of *critical diameter* and *ideal diameter* which are the bases for most procedures for estimating hardenability of steels from chemical composition. They defined the critical diameter D_C as the diameter of a steel bar that would just transform to 50% martensite at the center when quenched into a bath having a known *H* value (assumed to be constant). This critical diameter increases as *H* increases, until the ideal diameter is reached when *H* becomes infinite, an *ideal quench.* In such a quench the heat transfer from the surface to the quenching medium is infinitely rapid, so that the surface of the steel bar comes instantaneously to the temperature of the quenching medium and remains at that temperature for the duration of the quench. This ideal quench would produce the maximum possible depth of hardening, and the critical diameter would be the largest attainable, the ideal diameter, D_I.

Approximate *H* values for various quenching media are given in Table VII-5.

The relation between the critical diameter D_C and the ideal diameter D_I for various values of *H* is plotted in Fig. VII-26. The difference between an ideal quench and a quench into violently agitated brine is slight.

The Jominy end-quench test remains the most widely used method for determining hardenability.[2,3] The test specimen and method are shown in Fig. VII-27. After

[1] M.A. Grossmann, M. Asimow, and S.F. Urban, *Hardenability of Alloy Steels,* Am. Soc. Met., Metals Park, Ohio, 1939, p. 124.

[2] W.E. Jominy and A.L. Boegehold, *Trans. Am. Soc. Met.,* 26:574 (1938).

[3] C.A. Siebert, D.V. Doane, and D.H. Breen, *The Hardenability of Steels,* Am. Soc. Met., Metals Park, Ohio, 1977.

Table VII-5 Approximate values of *H* (severity of quench)[1]

Agitation	Oil	Water	Brine
None	0.25–0.30	0.9–1.0	2
Mild	0.30–0.35	1.0–1.1	2.0–2.2
Moderate	0.35–0.40	1.2–1.3	
Good	0.40–0.50	1.4–1.5	
Strong	0.50–0.80	1.6–2.0	
Violent	0.80–1.1	4.0	5.0

[1] H.E. McGannon, ed., *Making, Shaping and Treating of Steel,* U.S. Steel Corp., Pittsburgh, 1971, p. 1099.

austenitizing, the specimen is removed from the furnace quickly, dropped into the fixture, then the water jet is turned on, with the water striking only the end of the specimen. After the cylindrical specimen has cooled to room temperature, a flat is ground along the length, 0.38 mm deep, then Rockwell C hardness measurements are made at 1.6-mm intervals from the quenched end. Hardness is plotted vs. distance from the quenched end of the bar, as shown in Figs VII-38, VII-39, and VII-40. The point along the bar at which the structure becomes 50% martensite can be estimated

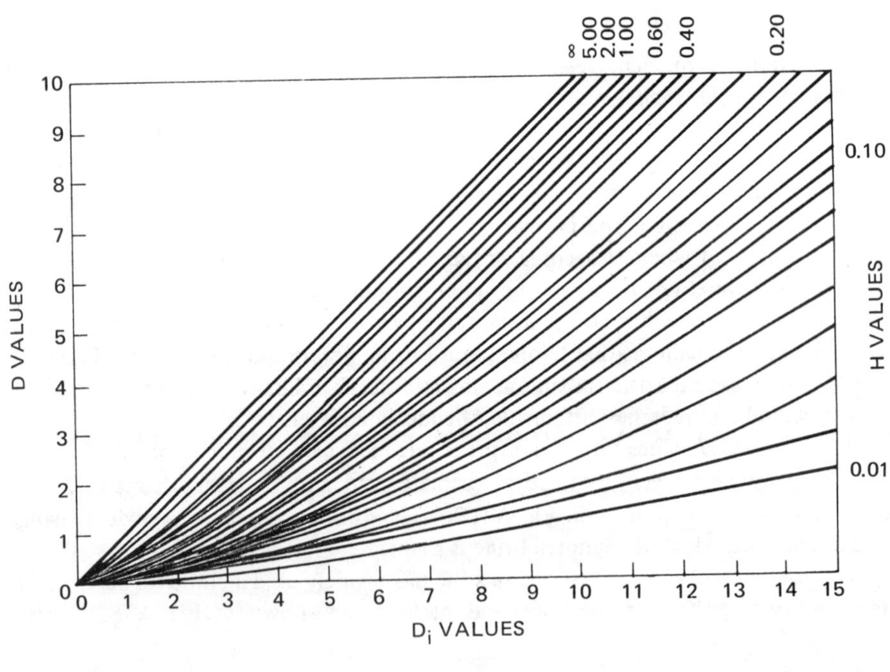

Figure VII-26. Relationships among ideal diameter, critical diameter, and severity of quench. (*Making, Shaping and Treating of Steel*, 9th ed., 1971, copyright United States Steel Corporation, by permission.)

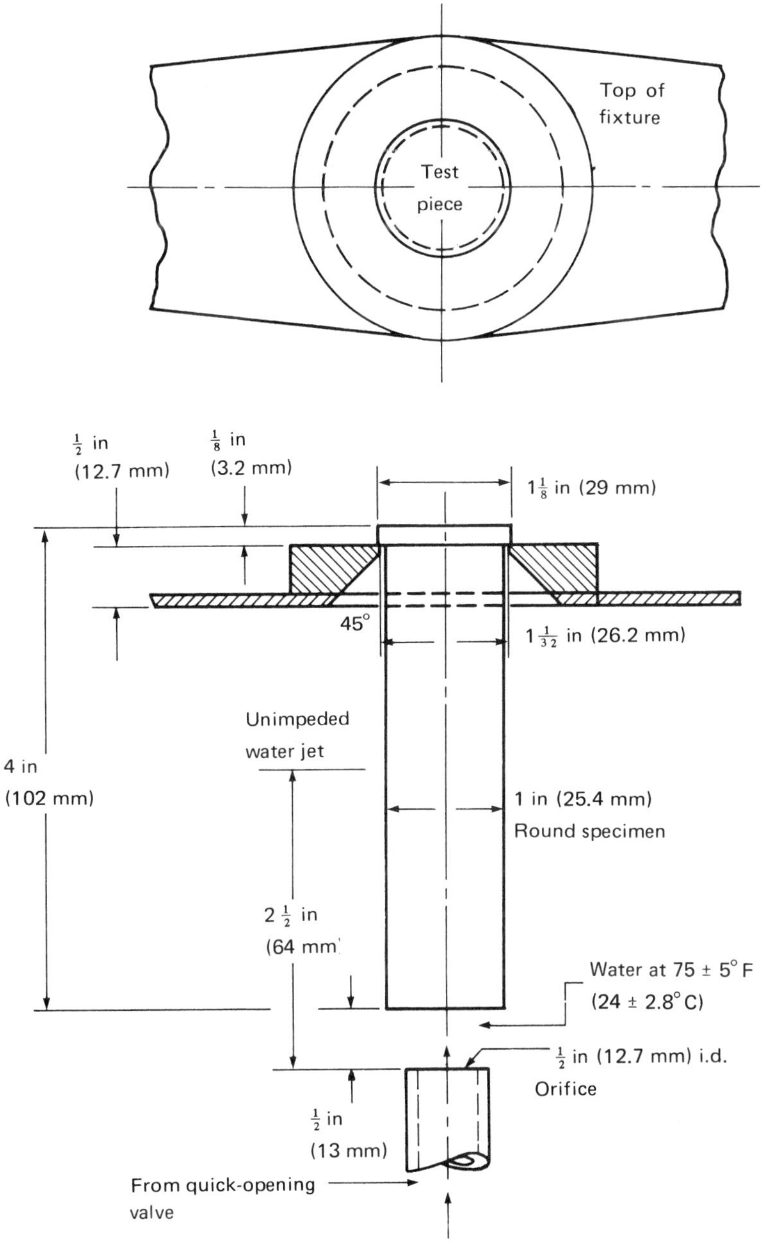

Figure VII-27. Standard form of Jominy end-quench test specimen and test arrangement.

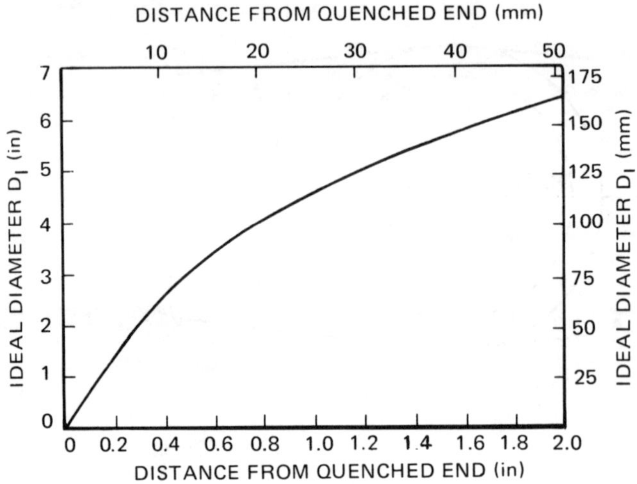

Figure VII-28. Relationship between end-quench distance and hardenability (D_I). (*Making, Shaping and Treating of Steel*, 9th ed., 1971, copyright United States Steel Corporation, by permission.)

as the point at which hardness decreases most rapidly, or determined more precisely by metallography. The end-quench distance to 50% martensite can be related to ideal diameter, as shown in Fig. VII-28.

Brown and James[1] have made the test more precise, with some penalty in speed. For Rockwell C measurements they substituted diamond pyramid hardness (DPH) and decreased the interval between hardness measurements to 0.64 mm. They also decreased the severity of the grinding operation by using fine cuts and grinding slurry. By such careful attention to detail, they were able to reduce the scatter in the hardness vs. distance plot, on successive tests from one heat of steel, to about 0.25 mm.

The Jominy specimen is customarily normalized before testing, to reduce the effects of prior structure. It is then austenitized, and close attention must be paid to the austenitizing temperature if results are to be reproducible. Specimens are usually held 30 min at the austenitizing temperature before quenching. The normalizing and austenitizing temperatures should be varied in accordance with the following schedule:

Carbon content	*Normalizing*	*Austenitizing*
Up to 0.25%C	925 ± 5°C	925 ± 5°C
0.26 to 0.36%C	900 ± 5°C	870 ± 5°C
0.37%C and over	870 ± 5°C	845 ± 5°C

It is essential that specimens be protected against decarburization during the treatments. This can be done by copper plating, then heating the specimens in a slightly carburizing atmosphere, as in a container filled with cast-iron chips.

[1] G.T. Brown and B.A. James, *Met. Trans.*, 4:2245 (1973).

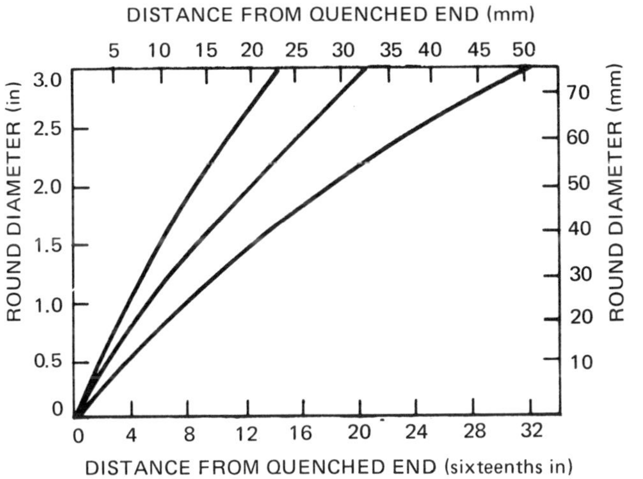

Figure VII-29. The 50% martensite position in end-quench bars and the center, $\frac{1}{2}$-radius, and $\frac{3}{4}$-radius positions of round bars oil-quenched from 845°C. (D.J. Carney, *Trans. Am. Soc. Met.*, 46 (1954).)

The end-quench hardenability can be used to approximate the hardness of steel parts having various sizes and shapes. This is based on the postulate that if cooling of two pieces of the same steel occurs at the same rate, their microstructures and hence their hardness will be the same. The relationships between the 50% martensite position and the $\frac{3}{4}$-radius, $\frac{1}{2}$-radius, and center positions of oil-quenched round bars are shown in Fig. VII-29; the equivalent round bar diameters for square bars and plates are given in Fig. VII-30.

The factors that affect the hardenability of steels are

1. austenite grain size
2. carbon content
3. alloy content

The austenite grain size is involved because transformation to ferrite or pearlite is nucleated at grain boundaries; therefore, the greater the grain boundary area per unit volume, the greater will be the density of nucleation sites and the lower the hardenability. In accordance with this, Grange[1] found a linear relation between the diameter of bars of shallow-hardening steels, hardenable to 90% martensite, and the inverse square root of grain diameter. From this he developed the chart shown in Fig. VII-31. Given a steel with the hardenable diameter indicated by the small circle, the hardenable diameter at any other grain size can be found by moving parallel to the diagonal lines, as indicated by the dashed line.

[1] R.A. Grange, *Met. Trans.*, 4:2231 (1973).

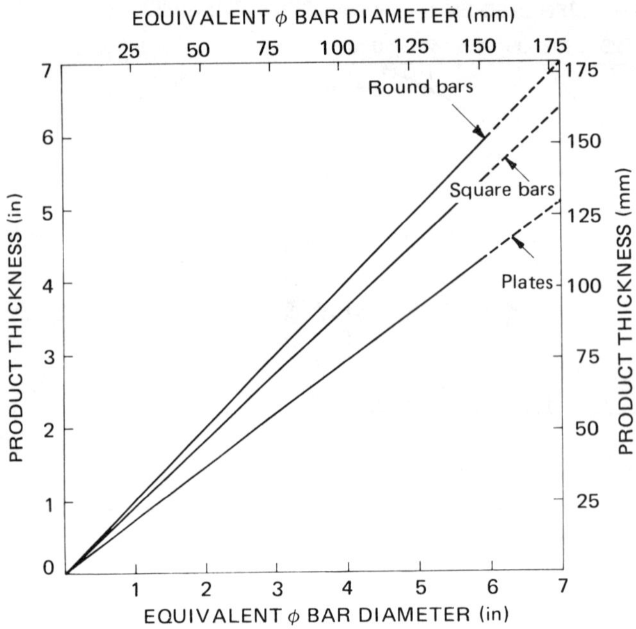

Figure VII-30. Equivalent round-bar diameters for square bars and plates. (U.S. Steel Corporation copyright 1970, by permission.)

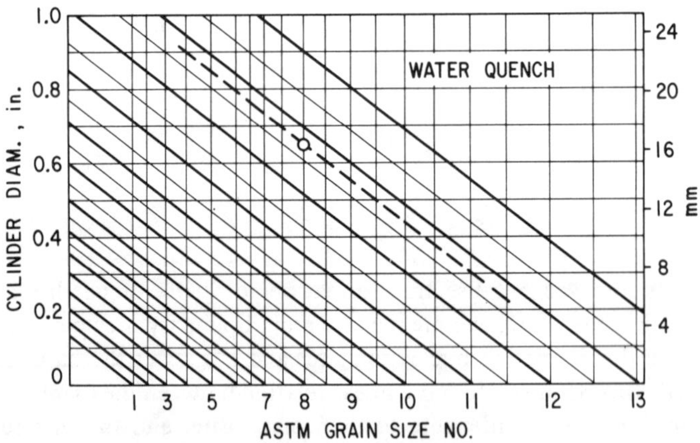

Figure VII-31. Chart for predicting hardenable diameter (90% martensite, water-quenched) for any grain size different from that at which hardenability was measured or estimated. (R.A. Grange, *Met. Trans.*, 4:2231 (1973), by permission.)

Prior austenite grain size of steels has been measured conventionally by the McQuaid-Ehn test.[1] In this procedure, a steel is carburized at 925°C for at least 8 h, then cooled slowly to allow formation of a carbide network in the austenite grain boundaries in the hypereutectoid carburized case. The comparatively high austenitizing temperature and the extended period at temperature make this an overly severe test for steels that usually might be austenitized at only 850°C.

A test that more clearly and reliably delineates austenite grains consists of austenitizing the steel at the recommended temperature, quenching, then tempering for 16 h at 510°C, near the optimum temperature for temper embrittlement.[2] Specimens are then polished and etched for one minute in a boiling, saturated aqueous solution of picric acid and finally lightly repolished to reveal a dark network of prior austenite grain boundaries on a light background. The procedure is much superior in delineation to the McQuaid-Ehn test, which should be limited to steels intended for carburized parts.

The large effect of small amounts of C in increasing the hardenable diameter is illustrated in Fig. VII-32. The hardenability effect of C reaches a maximum between 0.7 and 0.8 wt %.

The various methods for predicting hardenability of steels from chemical composition and austenite grain size are reviewed by Doane.[3] No one procedure is suitable for the entire range from low to high hardenability. For steels which harden during air-cooling, the Jominy end-quench test is inadequate because the cooling rate is too rapid and too constant at distances beyond 40 mm from the quenched end. In shallow-hardening steels, it is difficult to transform the quenched end of the Jominy bar to as much as 90% martensite.

The most widely used procedure for calculating hardenability from chemical composition is that developed by Grossmann.[4] In this technique, a steel is considered to have a base hardenability derived from carbon content and austenite grain size. The base hardenability is expressed as ideal diameter. This ideal diameter is then multiplied by a factor for each alloying element present, this factor varying with the effectiveness of the element in increasing hardenability and the concentration of the element. There are several compilations of these multiplying factors, differing in detail. Several important assumptions are made:

1. The hardenability due to C is the same in an alloy steel as in high-purity iron.
2. There is no substantial interaction between alloying elements.
3. When multiplying factors are used, a given concentration of an alloying element has the same effect when it increases D_I from 5 to 10 mm as when it increases D_I from 25 to 50 mm.

[1] H.W. McQuaid, *Metals Handbook* 1948 ed., *Am. Soc. Met.*, Metals Park, Ohio, p. 407.

[2] A.S. Keh and R.A. Grange, unpublished work, 1956.

[3] D.V. Doane, in *Hardenability Concepts with Applications to Steels*, AIME, Warrendale, Philadelphia, 1978, p. 351.

[4] M.A. Grossmann, *Trans. AIME*, 150:227 (1942).

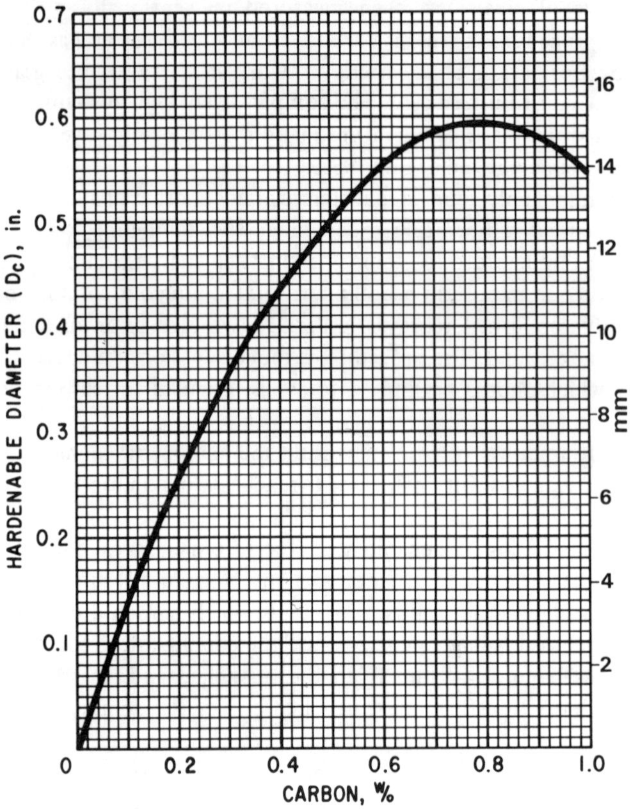

Figure VII-32. Hardenability of Fe–C alloys (90% martensite, water-quenched, grain size ASTM 4). (R.A. Grange, *Met. Trans.*, 4:2231 (1973), by permission.)

None of these assumptions is entirely valid and the third is especially doubtful, because the cooling rates in the two instances are quite different. Nevertheless, the Grossmann concepts proved highly useful. An example of their application is appropriate. What is the hardenability of an AISI 4140 steel, ASTM grain size 7, with the composition

C	Mn	Cr	Mo
0.41	0.82	0.91	0.19

The carbon content and grain size give an ideal diameter of 5.6 mm.[1] The multiplying factor for 0.82%Mn is 3.70; for 0.91%Cr, 2.90; and for 0.19%Mo, 1.70. The hardenability, expressed as an ideal diameter, D_I, for structures containing 50% martensite, is

$$D_I(4140) = 5.6 \text{ mm} \times 3.70 \times 2.90 \times 1.70$$
$$= 102 \text{ mm}$$

[1] H.E. McGannon, ed., *Making, Shaping and Treating of Steel,* 9th ed., U.S. Steel Corp., Pittsburgh, 1971, p. 1133.

Multiplying factors for alloying elements in steels containing from about 0.30 to 0.60%C are given in Table A-3, Appendix. Note that these factors are given as logarithms, so they can be added, then their sum converted to D_I. For lower carbon alloy steels, in the range from 0.15 to 0.30%C, multiplying factors taken from Fig. A-2, Appendix, are more accurate predictors of hardenability.

The Grossmann system has been in general use for more than 35 years. In that time, the average C content of heat treatable alloy steels has decreased from between 0.30 and 0.40 to nearer 0.20%, primarily because of concern with toughness. Also, there has been a trend toward lower cost, lower alloy, shallow-hardening steels that can be drastically quenched with less danger of cracking. The multiplying factor concept cannot be successfully extrapolated over the current range of C content and grain sizes. These deficiencies led Grange to develop a new procedure for determining the hardenability of shallow-hardening steels, which he termed the hot-brine or H-B test.[1]

The cooling rate of a steel specimen can be changed by varying the severity of the quenching medium; a simple way to do this is to change the temperature of the quenchant. Grange adopted a 6% solution of NaCl in water as a standard. The specimen used is shown in Fig. VII-33. It can be hot or cold rolled to 2.5 mm, then sheared to

[1] R.A. Grange, *Met. Trans.*, 4:2231 (1973).

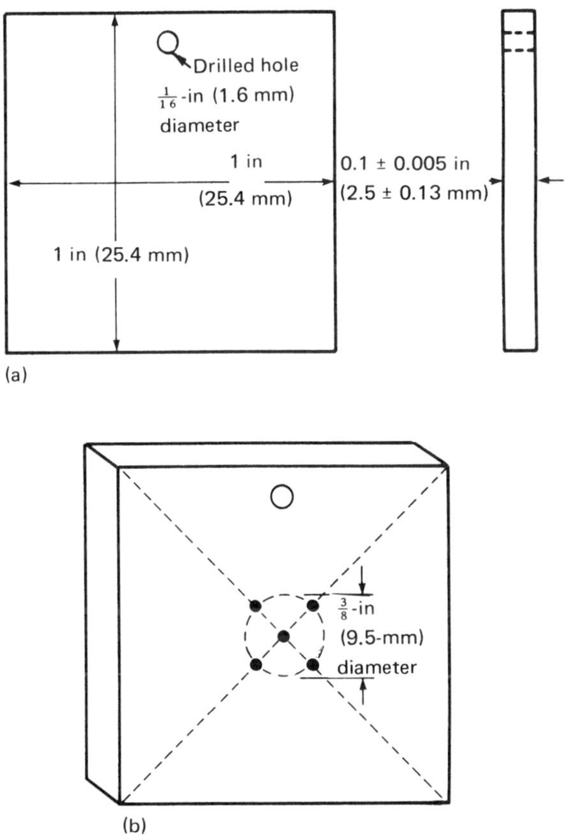

Figure VII-33. Hot-brine hardenability test specimen: (*a*) specimen dimensions, and (*b*) method of locating hardness impressions after heat treatment. (R.A. Grange, *Met. Trans.*, 4:2231 (1973), by permission.)

size. It is normalized at 925°C, austenitized at 845 to 870°C, then quenched into brine held at some preselected temperature. Hardness is measured at locations shown; then the piece is sectioned, mounted, polished, and etched to determine the martensite content. Hardness and martensite content are then correlated with brine temperature, with a typical result being given in Fig. VII-34. The criterion of hardenability chosen was 90% martensite, because in shallow-hardening steels the mechanical properties have deteriorated too much at 50% martensite. Examination of microstructure is more precise in determining the 90% martensite point than is hardness. The brine temperature corresponding to the 90% martensite point is then read. Results from the hot-brine test have been correlated with the diameter of water- and oil-quenched cylinders (Fig. VII-35). As an example, an AISI 1020 steel had a hot-brine temperature of 63°C, a cylinder diameter, water quenched, of 4.8 mm, and a cylinder diameter, oil-quenched, of 3.1 mm. The test is applicable to steels hardenable by water quenching to diameters from 2.5 to 23.0 mm.

The effect of grain size was examined, resulting in the plot of Fig. VII-31. The effect of C content is shown in Fig. VII-32. The concept of multiplying factors for Mn in shallow-hardening steel was found to be untenable. As Grange remarks, ". . . it is more realistic to consider an alloying element increasing the hardenable diameter

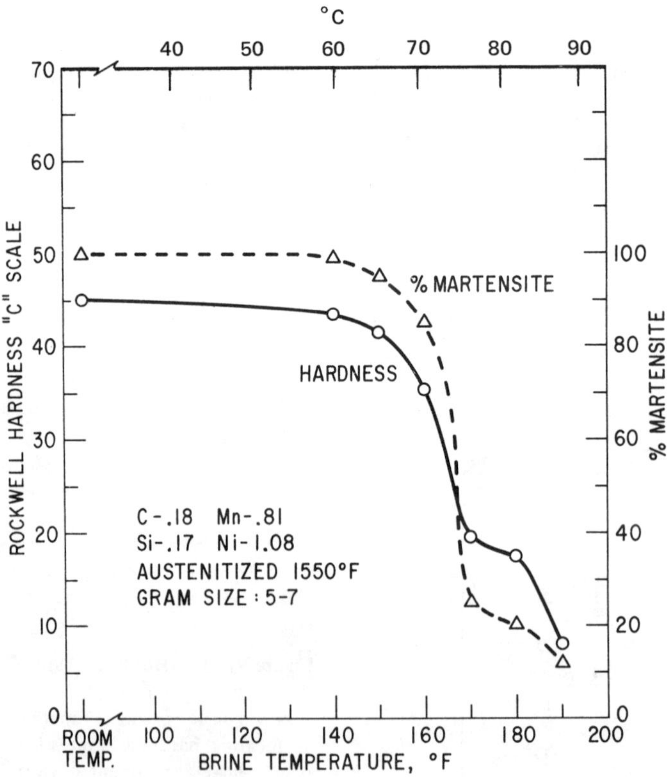

Figure VII-34. Typical results of the hot-brine hardenability test. (R.A. Grange, *Met. Trans.*, 4:2231 (1973), by permission.)

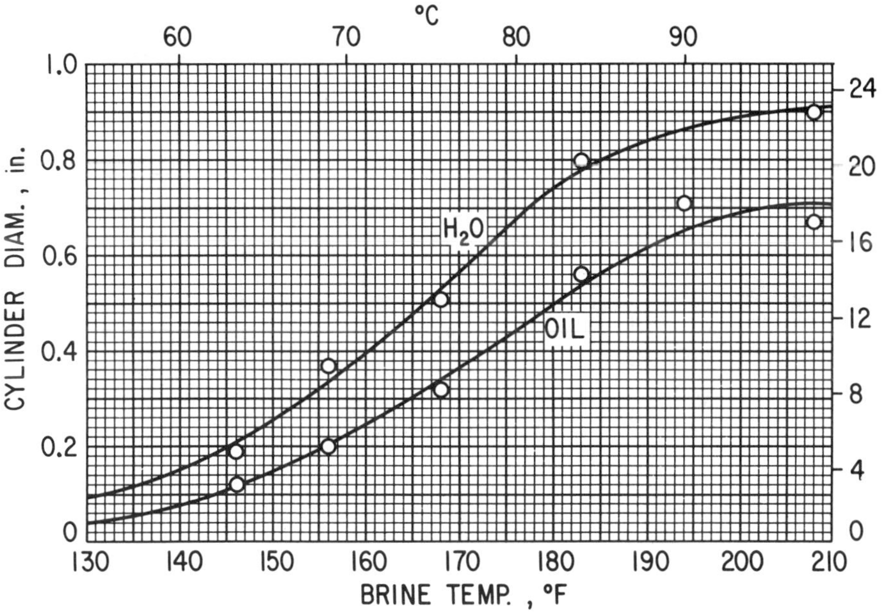

Figure VII-35. Correlation of the hot-brine hardenability test with water- and oil-quenched cylinders. (R.A. Grange, *Met. Trans.*, 4:2231 (1973), by permission.)

by the addition of a layer of constant thickness." This concept of ΔD was proved for Mn. A given concentration of Mn provides a constant increment of hardenable diameter for all C contents between zero and 1.0%. The hardenability effect of Mn was determined in Fe-0.2%C-Mn alloys, over a range of Mn contents from zero to 2.0%. Similar determinations were made for Si, P, S, Ni, Cr, Mo, V, Ti, Zr, and Cu. In estimating hardenability by Grange's technique it is assumed that the hardenable diameter D_H, for iron is zero. The values for ΔD appropriate to the steel composition are read from Fig. VII-36 and its supplement, Fig. VII-37. Ti and Zr were omitted because of their very low concentrations in commercial shallow-hardening steels and because their effects varied with C and N content. The data contained in Figs. VII-36 and 37 are for an ASTM grain size of 4. After the cylinder diameter hardenable to 90% martensite in the center is calculated, it must be corrected for the existing grain size by reference to Fig. VII-31. The equation is

$$D_H \text{ (90\% martensite)} = \Delta D_C + \Delta D_{Mn} + \Delta D_P + \Delta D_{Si} + \Delta D_{Cu} + \Delta D_{Ni} + \Delta D_{Cr} + \Delta D_{Mo} + \Delta D_V \qquad \text{(VII-18)}$$

By avoiding the concept of an ideal diameter, this procedure eliminates the embarrassment of assuming the severity of quench H to be independent of specimen cross section, when, as previously discussed, this is known to be incorrect.

Kirkaldy[1] has begun the long process of replacing empirical methods of calculating hardenability with calculations based on transformation kinetics and thermodynamics.

[1] J.S. Kirkaldy, *Met. Trans.*, 4:2327 (1973).

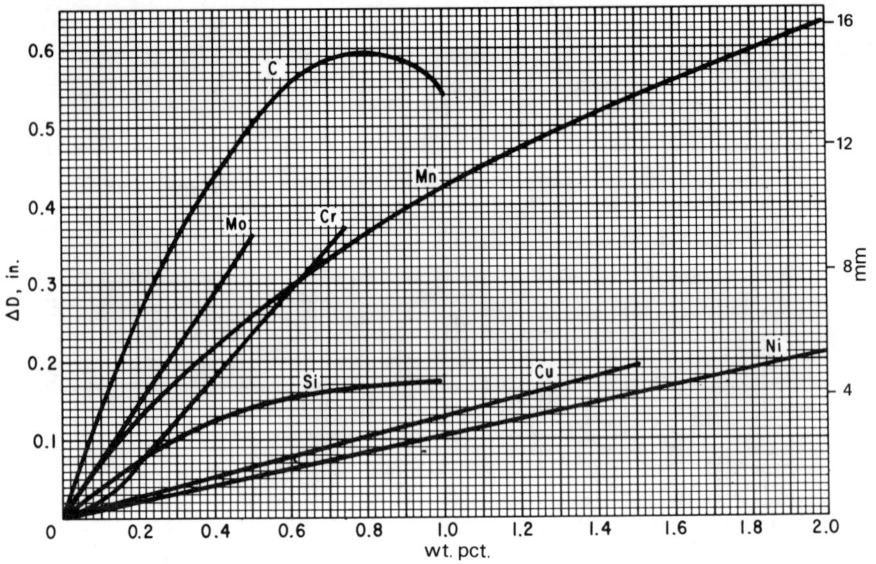

Figure VII-36. Change in hardenable diameter (ΔD) with C, Mn, Si, Cu, Ni, Cr, and Mo. (R.A. Grange, *Met. Trans.*, 4:2231 (1973), by permission.)

He has developed a computerized predictor of hardenability which, given an input of chemical composition, austenite grain size, and austenitizing temperature, will produce a Jominy end-quench curve. Because of proprietary restrictions, the exact blend of empiricism and fundamentals is not known.

VII-8. ISOTHERMAL TRANSFORMATION DIAGRAMS

Carbon and alloying elements affect the transformation of austenite in several ways. First, they alter the Ac_1 and Ac_3 temperatures. Andrews[1] suggests the equations

[1] K.W. Andrews, *J. Iron Steel Inst.*, 203:721 (1965).

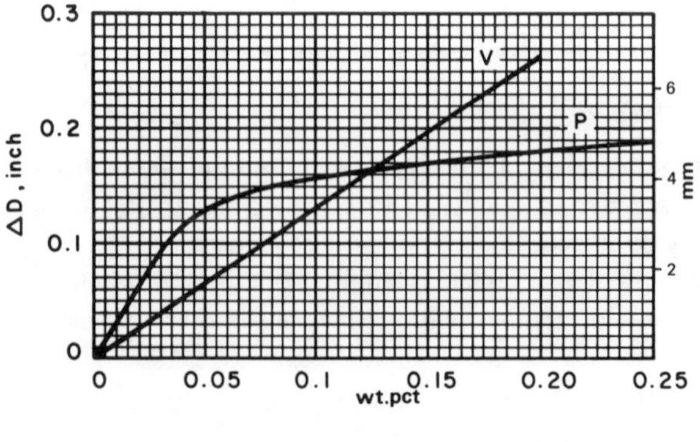

Figure VII-37. Change in hardenable diameter (ΔD) with V and P. (R.A. Grange, *Met. Trans.*, 4:2231 (1973).)

$$Ac_1\,(^\circ C) = 723 - 10.7(\%Mn) - 16.9(\%Ni) + 29.1(\%Si) + 16.9(\%Cr) + 290(\%As) + 6.38(\%W) \qquad \text{(VII-19)}$$

with a standard deviation of 11.5 for a sample of 196 steels. The same equation can be used for Ae_1.

$$Ac_3 = 910 - 203\sqrt{\%C} - 15.2(\%Ni) + 44.7(\%Si) + 104(\%V) + 31.5(\%Mo) + 13.1(\%W) \qquad \text{(VII-20)}$$

with a standard deviation for 155 steels of 16.7. The effects of other elements on Ac_3 are given by the additional factors $-[30(\%Mn) + 11(\%Cr) + 20(\%Cu) - 700(\%P) - 400(\%Al) - 120(\%As) - 400(\%Ti)]$. Although these additional factors are not statistically significant, the very large ferrite stabilizing effects of P, Al, and Ti are obvious.

By lowering the Ac_3 temperature the austenite stabilizing elements reduce the undercooling, and hence the driving force for transformation of austenite, at any lower temperature. This effect increases hardenability. On the same basis, ferrite stabilizing elements, because they *increase* undercooling, would be expected to decrease hardenability, but here another factor enters: most ferrite stabilizing elements tend to segregate between austenite transformation products to a greater extent than do the austenite stabilizers. The rate of austenite transformation is thereby delayed by diffusion of solutes. Other factors that may delay transformation are preprecipitation clustering of substitutional and interstital solutes and solute drag by the advancing γ-α interface boundary.

The technique of determining the rate of transformation of austenite to ferrite, pearlite, or bainite is thoroughly covered in elementary texts and will not be discussed further here. The connection between rate of transformation and hardenability is obvious: the slower the time-dependent processes of nucleation and growth, the greater the time available to reach the Ms temperature and the larger the cross-section that can be transformed to martensite. The most readily available compilation of isothermal transformation diagrams is one produced by the American Society for Metals,[1] largely a reprint of earlier collections by the United States Steel Corporation. From this compilation, diagrams for three steels and their associated end-quench hardenability are selected as examples. The steels have about the same C content but differ in alloying elements.

The first is a carbon steel, AISI 1054 (Fig. VII-38). The rate of transformation of austenite reaches a maximum at about 540°C, where the product is an aggregate of ferrite, pearlite, and bainite. The rate of transformation is so rapid that the times for the beginning and completion of transformation cannot be determined by the techniques employed, hence the dash lines at the "nose" of the diagram. Because of such rapid transformation, this steel has low hardenability; the 50% martensite point in the Jominy test occurs at about 3 mm from the quenched end, as indicated in the lower portion of Fig. VII-38.

[1] *Atlas of Isothermal Transformation and Cooling Transformation Diagrams,* Am. Soc. Met., Metals, Park, Ohio, 1977.

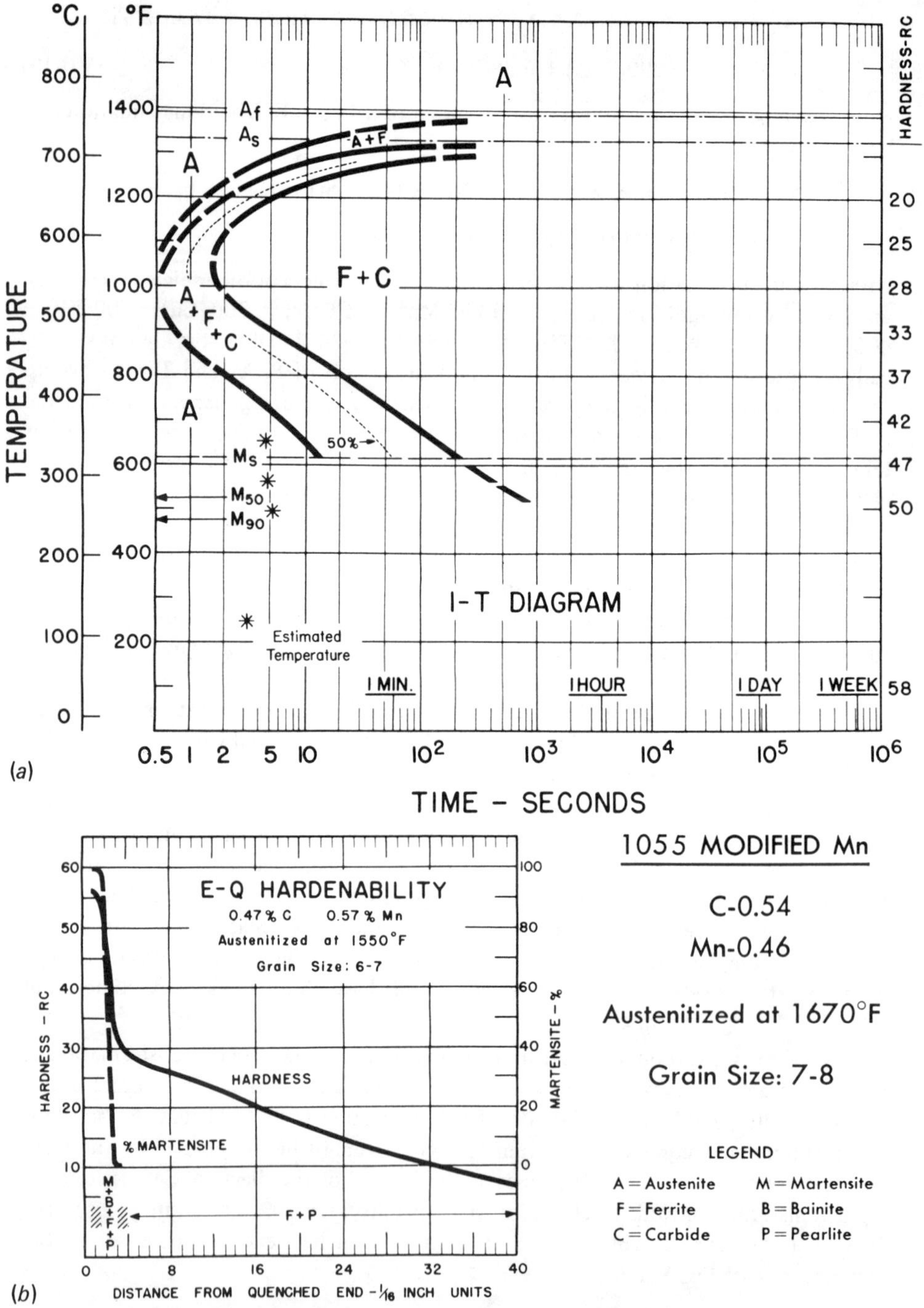

Figure VII-38. (*a*) Isothermal transformation diagram and (*b*) end-quench hardenability of modified AISI 1054 steel austenitized at 910°C. (*Atlas of Isothermal Transformation and Cooling Transformation Diagrams*, Am. Soc. Met., Metals Park, Ohio, 1977, by permission.)

The addition of a single alloying element, Ni, in the rather high concentration of 3.4%, does not change the shape of the isothermal transformation curves (Fig. VII-39), but it retards the beginning and ending of transformation over the entire temperature range between Ae_3 and Ms. The nose of the curve is at about 480°C. The hardenability is increased so that the 50% martensite point now occurs at about 9.5 mm from the quenched end of the Jominy bar. However, this increase in hardenability is purchased at the high cost of an addition of 3.4%Ni, which explains why this AISI 2340 steel is no longer a standard grade.

The addition of smaller amounts of three alloying elements, Ni, Cr, and Mo, to a steel containing 0.40%C produces drastic changes in the IT diagram and in hardenability (Fig. VII-40). The rates of transformation now have two maxima, one at about 650°C, the pearlite nose, and the other at about 440°C, the bainite nose. These are separated by a bay at about 540°C, where transformation is greatly delayed. Transformation over the entire temperature range between Ae_3 and M_s is delayed as compared to the two steels previously discussed, and, as a consequence, the hardenability is very high. A structure of 95% martensite is obtained at about 64 mm from the quenched end of the Jominy bar.

The isothermal transformation diagram provides the basis for the design of several types of heat treatment, as illustrated in Figs. VII-41 to VII-45. Full annealing and isothermal annealing (Figs. VII-41 and 42), in which the products are ferrite and pearlite, can be difficult and time-consuming for high-hardenability steels such as AISI 4340, as can be realized by reference to Fig. VII-40.

Conventional quenching and tempering (Fig. VII-43) depends upon cooling with sufficient rapidity to obtain a microstructure consisting largely or entirely of martensite, then tempering at some preselected temperature to obtain the desired combination of mechanical properties. It is obvious that there is a critical cooling rate that must be equalled or exceeded if the steel is to transform to martensite, and this cooling rate depends upon carbon content and alloy content. Kunitake and Ohtani[1] used time (in seconds) to cool from Ac_3 to 500°C in the derivation of three empirical formulas relating critical cooling rates to chemical composition. The maximum cooling time allowable if martensite is to be obtained, C_B, is given by

$$\log C_B = 3.725(\%C) + 0.046(\%Si) + 0.626(\%Mn) + 0.706(\%Cr) + 0.520(\%Mo) + 0.026(\%Ni) + 0.675(\%Cu) - 1.818 \qquad \text{(VII-21)}$$

Although more applicable to transformations during continuous cooling, two other formulas by Kunitake and Ohtani[1] give the maximum cooling time allowable for the formation of bainite, without ferrite or pearlite

$$\log C_F = 3.288(\%C) - 0.168(\%Si) + 1.068(\%Mn) + 1.266(\%Cr) + 2.087(\%Mo) + 0.300(\%Ni) + 0.626(\%Cu) - 1.931 \qquad \text{(VII-22)}$$

and the cooling rate which just suppresses pearlite, giving a structure of ferrite + martensite or ferrite + bainite

[1] T. Kunitake and H. Ohtani, *Sumitomo Search* (Nov. 1979), p. 18.

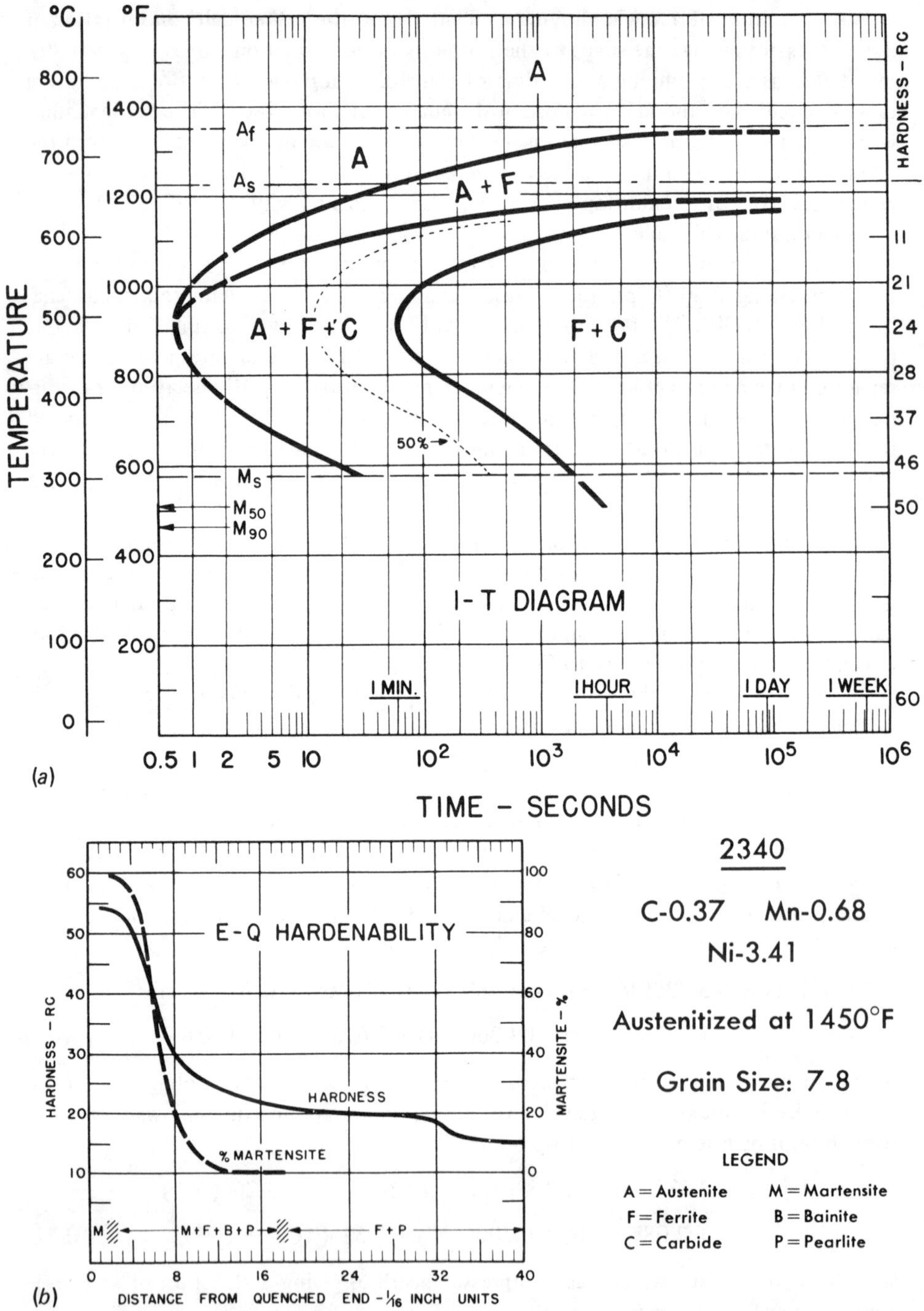

Figure VII-39. (*a*) Isothermal transformation diagram and (*b*) end-quench hardenability of AISI 2340 steel austenitized at 790°C. (*Atlas of Isothermal Transformation and Cooling Transformation Diagrams*, Am. Soc. Met., Metals Park, Ohio, 1977, by permission.)

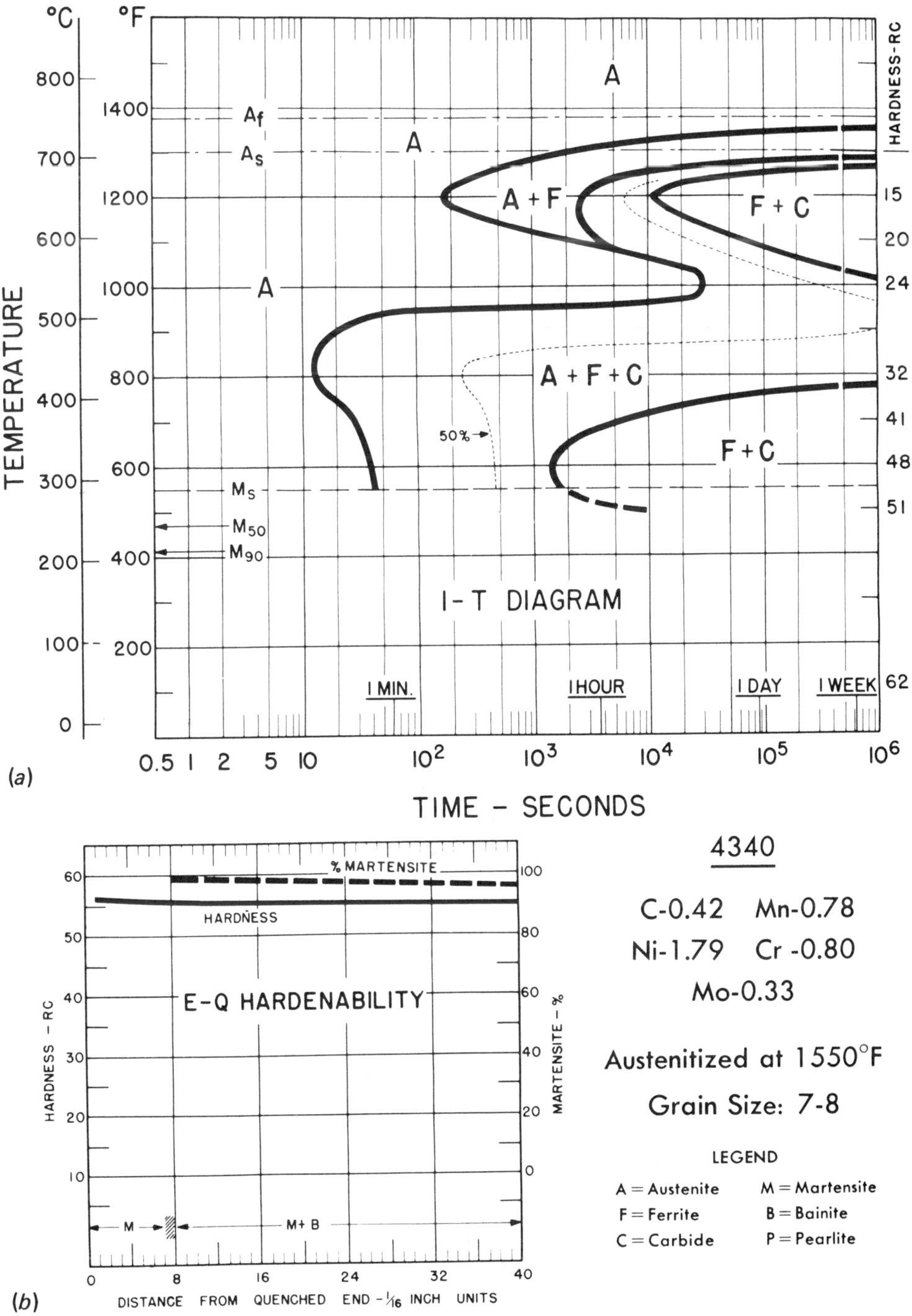

Figure VII-40. (*a*) Isothermal transformation diagram and (*b*) end-quench hardenability of AISI 4340 steel austenitized at 845°C. (*Atlas of Isothermal Transformation and Cooling Transformation Diagrams*, Am. Soc. Met., Metals Park, Ohio, 1977, by permission.)

$$\log C_P = 0.597(\%C) - 0.100(\%Si) + 1.395(\%Mn) + 1.295(\%Cr) + 3.720(\%Mo) + 0.395(\%Ni) + 0.398(\%Cu) - 0.869 \qquad \text{(VII-23)}$$

The latter could be useful in the development of dual-phase HSLA steels.

The processes called martempering and austempering have the same objective: to minimise distortion and cracking during the hardening of steel. In martempering (Fig. VII-44), the steel is cooled rapidly by quenching into a bath at a temperature just above the Ms, held just long enough for the center of the piece to come to the bath temperature, but not so long that bainite begins to form, then it is removed from the bath and cooled in air. This process eliminates most of the quenching stresses, as distinct from transformation stresses.

Austempering (Fig. VII-45) begins in a manner similar to martempering, but the steel is allowed to remain in the bath until the transformation to lower bainite is complete, after which the piece is air cooled. Isothermal transformation diagrams are essential for the proper control of martempering and austempering.

In addition to the United States Steel Corporation–American Society for Metals compilation of isothermal transformation diagrams previously mentioned, other collections have been published by the Iron and Steel Institute of Great Britain,[1] the Institute

[1] *Atlas of Isothermal Diagrams,* Iron and Steel Inst. Spec. Rept. No. 40, March 1949.

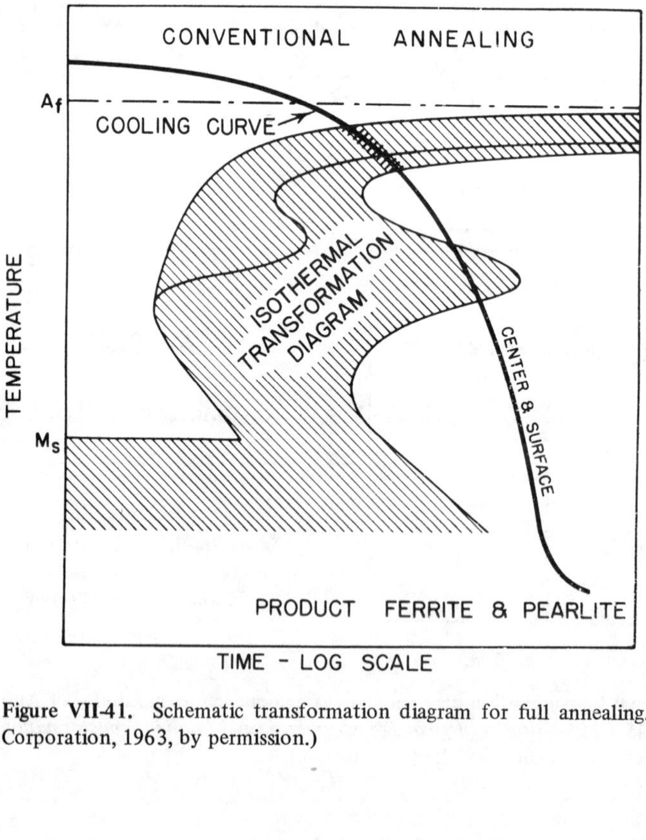

Figure VII-41. Schematic transformation diagram for full annealing. (*I–T Diagrams*, U.S. Steel Corporation, 1963, by permission.)

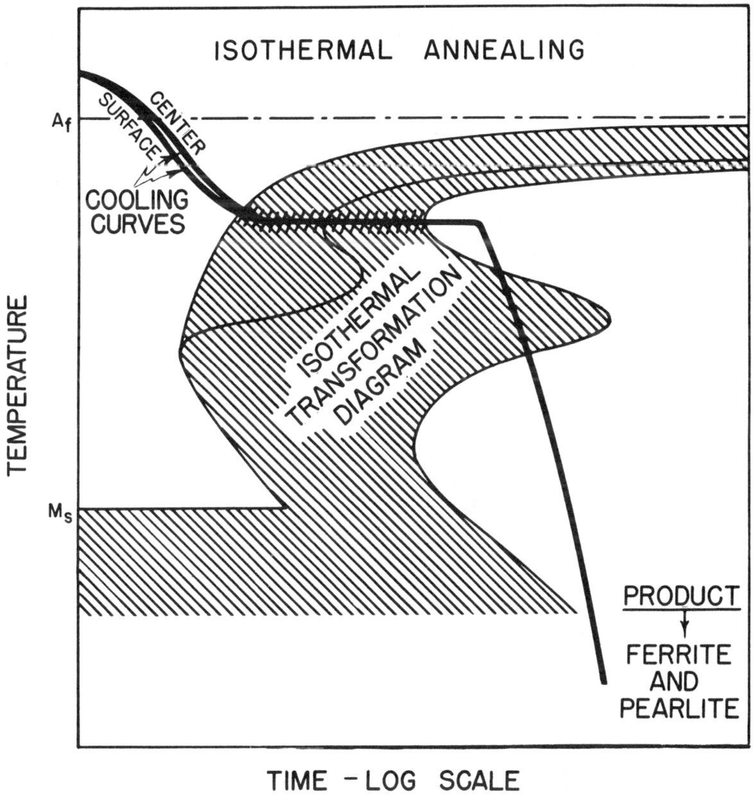

Figure VII-42. Schematic transformation diagram for isothermal annealing. (*I-T Diagrams*, U.S. Steel Corporation, 1963, by permission.)

de Recherches de la Siderurgie (IRSID) of France,[1] and by the Max Planck Institut für Eisenforschung of the Federal Republic of Germany.[2]

VII-9. CONTINUOUS COOLING TRANSFORMATION DIAGRAMS

In most industrial heat-treating operations a knowledge of the microstructure and hardness of the steel after continuous cooling from some temperature in the austenite region is highly desirable. Such information cannot be obtained directly from an

[1] G. Delbart, A. Constant, and A. Clerc, *Courbes de transformation des aciers de fabrication francais,* IRSID, St. Germain-en-Laye, 1956.

[2] F. Wever, A. Rose, W. Peter, W. Strassburg, and L. Rademacher, *Atlas zur Wärmbehandlung der Stähle,* Vol. I, 1961; A. Rose and H. Hougardy, vol. II, Verlag Stahleisen m.b.H., Dusseldorf, 1972.

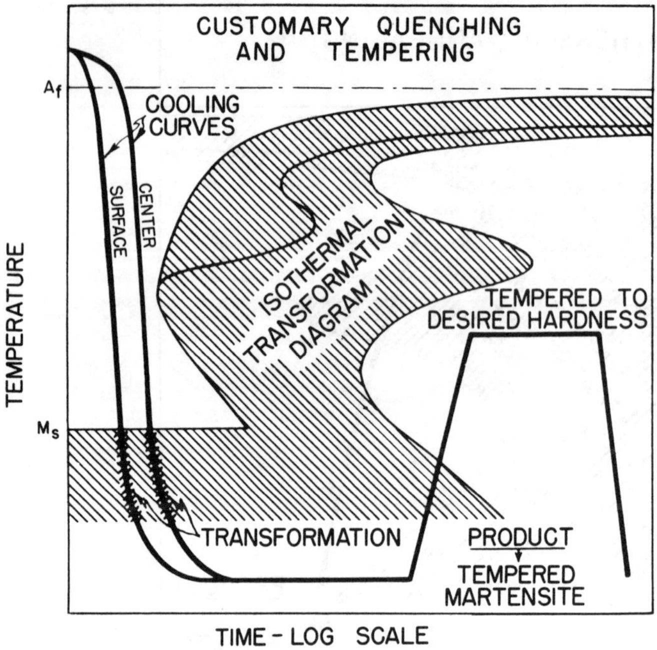

Figure VII-43. Schematic transformation diagram for quenching and tempering. (*I–T Diagrams*, U.S. Steel Corporation, 1963, by permission.)

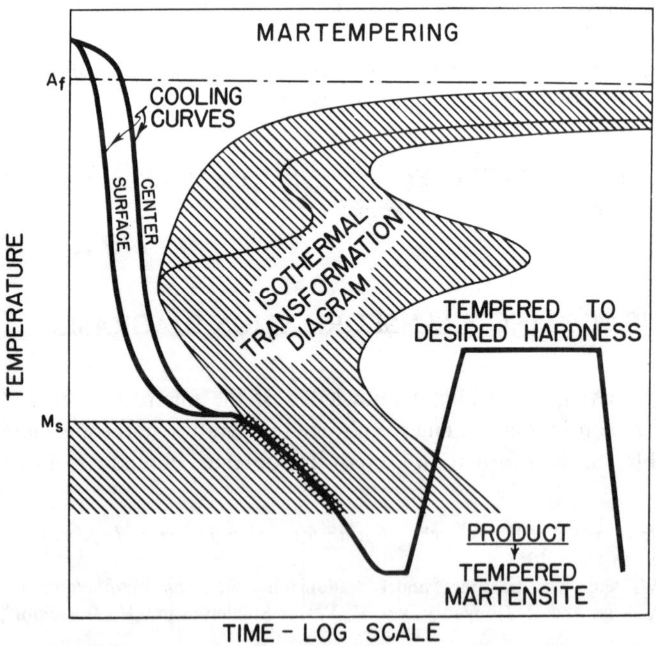

Figure VII-44. Schematic transformation diagram for martempering. (*I–T Diagrams*, U.S. Steel Corporation, 1963, by permission.)

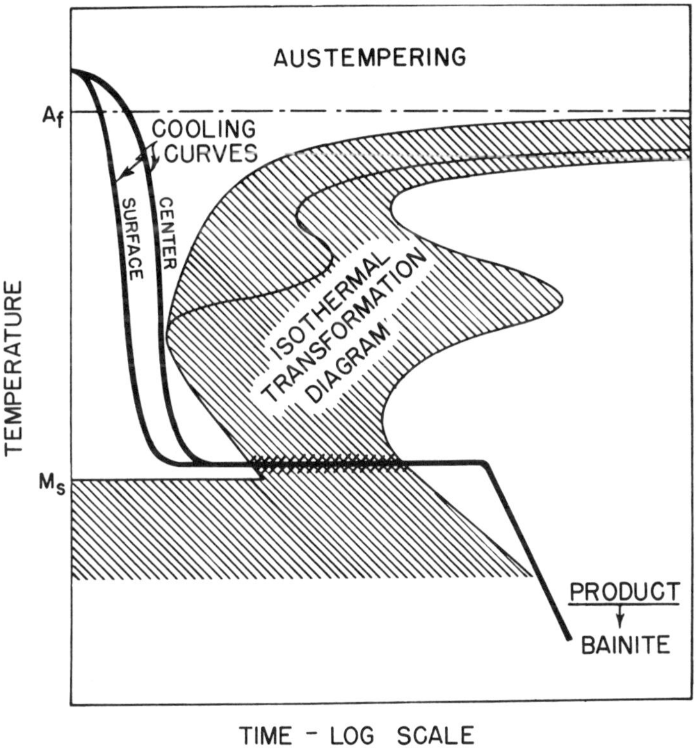

Figure VII-45. Schematic transformation diagram for austempering. (*I-T Diagrams*, U.S. Steel Corporation, 1963, by permission.)

isothermal transformation diagram, so it has been necessary to develop continuous cooling transformation diagrams. These are determined by dilatometry, hardness measurements, and examination of microstructures. In dilatometry, temperature and specimen length are recorded automatically. When transformation of austenite begins, the length of the specimen increases and the start of transformation can be determined quite accurately, often between ±5°C. The completion of transformation during any given cooling schedule can be determined by an inflection in the temperature-length curves, but this determination is less precise than that for the start of transformation. The techniques are described in more detail by Eldis.[1] After cooling to ambient temperature, the hardness and constituents of the microstructure are determined.

The data obtained can be presented in several ways; most frequently a plot of temperature on the ordinate and time on a logarithmic scale on the abscissa is employed. However, the zero of time can be taken as the start of cooling from the austenitizing

[1] G.T. Eddis, *Hardenability Concepts with Applications to Steel*, AIME, Warrendale, Pennsylvania, 1978, p. 126.

temperature[1] or as the time at which the temperature of the specimen reaches the Ac_3.[2] A third alternative is to plot temperature on a linear scale versus the time required to cool between a fixed temperature interval, 800 to 500°C for example, on a logarithmic scale, as done by Economopoulus et al.[3] A fourth method employs as the abscissa the size of the piece being cooled.[4] The second and fourth procedures are probably those that will be encountered most frequently, so an example of each is given.

Figure VII-46 is the continuous cooling transformation diagram of a steel near AISI 4140 in composition. In using this diagram to predict the microstructure and hardness of a continuously cooled steel part, it is necessary to know the cooling curve of the part, or the time required to cool the part from Ac_3 to 500°C or to a temperature half-way between Ac_3 and ambient temperature.

Figure VII-47 is also a continuous cooling transformation diagram of a steel near AISI 4140 in composition, but now the abscissa is the diameter of bars air-cooled, oil-quenched, or water-quenched from above Ac_3. For the user, this type of diagram has the great virtue of simplicity. For a given diameter of bar and a given quenching procedure, one need only follow the appropriate vertical line on the diagram to determine the approximate fraction of each microconstituent and the hardness at the center of the bar. The variable that is not accounted for is agitation during the quench. Also, in any transformation of austenite, whether accomplished isothermally or by continuous cooling, it is necessary to consider whether austenitizing temperature and austenite grain size in the part are comparable to those used in developing the diagram.

Continuous cooling transformation diagrams of the type shown in Fig. VII-46 have greater flexibility than those of the type shown in Fig. VII-47, but are much less convenient to use.

VII-10. HARDENABILITY BANDS

Steels can be ordered to hardenability limits (*H* steels), in which instance the chemical composition ranges are slightly greater than in comparable standard AISI grades, to allow the producers greater flexibility in obtaining hardenability within the specified limits. These hardenability limits–hardenability bands–are published in the SAE Handbook[5] and elsewhere. Hardenability bands of steels can be compared and a rational choice made on the basis of cost to obtain the desired minimum hardenability. The width of a hardenability band for a given steel is arrived at by a compromise

[1] G. Delbart, A. Constant, and A. Clerc, *Courbes de transformation des aciers de fabrication francais,* IRSID, St. Germain-en-Laye, 1956.

[2] W.W. Cias, *Austenite Transformation Kinetics of Ferrous Alloys,* Climax Molybdenum Co., Greenwich, Connecticut, 1978.

[3] M. Economopoulos, N. Lambert, and L. Habraken, *Diagrammes de Transformation des Aciers Fabriques dans le Benelux,* CNRM, Liege, 1967.

[4] M. Atkins, *Atlas of Continuous Cooling Transformation Diagrams for Engineering Steels,* British Steel Corp., Sheffield, England, 1977.

[5] SAE Handbook, 1979, Society of Automotive Engineers, Warrendale, Philadelphia, p. 219.

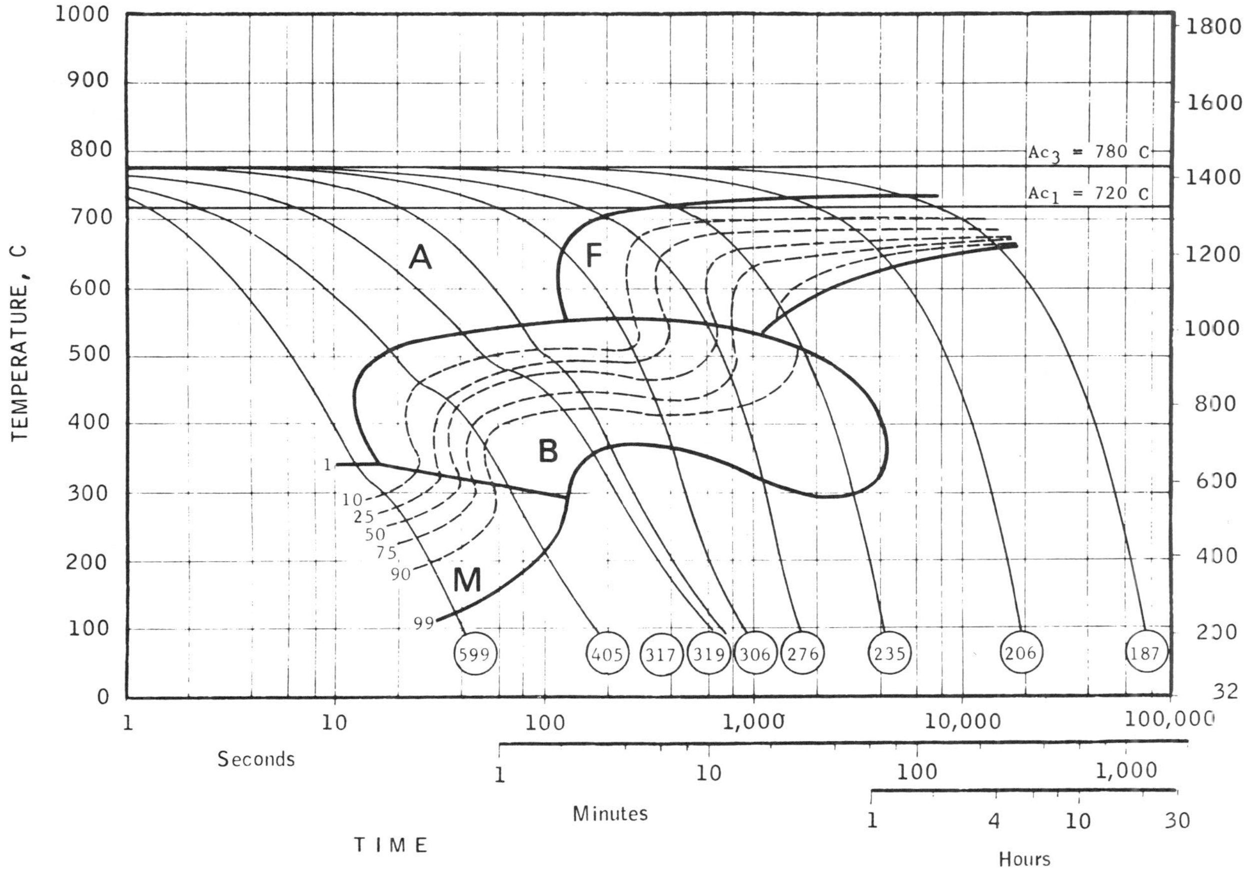

Figure VII-46. Continuous cooling transformation diagram of a Si–Mn–Cr–Mo steel. Numbers in circles at ends of cooling curves are hardness (HV) of transformed specimens. (W.W. Cias, *Austenite Transformation Kinetics of Ferrous Alloys,* Climax Molybdenum Co., Greenwich, Connecticut, 1978, by permission.)

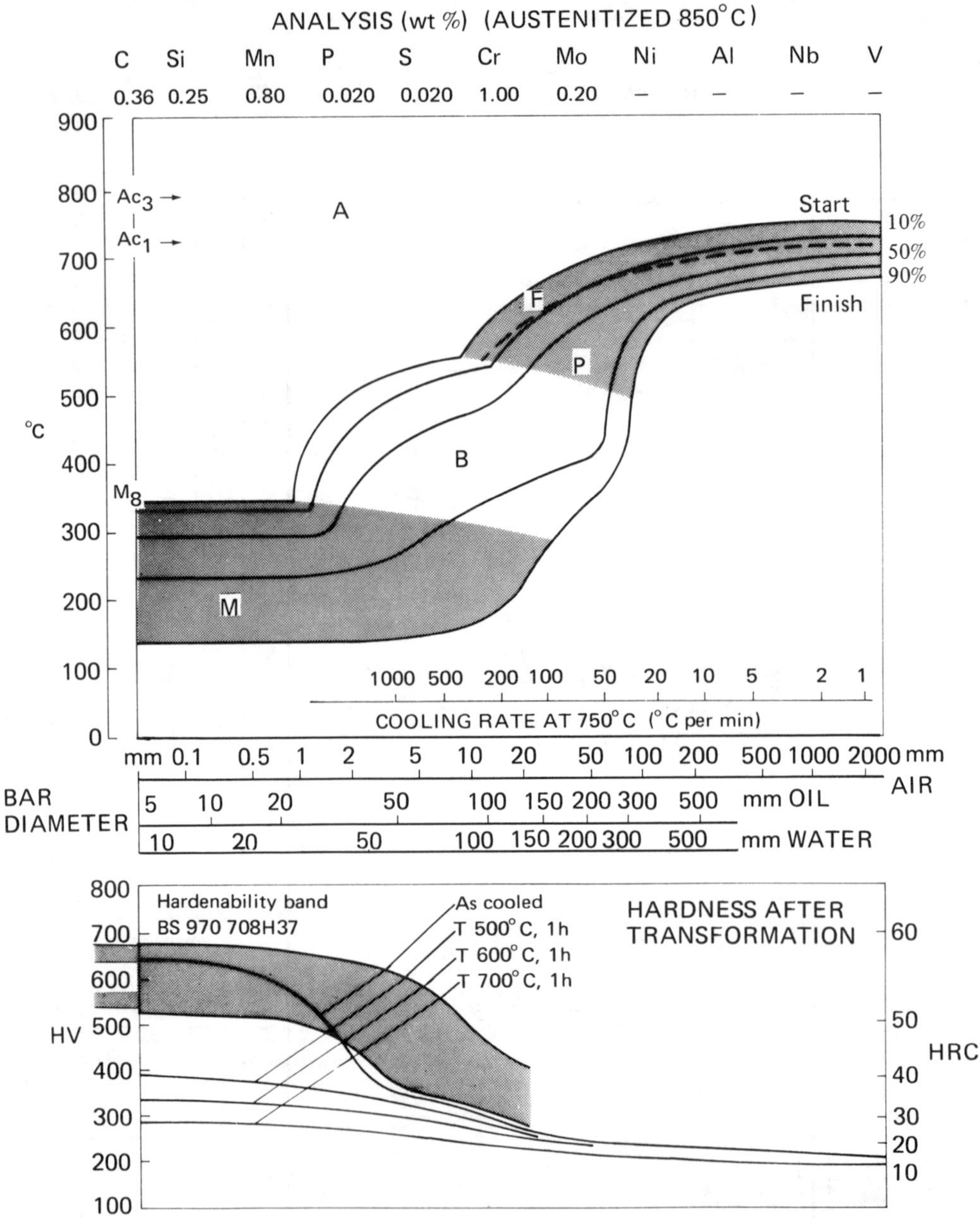

Figure VII-47. Continuous cooling transformation diagram of a Si–Mn–Cr–Mo steel, with hardness after transformation and after tempering. (M. Atkins, *Atlas of Continuous Cooling Transformation Diagrams for Engineering Steels*, British Steel Corporation, Sheffield, England, 1977, by permission.)

between the interests of the user, who would prefer a very narrow band so that the finished products will have uniform mechanical properties, and the steel producer, who would prefer a more generous width to increase the fraction of heats that meet the specification.

Figure VII-48 summarizes the procedures used in specifying a steel according to the hardenability band. For each *H* steel, the hardness limits at each Jominy distance are tabulated (Fig. VII-49) and specifications are written in terms of these tabulations. The minimum and maximum hardness at any desired point in the Jominy curve can be specified, as at points *A–A* in Fig. VII-48, corresponding to the section size of interest to the user. A second alternative is to specify the minimum and maximum Jominy distance at which a desired hardness is attained, as at points *B–B.* A third procedure is to specify two maximum hardness values at two distances, as at points *C–C,* or two minimum hardnesses, as at points *D–D.* If the entire hardenability band is specified, it should be described by the hardness at $\frac{1}{16}$ in, $\frac{1}{8}$ in, $\frac{1}{4}$ in, $\frac{1}{2}$ in, $\frac{3}{4}$ in, 1 in, $1\frac{1}{4}$ in, $1\frac{1}{2}$ in, $1\frac{3}{4}$ in, and 2 in (or the equivalent distances in mm).

For each heat-treated steel part it is necessary to determine the minimum hardenability required for the proper functioning of the part. No more hardenability should be specified than is necessary, for increasing hardenability increases costs and may increase the incidence of quench cracking. There are some rough guidelines that can be applied.[1] For highly stressed parts, a minimim of 80% martensite at the center of the largest section is generally required. For less highly stressed parts, 50% martensite may suffice. For most automotive parts, 80% martensite at the $\frac{3}{4}$ radius position is enough. The state of stress in each part should be known. If stress is applied in bending so that the outer fibers are the most highly stressed, the steel can have lower hardenability than if the entire section is stressed in tension.

The difficulty of quenching large sections to a high volume fraction of martensite, even with steels of very high hardenability, is illustrated by the fact that AISI 4340 can be oil-quenched to 80% martensite at the center of an 85.7-mm round bar, but if 95% martensite is required the diameter of the bar is reduced to 51 mm.[1]

Tables VII-6 and VII-7 list some steels that are suitable for oil-quenched moderately stressed and highly stressed parts.[1] These are ranked roughly in order of increasing cost. When one steel is substituted for another, it may be shifted to the left in the table, to a smaller diameter, or upward to a lower strength, as indicated by the arrows.

VII-11. THE BORON EFFECT

Boron is added to steel for only one reason—to increase hardenability. Boron-treated steels have certain peculiar characteristics:

1. Boron increases hardenability in hypoeutectoid steel, has no effect on eutectoid steel, and decreases hardenability in hypereutectoid steel.
2. As the austenite grain size becomes finer, the hardenability effect of B increases.

[1] *ASTM A400–69, 1979 ASTM standards,* Part 5, ASTM, Philadelphia.

DIAMETERS OF ROUNDS WITH SAME AS QUENCHED HARDNESS									LOCATION IN ROUND	QUENCH
3.8									SURFACE	MILD
1.1	2.0	2.9	3.8	4.8	5.8	6.7			3/4 RADIUS FROM CENTER	WATER
0.7	1.2	1.6	2.0	2.4	2.8	3.2	3.6	3.9	CENTER	QUENCH
0.8	1.8	2.5	3.0	3.4	3.8				SURFACE	MILD
0.5	1.0	1.6	2.0	2.4	2.8	3.2	3.6	4.0	3/4 RADIUS FROM CENTER	OIL
0.2	0.6	1.0	1.4	1.7	2.0	2.4	2.8	3.1	CENTER	QUENCH

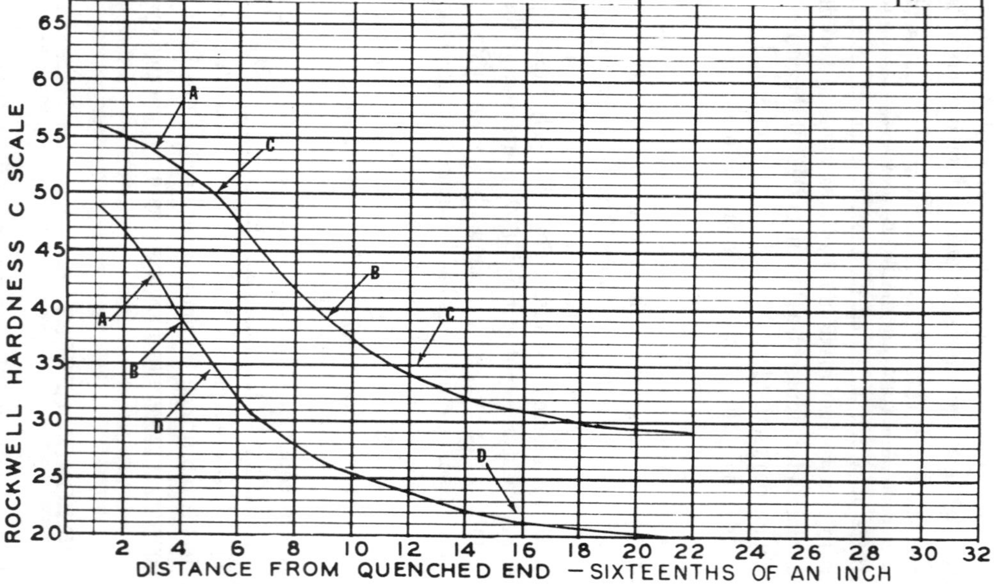

Figure VII-48. Alternate methods of specifying hardenability requirements. (AISI Steel Products Manual, *Alloy Steel: Semifinished; Hot Rolled and Cold Finished Bars,* AISI, Washington, D.C., August 1970, by permission.)

HARDNESS LIMITS FOR SPECIFICATION PURPOSES		
"J" DISTANCE SIXTEENTHS OF AN INCH	4140 H MAX.	4140 H MIN.
1	60	53
2	60	53
3	60	52
4	59	51
5	59	51
6	58	50
7	58	48
8	57	47
9	57	44
10	56	42
11	56	40
12	55	39
13	55	38
14	54	37
15	54	36
16	53	35
18	52	34
20	51	33
22	49	33
24	48	32
26	47	32
28	46	31
30	45	31
32	44	30

HEAT TREATING: TEMPERATURES RECOMMENDED BY SAE

*NORMALIZE 1600 °F

AUSTENITIZE 1550 °F

*For forged or rolled specimens only.

HARDENABILITY BAND 4140 H

C	Mn	Si		Cr	Mo	
.37 / .44	.65 / 1.10	.15 / .30		.75 / 1.20	.15 / .25	

DIAMETERS OF ROUNDS WITH SAME AS QUENCHED HARDNESS									LOCATION IN ROUND	QUENCH
3.8									SURFACE	MILD WATER QUENCH
1.1	2.0	2.9	3.8	4.8	5.8	6.7			3/4 RADIUS FROM CENTER	MILD WATER QUENCH
0.7	1.2	1.6	2.0	2.4	2.8	3.2	3.6	3.9	CENTER	MILD WATER QUENCH
0.8	1.8	2.5	3.0	3.4	3.8				SURFACE	MILD OIL QUENCH
0.5	1.0	1.6	2.0	2.4	2.8	3.2	3.6	4.0	3/4 RADIUS FROM CENTER	MILD OIL QUENCH
0.2	0.6	1.0	1.4	1.7	2.0	2.4	2.8	3.1	CENTER	MILD OIL QUENCH

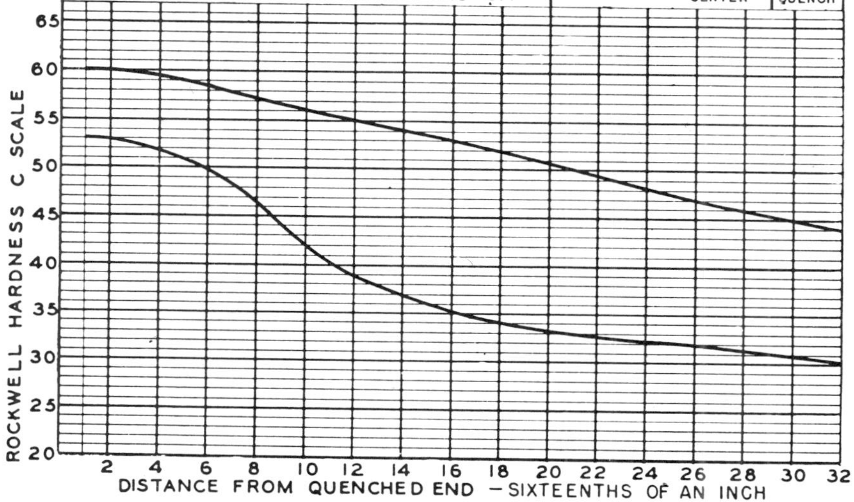

Figure VII-49. Tabulation and graphical presentation of the hardenability band of AISI 4140H. (AISI Steel Products Manual, *Alloy Steel: Semifinished; Hot Rolled and Cold Finished Bars,* AISI, Washington, D.C., August 1970, by permission.)

Table VII-6 H-Steels for oil-quenched moderately stressed parts (quenched to a min. of 50% martensite)[1]

Yield strength (MPa)	Tensile strength (MPa)	Hardness after temper R_C (Bhn)	Hardness after quenching 50% martensite minimum R_C (Bhn)	Round sections to 12 mm	Over 12 to 25 mm	Over 25 to 37 mm	Over 37 to 50 mm	Over 50 to 62 mm	Over 63 to 75 mm	Over 75 to 87 mm	Over 87 to 100 mm
				50% Martensite Full radius to center		$\frac{1}{2}$ Radius		$\frac{3}{4}$ Radius			
				Reference point 5.5 mm	9.5 mm	12 mm	16 mm	17 mm	21 mm	24 mm	29 mm
620 to 860	794 to 938	23 (243) to 30 (286)	42 (390)	1330H 5132H 4130H 8630H	8737H	50B40H 8642H 94B30H 8740H 3140H	4140H 94B40H		4142H		
860 to 1035	938 to 1104	30 (286) to 36 (336)	44 (409)	1335H 4042H 4047H 5135H	4135H 8640H 94B30H 3740H 3140H	50B44H 5147H 4137H 8645H 8742H		4142H	4145H	4147H 86B45H 9840H	4337H 4340H
1035 to 1172	1104 to 1296	36 (336) to 41 (381)	48 (455)	1340H 5140H 4135H	5150H 50B40H 4137H	5160H 50B50H 4140H	51B60H 8655H	4145H 9840H	4147H 86B45H 4337H	4150H 4340H	

				8637H 94B30H 3140H	8642H 8645H 8742H	94B40H 6150H					
1172 to 1275	1296 to 1530	41 (381) to 46 (432)	51 (496)	5145H 50B40H 50B40H 4063H 4140H 8640H 8642H 8645H 8740H 8742H	5155H 59B44H 5147H 94B40H 6150H	81B45H 4142H 4145H 8650H 8655H 4337H	86B45H 9840H	4147H 8660H 4340H	4150H		9850H E4340H
Over 1275	Over 1530	46 Min. (432)	55 (560)	5150H 5155H 50B44H 5147H 9260H 81B45H 8650H 86B45H 6150H	5160H 50B50H 9262H 4147H 8655H	50B60H 51B60H 8660H	4150H			9850H	

[1] *ASTMA 400–69, 1979 ASTM Standards,* Part 5, p. 310–314, Table 2.

Table VII-7 H-Steels for oil-quenched highly stressed parts (quenched to a min. of 80% martensite[1]

Yield strength (MPa)	Tensile strength (MPa)	Hardness after temper R_C (Bhn)	Hardness after quenching 80% martensite minimum R_C (Bhn)	Round sections to 12 mm	Over 12 to 25 mm	Over 25 to 37 mm	Over 37 to 50 mm	Over 50 to 62 mm	Over 62 to 75 mm	Over 75 to 87 mm	Over 87 to 100 mm
				80% Martensite Full radius to center		$\frac{1}{2}$ Radius		$\frac{3}{4}$ Radius			
				Reference point 5.5 mm	9.5 mm	12 mm	16 mm	17 mm	21 mm	24 mm	29 mm
620 to 860	794 to 938	23 (243) to 30 (286)	42 (390)	1330H 5132H 4130H 8630H		94B30H					
860 to 1035	938 to 1104	30 (286) to 36 (336)	44 (409)	1335H 5135H	4135H 8640H 94B30H 8740H	4137H		4142H	9840H		4337H 4340H
1035 to 1172	1104 to 1296	36 (336) to 41 (381)	48 (455)	1340H 5140H 4135H	50B40H 4137H 8642H	4140H 94B40H		4145H 9840H	86B45H 4337H	4340H	

				8637H 94B30H 3140H	8645H 8742H						
1172 to 1275	1296 to 1530	41 (381) to 46 (432)	51 (496)	50B46H 5145H 50B40H 4140H 8640H 8642H 8645H 8740H 8742H	5155H 50B44H 5147H 94B40H 6150H	81B45H 4142H 4145H 8650H 8655H 4337H	86B45H 9840H	4147H 4340H	4150H		9850H E4340H
Over 1275	Over 1530	46 Min. (432)	55 (560)	5150H 5155H 50B44H 5147H 9260H 81B45H 8650H 86B45H 6150H	5160H 50B50H 9262H 4147H 8655H	50B60H 51B60H 8660H	4150H			9850H	

[1] *ASTM A 400–69, 1979 ASTM Standards,* Part 5, p. 310–314, Table 4.

3. Austenitizing at high temperature reduces the hardenability effect of B.
4. For the maximum B effect, the concentration should be in the range 0.0005 to 0.003 wt %.
5. Increasing hardenability by adding B does not decrease the Ms temperature.

These arise because of the manner in which B produces its effect and the interactions of B with other elements in steel, especially C and N. To be effective, B must be in solid solution in austenite. The solubility of B in austenite, because it is very low, has been a matter of dispute for many years. The best estimate is that the solubility at 912°C is about 0.001%, increasing with temperature to a maximum of about 0.005% at the eutectic.[1] In alpha Fe, the solubility is essentially zero, but is influenced by the impurities present.[2] Because of its low solubility in austenite, B can be highly concentrated in grain boundaries (Fig. III-17).

Kapadia[3] has reviewed the data on the effect of B on the hardenability of steel. Boron segregated to austenite grain boundaries suppresses the nucleation of proeutectoid ferrite, but there are at least two opinions as to how this is accomplished. In one of these, it is held that since B reduces the free energy of austenite grain boundaries[4,5] it thereby reduces the effectiveness of such boundaries as nucleation sites for ferrite or upper bainite. In the other,[6,7] the critical factor is considered to be not B atoms segregated at the boundary, but the formation of particles of the fcc borocarbide, $Fe_{23}(C, B)_6$, at austenite boundaries. The borocarbide forms a coherent boundary with one austenite grain and a noncoherent boundary with the other. The coherent boundary is not a suitable site for ferrite nucleation. Because of the experimental difficulty of determining when B is present as a segregated solute and when it is present as very fine precipitate particles, the resolution of this question may require some time. It is possible that both effects may operate to increase hardenability.

It has been recognized for some time that increasing the C content of a steel decreases the hardenability effect of B. In a systematic study of this phenomenon, Grange and Mitchell[8] examined the hardenability effect of B in a matched series of AISI 8600 steels, that is, 8620; 86B20; 8650; 86B50; 8680; and 86B80.

The base composition was:

Mn	Ni	Cr	Mo	Al	B
0.75	0.60	0.50	0.21	0.07	0.0025

[1] A. Brown, J.D. Garnish and R.W.K. Honeycombe, *Met. Sci.*, 8:317 (1974).
[2] A. Lucci, G. Della Gatta, and G. Venturello, *Met. Sci. J.*, 3:14 (1969).
[3] B.M. Kapadia, *Hardenability Concepts with Applications to Steel,* AIME, Warrendale, Pennsylvania, 1978, p. 448.
[4] D.A. Mortimer and M.G. Nicholas, *Met. Sci.*, 10:326 (1976).
[5] J.E. Morral and J.B. Cameron, *Met. Trans. A.*, 8A:1817 (1977).
[6] Ph. Maitrepierre, D. Thivellier, J. Rofes-Vernis, D. Rousseau, and R. Tricot, *Hardenability Concepts with Applications to Steel,* AIME, Warrendale, Pennsylvania, 1978, p. 421.
[7] R.C. Sharma and G.R. Purdy, *Met. Trans.*, 5:939 (1974).
[8] R.A. Grange and J.B. Mitchell, *Trans. Am. Soc. Met.*, 53:157 (1961).

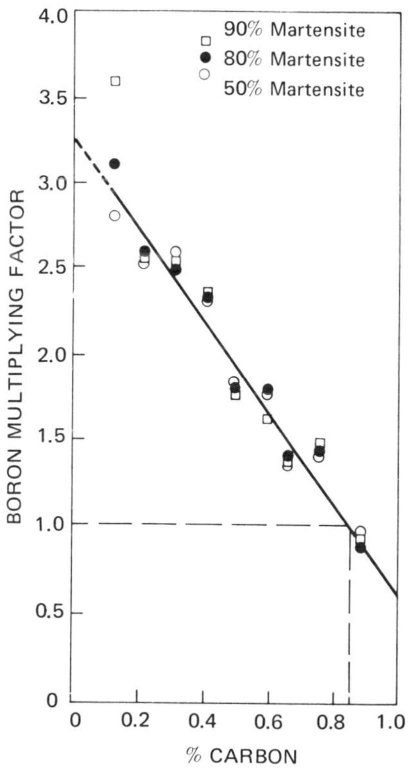

Figure VII-50. Effect of carbon on boron multiplying factor in 0.8%Mn steels. (D.T. Llewellyn and W.T. Cook, *Met. Technol.*, 1:517 (1974), by permission.)

Only the C content was varied. In agreement with the work of others, they found that the hardenability effect of B decreases with increasing C content (Fig. VII-50).

As the C content is increased, more borocarbide $Fe_{23}(C, B)_6$ precipitates from austenite, thereby decreasing the concentration of B remaining in solution. The borocarbide precipitates at austenite grain boundaries, depleting B in the regions where it is most effective. The borocarbide particles above some critical size serve as nucleation sites for ferrite and pearlite.

Further evidence regarding the relation between C content, B content, and hardenability, in the absence of alloying elements, was provided by Grange and Hribal[1] (Fig. VII-51). The largest effect of B on hardenability occurs at low C contents, ~0.10%.

When steels contain no B, the rate of subsequent isothermal transformation decreases with increasing austenitizing temperature because of the effect of larger austenite grains in delaying transformation. When B is added, the rate of transformation is *increased* after austenitizing in the range 845–985°C. This effect is shown in Fig. VII-52. As the austenitizing temperature is raised, the amount of B dissolved increases; but the higher the concentration of B in solution, the more readily the borocarbide precipitates in either stable or metastable austenite.

[1] R.A. Grange and C.R. Hribal, *U.S. Steel Corporation Technical Report 40.061(1)*, June 1972.

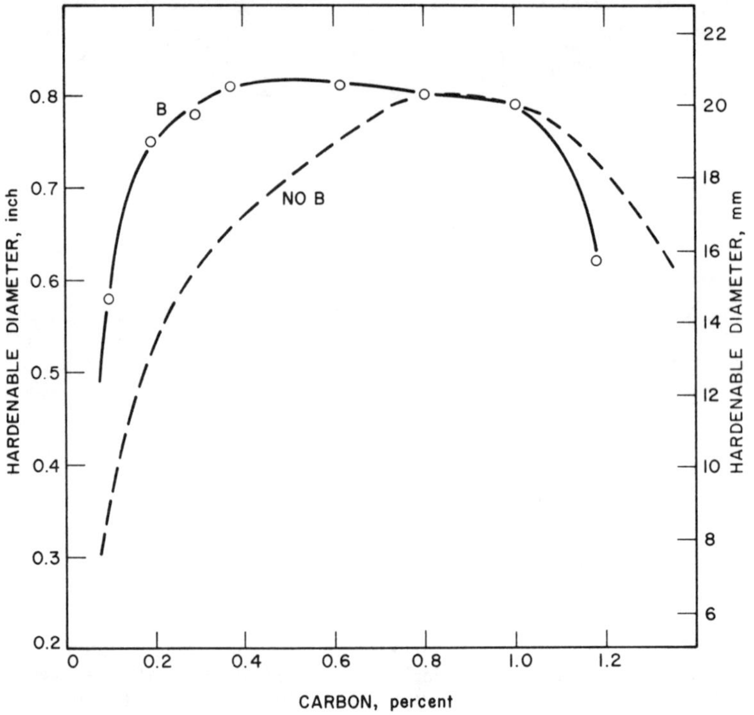

Figure VII-51. Hardenable diameter (90% martensite at center of Fe–0.5%Mn alloys, as-water-quenched, with and without boron at an austenite grain size of ASTM 4). (R.A. Grange and C.R. Hribal, U.S. Steel Corporation Tech. Rept 40.061 (1), June 1972, by permission.)

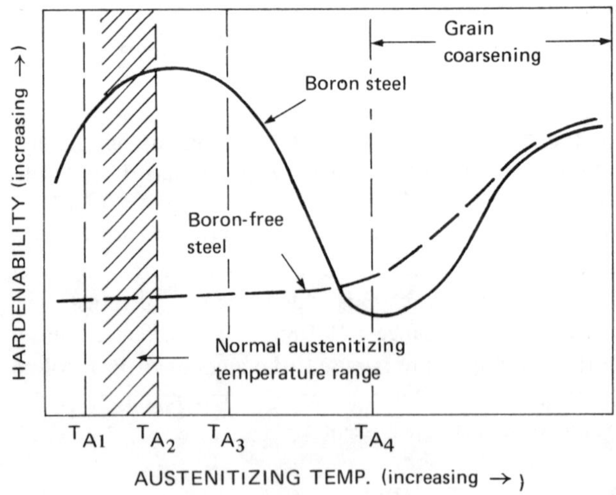

Figure VII-52. Schematic representation of effect of austenitizing temperature in a boron steel and a corresponding boron-free steel. (R.A. Grange and J.B. Mitchell, *Trans. Am. Soc. Met.*, 53:157 (1961).)

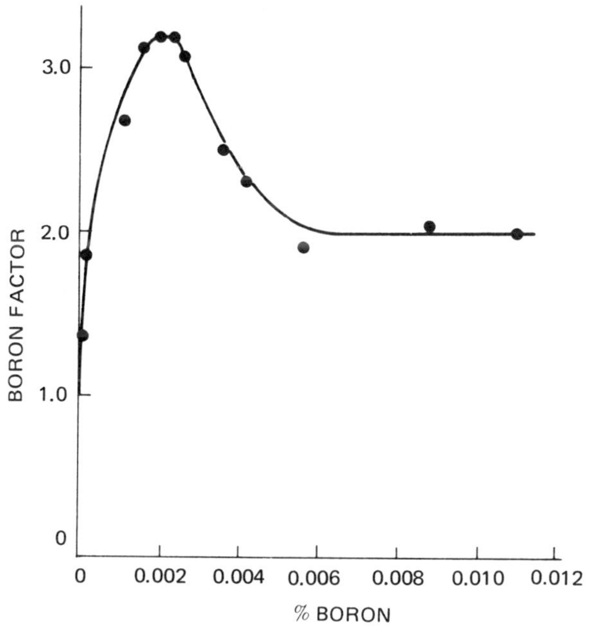

Figure VII-53. Boron hardenability factors as a function of boron content in a 0.20%C–0.55%Mo steel. (G.F. Melloy, P.R. Slimmon, and P.P. Podgurski, *Met. Trans.*, 4:2279 (1973), by permission.)

The optimum B content, near 0.002 wt %, for increasing hardenability in 0.20%C–0.65%Mn–0.55%Mo steel is shown in Fig. VII-53. The "boron factor" referred to is defined by

$$B_F = \frac{D_I \text{ actual}}{D_I \text{ no boron}}$$

Boron contents above about 0.003% lead to a loss in hardenability, and B in excess of about 0.004% causes a loss in toughness through precipitation of Fe_2B in austenite grain boundaries.

The effect of B on isothermal transformation and hardenability of AISI 8620 steel is shown in Fig. VII-54. Boron delays the start of transformation to ferrite, pearlite, or bainite, but at transformation temperatures above about 500°C it has little effect on the time required for completion of transformation.

As the austenite grain size becomes finer, the hardenability effect of B becomes greater. Therefore, B offsets losses in hardenability due to low C content and fine grain size. Also, because B does not lower the Ms temperature, more severe quenching can be used without danger of quench cracking. As the Ms temperature is reduced, the tendency for any particular steel part to crack on quenching increases.

Because of the strong tendency of B to combine with N, any steel to which B is added must be treated with other strong nitride-forming elements, such as Al, Ti, and Zr. Boron is customarily added as one component of complex ferroalloys which also

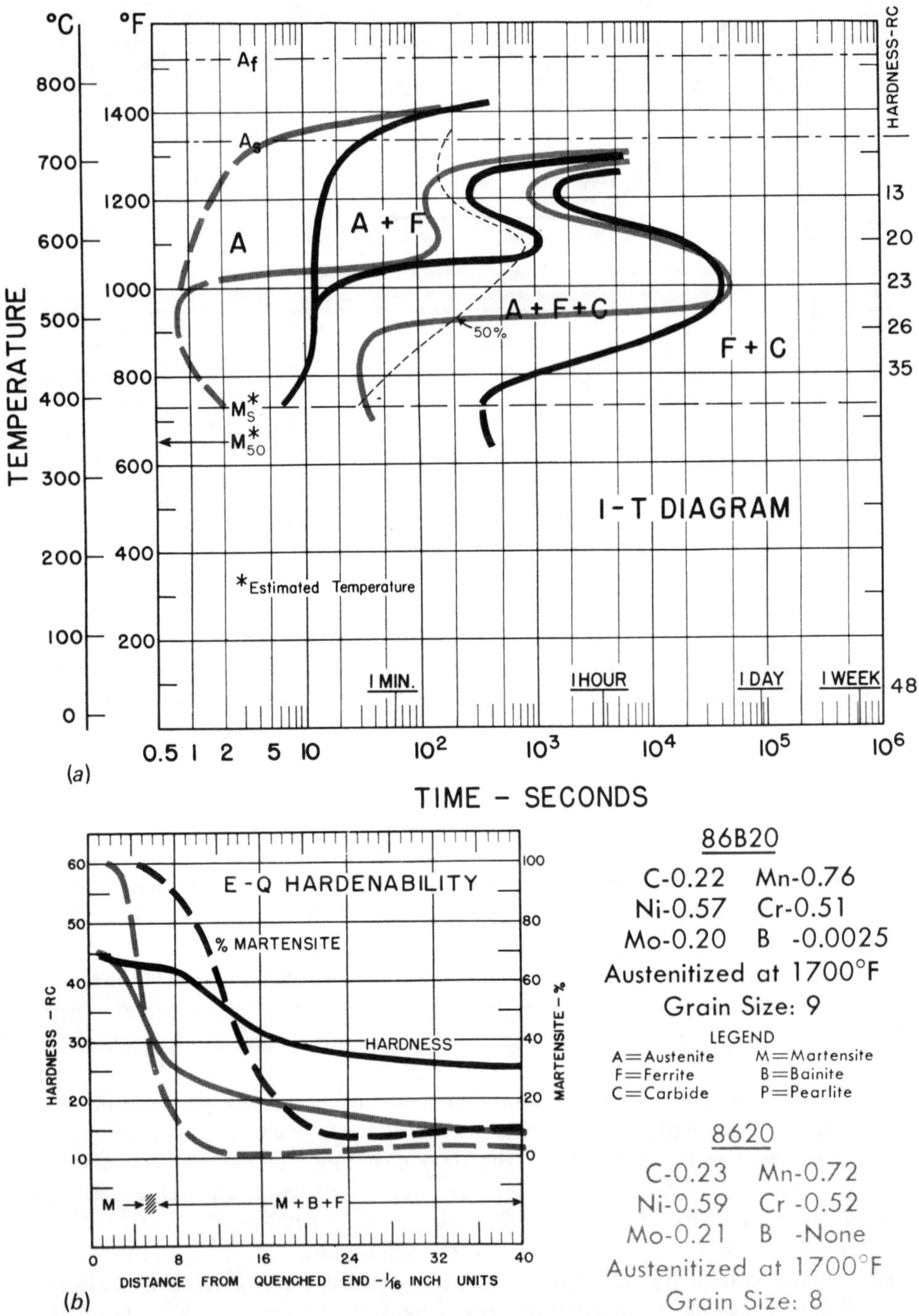

Figure VII-54. The effect of boron on (*a*) isothermal transformation and (*b*) hardenability of AISI 8620 steel. (R.A. Grange and J.B. Mitchell, *Trans. Am. Soc. Met.*, 53:157 (1961), by permission.)

contain Al, Mn, Si, Ti, and Zr. For the solubility product of BN in austenite, Fountain and Chipman[1] give

$$\log\,[\%B]\,[\%N] = \frac{-13{,}970}{T} + 5.24 \qquad \text{(VII-24)}$$

This solubility product should be compared with that of AlN, Table IV-1.

The largest use of B is probably in C–Mn steels containing from 0.15 to 0.40%C and 0.80 to 1.65%Mn.[2] These steels can replace alloy grades with large savings in cost. As compared to B-free steels of equal hardenability, these C–Mn–B grades have superior cold-forming properties, equal or better machinability, and less tendency to crack on quenching.

[1] R.W. Fountain and J. Chipman, *Trans. AIME,* 224:599 (1962).
[2] R.F. Kern, *Met. Prog.,* 102:3–90 (1973).

CHAPTER

EIGHT

THERMOMECHANICAL TREATMENT OF STEELS

A brief paper by Lips and Van Zuilen in 1954[1] generated great interest in the prospect of improving the properties of steels by various combinations of thermal and mechanical treatments. For several reasons, these processes have not been widely adopted by industry, but they are deserving of a review and to be kept in mind for possible future application as needs arise.

Thermomechanical treatments can be classified according to their timing with respect to the transformation of austenite during heat treatment,[2] as shown in Table VIII-1.

In classification Ia, austenite is deformed in the temperature range in which it is stable, just above Ae_3 (Fig. VIII-1); or additions can be made to the steel that inhibit recrystallization, such as V, Nb, or Ti, so that rolling can be done at a higher temperature without recrystallization occurring. The transformation takes place in deformed, unrecrystallized austenite. If sizeable changes in properties are to be obtained the deformation must be severe. The result is that dimensions of the austenite grains are greatly reduced in one orthogonal direction (normal to the rolling plane), unchanged in another, and greatly elongated in the rolling direction. After transformation the property changes are directional; the martensite is textured.

If austenite is severely deformed by isothermal hot rolling and does not recrystallize, it takes up the copper texture, $\{124\}\langle 112\rangle$ to $\{146\}\langle 112\rangle$. When this strongly textured austenite transforms to martensite, the martensite adopts a strong (112) + (111) texture which does not change during subsequent tempering at a low temperature, 175°C for example. A martensite texture with (110) in the rolling plane can be produced by recrystallizing the isothermally rolled austenite to form a well-defined cube texture. A martensite texture near (111) can be formed by isothermal cross-rolling of austenite,

[1] E.M.H. Lips and H. Van Zuilen, *Met. Prog.*, 66:103 (1954).

[2] E.B. Kula, in *Strengthening Mechanisms,* Syracuse Univ. Press, 1966, p. 83.

Table VIII-1 Classification of thermomechanical treatments

I.	Supercritical TMT (above Ae_3)	
	a.	Without recrystallization (hot-cold work, high-temperature TMT)
	b.	With recrystallization (controlled hot rolling, high-temperature TMT)
II.	Intercritical TMT (between Ae_3 and Ae_1)	
III.	Subcritical TMT (below Ae_1)	
	a.	Prior to transformation (ausforming, ausworking, ausrolling, low-temperature TMT)
	b.	During transformation (isoforming)
	c.	After transformation
IV.	Thermomechanical annealing	

followed by quenching before recrystallization occurs. All of these textures in the martensite are generated through the Kurdjumov–Sachs relation between austenite and ferrite (martensite)[1,2]

$$\{111\}_\gamma \parallel \{110\}_\alpha \quad \text{or} \quad \{110\}_\gamma \parallel \{111\}_\alpha$$
$$\langle 111\rangle_\gamma \parallel \langle 110\rangle_\alpha \qquad\quad \langle 110\rangle_\gamma \parallel \{111\}_\alpha$$

[1] H. Hu, G.R. Speich, and R.L. Miller, *AMMRC Report CTR 76-22,* July 1976.
[2] H. Hu, *AMMRC Report CTR 77-19,* July 1977.

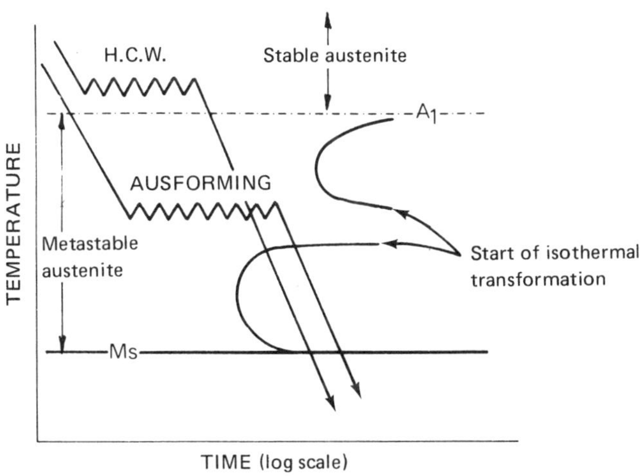

Figure VIII-1. Schematic isothermal transformation diagram with thermal cycles for hot–cold working and ausforming operations. (V.F. Zackay, *J. Iron Steel Inst.*, 207:894 (1969), by permission.)

The intensity of the (112) + (111) texture in martensite transformed from textured austenite can be as great as nine times random. The presence of this pronounced texture in quenched and tempered light armor plate can increase the ballistic limit by about 15%, with the ballistic limit increasing linearly with texture intensity. The ballistic limit is the maximum impact velocity of a standard projectile that the plate can withstand without penetration. This improvement is attributed to a high modulus of elasticity in a direction normal to the plate surface and to an increased yield strength in compression in the through-thickness direction.[1]

In classification Ib, the austenite recrystallizes to a fine, equiaxed structure. The grain size is smaller than can be attained by conventional austenitizing, quenching, and tempering. The mechanical properties after transformation are isotropic, except as affected by inclusions. After either Ia or Ib treatments the entire stress-strain curve is raised, with little or no penalty in ductility, because of the refinement of austenite grains.

In treatment II, the steel is deformed in the ferrite-plus-austenite region, which limits the process to hypoeutectoid steels. Such deformation followed by quenching yields a structure of martensite fibers in ferrite, as previously discussed in Chap. V. The properties will be highly directional, with high toughness in directions normal to the fibers.

In IIIa the deformation is done at temperatures below A_1, as indicated in Fig. VIII-1. It is necessary that the steel transform slowly at the deformation temperature. Medium-carbon steels containing several percent Cr are suitable, as they tend to have a deep 'bay' between the pearlite and bainite noses of the pertinent IT diagram. In this ausforming process the yield and tensile strengths of the quenched and tempered martensite are proportional to the strain imparted to the austenite, as indicated in Fig. VIII-2 for an H11 tool steel. Strengthening is usually in the range 4 to 9 MPa for each 1% of deformation. The higher the initial yield strength, the greater is the strengthening. In the absence of C the strengthening obtained by ausforming is small. Some minimum C content, perhaps about 0.05 to 0.10%, is necessary, but greater additions produce little effect. The presence of strong carbide-forming elements promotes increased strength.

The mechanisms of strengthening in ausforming have given rise to much controversy, and the relative importance of several processes is still in doubt. These changes in martensite structure as a result of prior plastic deformation of unstable austenite are:

1. Refinement of martensite packets or platelets
2. Changes in the size, amount, and distribution of carbides
3. Possible changes in dislocation density in the martensite
4. Development of texture in the martensite

The first change is undeniable. Because of the reduction of size of austenite grains

[1] A.K. Ghosh and N.E. Paton, *AMMRC Report TR78-40,* September 1978.

in a direction normal to the rolling plane, the structural units of martensite are also reduced in that direction.

Precipitation of carbides occurs in deformed austenite before transformation, particularly when strong carbide formers such as Mo are present. The martensite is dispersion-strengthened as soon as it is formed, and the fine, preexisting carbides also serve to refine the structure of the martensite.

The proposed strengthening mechanism that has generated the most controversy is the inheritance of defects by martensite from deformed austenite. If this occurred in a major way it would cause the strengthening of C-free martensite, but as previously mentioned (Chap. VII-3), such strengthening is small.

Ausforming, in at least some instances, improves the ductility of quenched and tempered steels. This probably occurs because the strain-induced precipitation of carbides in untransformed austenite reduces the C content of the martensite. Ausformed steels are also more resistant to tempering than are conventionally treated steels, but the difference between them decreases as the tempering temperature increases.

In the type IIIb process, the steel is deformed as it is transforming to pearlite or bainite, as illustrated in Fig. VIII-3. This is the so-called isoforming process. The product, which can be fine pearlite or bainite, must have properties that are highly anisotropic. The process also can be applied to metastable austenitic stainless steels, such as AISI 301 and 304, strained at a temperature where martensite forms during the straining.

Type IIIc processing can be applied to steels that have pearlitic, bainitic, or marten-

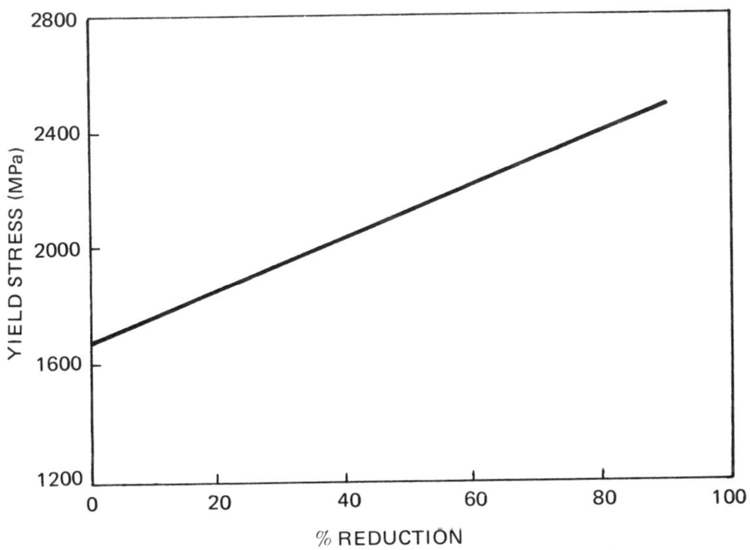

Figure VIII-2. Strengthening of H-11 steel (0.4%C–5.0%Cr–1.3%Mo) by ausforming. (V.F. Zackay and W.M. Justusson, *Iron and Steel*, 76:14 (1962), by permission.)

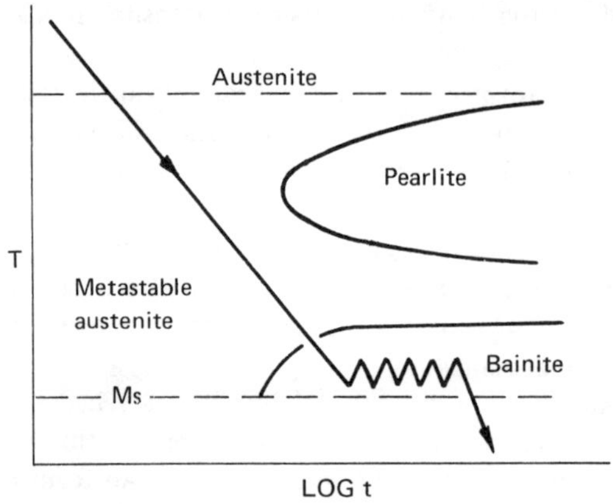

Figure VIII-3. Schematic isothermal transformation diagram with thermal cycle for the isoforming process.

sitic structures. It amounts to static or dynamic strain aging. The yield and tensile strengths increase rapidly with plastic strain and only small strains can be used. The principal penalty is a severe loss in ductility and a decrease of uniform strain to zero, or close to it. Stephenson and Cohen[1] strained quenched-and-tempered AISI 4340 steel 3%, then aged specimens at various temperatures. Maximum strengthening was obtained by aging at 150°C. Greater deformations, up to about 10%, can be imparted by cold drawing than by uniaxial tension or compression. Because of the very rapid work hardening of martensite (Fig. VII-7), very large increases in yield strength can be obtained. The drawing process can be done at temperatures in the range 100 to 300°C where the rate of work hardening is even more rapid than at room temperature.

Astonishing levels of strength can be obtained by a combination of ausforming and strain aging, as indicated in Table VIII-2. Needless to say, ductility after treatment 4 was poor, although not entirely lacking; witness the difference between the yield and tensile strengths.

[1] E.T. Stephenson and M. Cohen, *Trans. Am. Soc. Met.*, 54:72 (1961).

Table VIII-2 Effects of combined thermomechanical treatments[1] (0.6%C–5%Ni)

Treatment	0.2% Offset yield strength (MPa)	Tensile strength (MPa)
1. Conventional, quenched, tempered 120°C	1800	2280
2. Hot–cold worked, quenched, tempered 120°C	2720	3070
3. 1 + 3% strain, aged 120°C	2450	2520
4. Hot–cold worked, quenched, tempered strain 3%, aged 120°C	3590	3630

[1] S.J. Matas, M. Hill, and H.P. Munger, *Mechanical Working of Steel 1*, Gordon & Breach, New York, 1964, p. 143.

In contrast, type IV thermomechanical processing has as its aim the sharp *reduction* of strength. Alloy steels are often difficult to anneal, especially if they are high-C varieties. They must be softened to a level at which economical machining is possible, but this may require days at a temperature just below A_1 and the difficulties are compounded if the steel contains substantial amounts of Mn or Ni which reduce A_1. On the basis that deformation of a metastable structure hastens the approach to equilibrium, Grange[1] developed a thermomechanical annealing process which consists of heating the steel to a temperature just below A_3, then deforming (rolling) it as it cools, with the last rolling pass being made just below A_1. The resulting structure is a dispersion of spheroidal carbides in a matrix of ferrite, and this structure can be achieved in a matter of minutes.

These thermomechanical processes for strengthening have not been adopted by industry to any appreciable extent, for the reasons that follow:

1. For most of them, and especially for ausforming, large reductions must be made, with any amounts less than about 50% being ineffective. This means that large sections cannot be treated. Also, there is a danger that the steel will crack.
2. Because of heavy deformation of strong steels at relatively low temperatures, the loads on the rolling, forging, extruding, or drawing equipment are very high.
3. The steel must be used in nearly the simple shape in which it is formed, for machining must be done in the hardened condition.
4. Joining by any process is difficult and welding is out of the question.
5. The resulting mechanical properties are anisotropic, but this is not always a disadvantage.
6. Thermomechanical processing must compete in cost with alloying as a means for obtaining the desired strength and toughness.

Nevertheless, some thermomechanical processes may be near to practical utilization—for example, the improvement of light armor plate previously cited.

[1] R.A. Grange, *U.S. Patent 3,459,599,* Aug. 5, 1969.

CHAPTER

NINE

STEELS OF VERY HIGH STRENGTH: FRACTURE PREVENTION

IX-1. MODIFICATIONS OF STANDARD ALLOY STEELS

Standard alloy steels of high hardenability, such as AISI 4130, 4140, 4330, or 4340 (Table VII-1) can be quenched and tempered to high yield strengths, in excess of 1500 MPa. However, with such high strengths these steels are deficient in notch toughness. To improve this property, it is usually desirable to increase the tempering temperature, but this leads to a reduction in the yield strength and to the occurrence of "500°F embrittlement" (Chap. VII-6). Both effects can be counteracted by the addition of between 1 and 2% silicon. As previously discussed (Chap. IV), silicon strongly inhibits the growth of ϵ carbide and the formation of Fe_3C during tempering, allowing the steel to be tempered in the range 260–315°C, thereby increasing the relief of thermal and transformation stresses while avoiding embrittlement. Two steels containing higher than normal silicon, Crucible Hy-Tuf and INCO 300-M, are listed in Table IX-1. Such steels are used in critical applications such as aircraft landing gear. Figure IX-1 shows that tempering 300-M at 300°C produces maxima in both yield strength and impact strength at room temperature.

A second approach to achieving high toughness at high strength is through austenite grain refinement, which leads to refinement of the microstructure after quenching and tempering. Austenite grain growth can be inhibited by adding vanadium, as in AMS 6434 or Ladish D6AC (Table IX-1). Grain growth inhibition comes from finely divided vanadium carbides.

Table IX-1 Percent composition of steels of very high strength

Steel	C	Mn	Si	P	S	Cr	Ni	Mo	Other
Modified standard alloy steels									
Crucible Hy-Tuf	0.23–0.28	1.20–1.50	1.30–1.70	0.04 max	0.04 max	0.20–0.40	1.65–2.00	0.35–0.45	
INCO 300-M	0.35–0.45	0.65–0.90	1.45–1.80	0.025 max	0.025 max	0.70–0.95	1.65–2.00	0.30–0.45	0.05 V min
AMS 6434	0.31–0.38	0.60–0.80	0.20–0.35	0.025 max	0.025 max	0.65–0.90	1.65–2.00	0.30–0.40	0.17–0.23 V
Ladish D6AC	0.42–0.48	0.60–0.90	0.15–0.30	0.025 max	0.025 max	0.90–1.20	0.40–0.70	0.90–1.10	0.05–0.10 V
High-strength plate steels									
HY-80	0.18 max	0.10–0.40	0.15–0.35	0.025 max	0.025 max	1.00–1.80	2.00–3.25	0.20–0.60	
HY-100	0.20 max	0.10–0.40	0.15–0.35	0.025 max	0.025 max	1.00–1.80	2.25–3.50	0.20–0.60	
HY-130	0.12 max	0.60–0.90	0.20–0.35	0.010 max	0.015 max	0.40–0.70	4.75–5.25	0.30–0.65	0.05–0.10 V
Tool steels used in structural applications									
H-11	0.35					5.00		1.50	0.40 V
H-13	0.35					5.00		1.50	1.00 V
M-50	0.80	0.25	0.20			4.00		4.25	1.00 V
T1	0.70	0.30	0.20			4.00		0.30	18.0 W, 1.2 V
High-strength carburizing steels									
CBS-600	0.16–0.22	0.40–0.70	0.90–1.25	0.025 max	0.025 max	1.25–1.65		0.90–1.10	0.03 min Al
CBS-1000M	0.10–0.16	0.40–0.60	0.40–0.60	0.015 max	0.015 max	0.90–1.20	2.75–3.25	4.00–5.00	0.20–0.50 V
EX00053	0.10	0.35	1.00			1.00	2.00	3.25	2.00 Cu
X2-CVM	0.15	0.25	1.00	0.015 max	0.015 max	5.00		1.4	1.28 W, 0.40 V

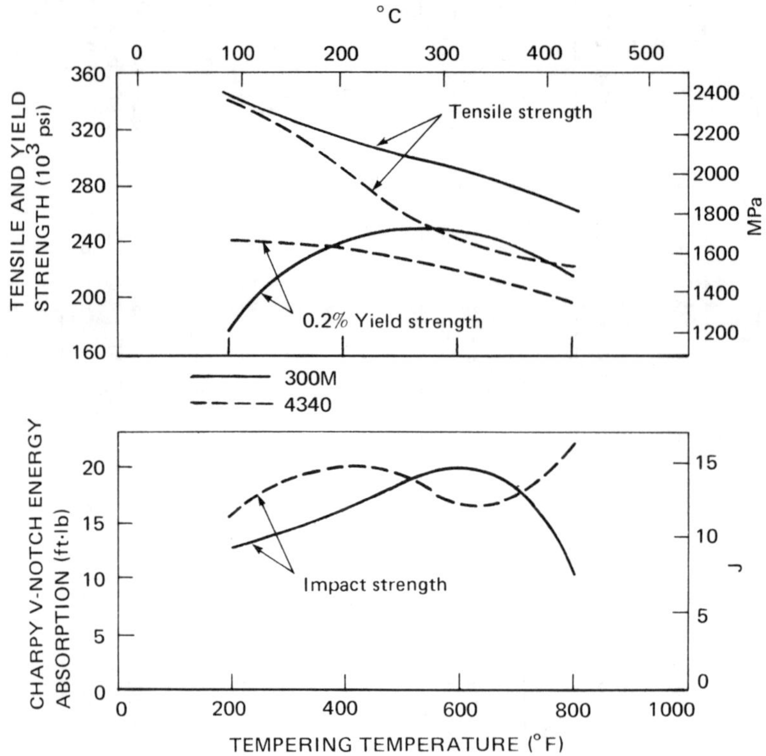

Figure IX-1. Effects of tempering temperature on the tensile and impact properties of 25-mm round bars of two medium-carbon low-alloy steels oil-quenched from 855°C. (J.W. Sands and O.O. Miller, *Mater. Methods*, 43:94 (3) (1956), by permission.)

IX-2. HIGH-STRENGTH PLATE STEELS

Recent years have seen a large increase in the demand for high-strength, high-toughness steel plates for submarine hulls, nuclear reactor vessels, high-pressure gas storage vessels, and solid-propellant rocket motor cases. Such plates, of course, must be weldable, so their development must include the simultaneous development of welding rod materials and welding techniques. The compositions of HY-80, HY-100, and HY-130 are listed in Table IX-1. The numerals in these designations indicate the minimum tensile yield strength in ksi; these are equivalent to 552, 690, and 897 MPa, respectively. To achieve such strengths requires a quenched and tempered structure. For maximum toughness, HY-100 and HY-130 must be quenched to a minimum of 80% martensite, which limits plate thicknesses to 75 mm for HY-100 and 100 mm for HY-130. All three of these steels must be tempered at a minimum of 565°C. The energy absorbed at −85°C by longitudinal Charpy V-notch impact specimens cut from the thickest plates is a minimum of 22 J for HY-80 and HY-100. For HY-130, the minimum is 44 J at −18°C.

In Table IX-1, note that as the yield strength of these three plate steels is increased, the nickel content is also increased, and the allowable sulfur and phosphorus content is reduced. More will be said of this later. The yield strength of HY-130 rises rapidly as carbon content is increased, but this is accompanied by a sharp drop in V-notch energy absorption (Fig. IX-2). This figure makes it clear that a yield strength goal must be reached with the minimum carbon content if toughness is an essential requirement.

IX-3. TOOL STEELS USED IN STRUCTURAL APPLICATIONS

The H11 and H13 tool steels (Table IX-1), together with similar 5%Cr-base steels, have certain attractive features as high-strength structural materials. They are air hardening, and therefore their mechanical properties are less sensitive to section size than are those of steels that must be quenched. Because they are air hardening they suffer little distortion during heat treatment. Their molybdenum and vanadium contents produce secondary hardening and resistance to tempering, by the formation of Mo_2C and V_4C_3. Therefore, high tempering temperatures can be employed, reducing the residual transformation stresses, and the range of service temperatures is increased. The

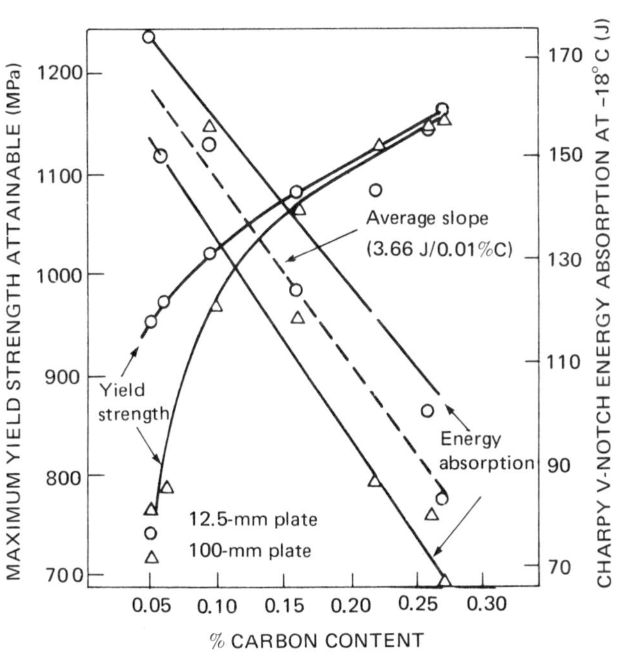

Figure IX-2. Effect of carbon content on maximum yield strength and notch toughness of 5%Ni-Cr-Mo-V steel. (L.F. Porter, S.J. Manganello, D.S. Dabkowski, and J.H. Gross, *Met. Eng. Q.*, 6:17 (Aug. 1966), by permission.)

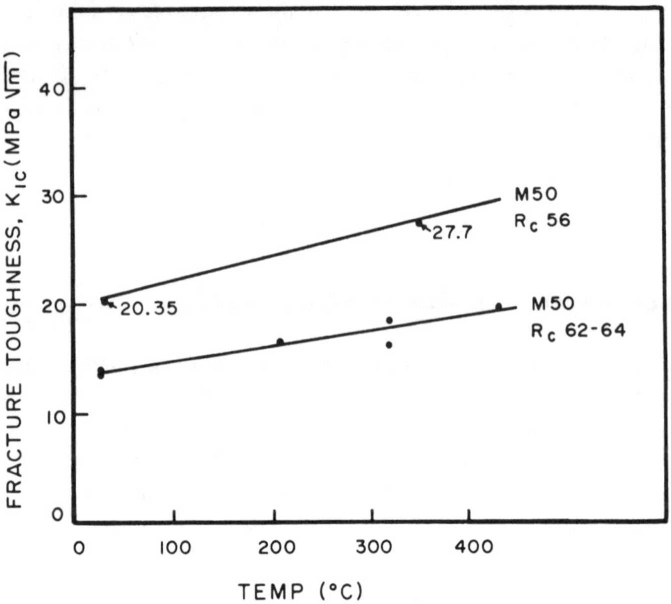

Figure IX-3. Fracture toughness of M50 bearing steel. (Courtesy R.R. Hendrickson, Terra Tek, Inc., 1979.)

weak point of these steels is their low toughness; even after tempering at 600°C the energy absorbed by a V-notch Charpy impact specimen at room temperature may be only about 15 J.

Higher temperatures in aircraft engines, and the development of lubricants to withstand these temperatures, has meant that the standard bearing steel, AISI 52100, no longer has sufficient resistance to softening. In critical high-temperature bearings it has been replaced by M50 or T1 tool steels (Table IX-1). These latter two steels are made by vacuum induction melting and vacuum arc remelting, or by electroslag remelting. The function of these expensive processes is to reduce the content of nonmetallic inclusions. M50 steel is austenitized at about 1100°C, air cooled, then repetitively tempered in the range 520-540°C. T1 steel is austenitized at about 1250°C, air cooled, then repetitively tempered in the range 540-560°C. These treatments produce a hardness of 62 to 64 R_C. For bearings, it is important that retained austenite be minimized.

At these hardness levels, fracture toughness (K_{IC}) at room temperature is necessarily low, 14 to 20 $MPa \cdot \sqrt{m}$.[1,2] In applications such as bearings for bits for geothermal drilling where wear resistance can be sacrificed for toughness, M50 can be tempered to a hardness of R_C 56 and the fracture toughness rises to slightly above 20 $MPa \cdot \sqrt{m}$ at room temperature. The toughness rises slightly with service temperature (Fig. IX-3).

[1] L.M. Barker and W.C. Leslie, *Fracture 77,* 2:305 (1977).
[2] J.A. Rescalvo and B.L. Averbach, *Met. Trans.,* 10A:1265 (1979).

XI-4. HIGH-STRENGTH CARBURIZING STEELS

Applications of carburized steel parts demanding the retention of high core strength and toughness to temperatures of 250°C and higher has led to the development of new types of steels for carburizing. Four such steels are listed in Table IX-1. The least expensive of the four, Timken CBS-600, is processed by carburizing at 925°C, oil quenching, tempering at 650°C, austenitizing at 830°C, oil quenching, then double tempering at 315°C. Following this, the hardness of the case at a service temperature of 315°C is about 53 R_C and the hardness of the core is 38 R_C. The core hardness is constant from room temperature to 315°C. Resistance to tempering is gained by increased silicon, chromium, and molybdenum contents. For this steel, or for any steel containing appreciable amounts of silicon or chromium, it is advisable to oxidize the part in air at about 950°C before carburizing, to avoid the formation of films of oxides of chromium and silicon that can be nearly impervious to carbon.

CBS-1000M retains useful strength in the case and core to a temperature of 430°C, by virtue of its high molybdenum and vanadium content, so a gain of about 115°C in service temperature is purchased at quite a high cost. The best high-temperature properties in this steel are obtained by austenitizing at 1095°C (far above the carburizing temperature of 940°C), oil quenching, refrigerating to transform some retained austenite, then triple tempering at 535°C at the peak of secondary hardening.

The steel EX00053 uses less molybdenum than CBS-1000M and substitutes copper, which produces a precipitation-hardening reaction. The core hardness of this steel remains essentially constant at about 35 R_C from room temperature to about 480°C, but the case hardness falls to about 43 R_C at that temperature.

The last of these high-strength carburizing steels, Vasco X2-CVM, is essentially an H12 hot-work tool steel (Table XIII-1) with carbon content reduced to about 0.15% from the normal 0.35%. Because of this reduced carbon content and the content of chromium, molybdenum, and tungsten, the steel must be heated to 1065°C to reduce the amount of primary ferrite to 5% or less, which is required for high toughness.[1] Tempering for maximum toughness should be done at 315°C, which is low for a steel with a secondary hardening peak, but desirable for maximum toughness of the core. This steel can maintain a hardness of 50 R_C in the case at 420°C and 38 R_C in the core.

This is a new field and it may be some time before economical carburizing steel compositions are standardized for severe applications such as helicopter gears and geothermal drill bits.

IX-5. MARAGING STEELS

Maraging steels first appeared in 1960 as a result of alloying studies by Bieber of the International Nickel Company. Since that time a reasonably good understanding of

[1] P.J. Fopiano, S.A. Oliver, and E.B. Kula, *AMMRC-TR 77-8,* March 1977.

Table IX-2 Typical compositions of maraging and TRIP steels

Steel	C	Mn	Si	Cr	Ni	Mo	Other
Maraging steels							
200	0.03 max	0.10 max	0.10 max		18.0	3.25	8.5Co, 0.20Ti, 0.10Al
250	0.03 max	0.10 max	0.10 max		18.0	4.90	8.0Co, 0.40Ti, 0.10Al
300	0.03 max	0.10 max	0.10 max		18.0	4.90	9.0Co, 0.65Ti, 0.10Al
350	0.01 max	0.10 max	0.10 max		17.5	3.75	12.5Co, 1.80Ti, 0.15Al
TRIP steels							
	0.25	2.5	2.0	9.0	9.0	5.0	
	0.25	1.7	2.0	10.0	8.8	5.5	
	2.0	2.0	2.0	13.5	8.8	3.0	

the hardening mechanisms in these steels has been developed.[1] Hardening depends upon a substantially complete transformation of austenite to martensite, followed by precipitation of intermetallic compounds during isothermal aging. Four standard compositions are listed in Table IX-2. The as-quenched Fe–Ni martensite is of the lath type, containing a high density of dislocations. Its function is to provide a very large number of nucleation sites for precipitation of intermetallic compounds. The martensite before aging is comparatively soft, about 30 R_C, so the steel can be formed before aging. Some care must be exercised in deformation, for the rate of work hardening is low and the uniform elongation of the martensite is slight, only 1 to 3%.

The basic composition is essentially 18%Ni–8%Co–5%Mo. The yield strength is varied between 1375 and 2410 MPa by raising the titanium content from 0.20% to 1.4%. To ensure high toughness, the content of the residual elements P, S, C, and N must be kept as low as possible,[2] which entails vacuum melting. The processing involves ingot heating to about 1250°C, with hot rolling being finished at about 920°C. The steels are subsequently austenitized at 850°C, cooled to room temperature, then aged 3 h at 480°C.

The principal hardening agent is orthorhombic Ni_3Mo, which precipitates as rods. This is accompanied by formation of spherical particles of tetragonal FeTi. Particle sizes of both are near 10 nm.[3] The strengthening effect of cobalt apparently arises through short-range ordering. Nickel is rejected from the ordered Fe–Co regions, and this affects the size of the Ni_3Mo particles by changing the local solubility of molybdenum.

[1] S. Floreen, *Met. Rev.*, 13:115 (1968).
[2] G.J. Spaeder, *Met. Trans.*, 1:2011 (1970).
[3] H.J. Rack and D. Kalish, *Met. Trans.*, 2:3011 (1971).

The regions depleted in Ni serve as nucleation sites for precipitation of FeTi. Overaging occurs by recovery of the martensite dislocation structures and formation of austenite.

Toughness suffers by precipitation or segregation at prior austenite grain boundaries. The main culprits are Ti(C, N) and Ti_2S.[1,2] Embrittled material can be restored to ductility by a solution treatment at above 980°C, followed by rapid cooling.

The maraging steels offer a remarkable combination of strength and ductility, with yield strengths as high as 2400 MPa with total elongation of 6%. They can be welded readily before solution treating and aging. Unfortunately, their cost has become almost prohibitive and attempts to develop more economical alloys have not been successful.

IX-6. TRIP STEELS

TRIP, an acronym for TRansformation-Induced Plasticity, is the name given to experimental compositions of steels that can be greatly strengthened by strain-induced transformation of austenite,[3,4] which also improves ductility. It has long been known[5] that if a metastable austenitic stainless steel such as Type 301 (Table XII-6) is plastically deformed at room temperature, some of the austenite in the most severely strained portions of the specimen will transform to martensite; and such transformation, by locally increasing the rate of work hardening, inhibits necking and increases the uniform elongation. TRIP steels employ the phenomenon, but in austenite that has been previously strengthened by thermomechanical processing (akin to ausforming) or by thermal cycling to produce a high density of dislocations in the austenite. For example, a processing sequence could be:

1. Solution anneal at 1200°C, quench
2. Strain to 80% reduction of thickness at 450°C (above M_D) to deform the austenite and to produce a fine carbide precipitate
3. Cold reduce at room temperature (below M_D) to transform some austenite to martensite

A variety of compositions (Table IX-2) and processing steps can be employed. The excellent combinations of yield strength and total elongation obtainable are shown in Fig. IX-4. However, because of the cost of the alloys, the difficulty of large plastic deformation at low temperature, and problems with joining, these steels will only be used in special applications of flat-rolled products or wire.

[1] R.L. Cairns and C.J. Novak, *Met. Trans.*, 2:1837 (1971).
[2] D. Kalish and H.J. Rack, *Met. Trans.*, 2:2665 (1971).
[3] V.F. Zackay, E.R. Parker, D. Fahr, and R. Busch, *Trans. Am. Soc. Met.*, 60:252 (1967).
[4] W.W. Gerberich, P.L. Hemmings, M.D. Merz, and V.F. Zackay, *Trans. Am. Soc. Met.*, 61:843 (1968).
[5] J.P. Bressanelli and A. Moskowitz, *Trans. Am. Soc. Met.*, 59:223 (1966).

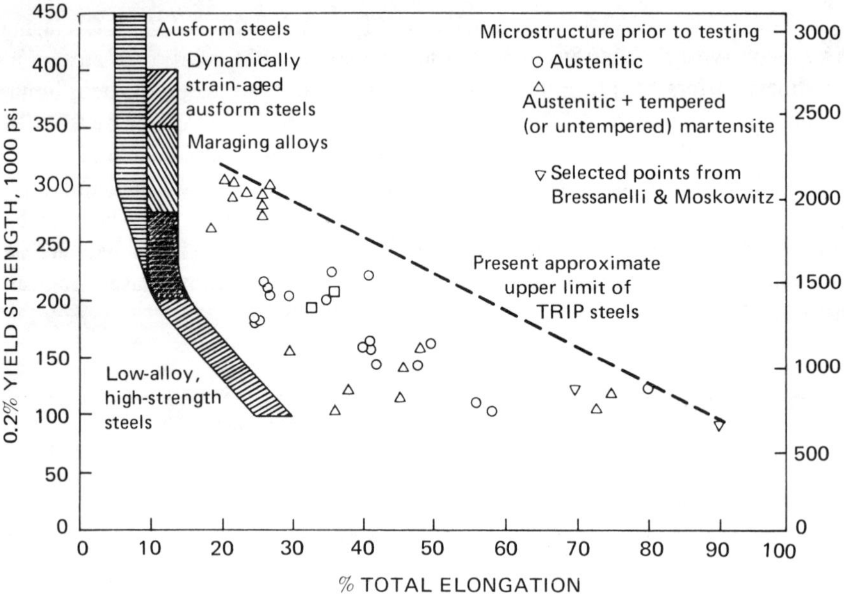

Figure IX-4. Ranges of yield strength and total elongation at failure characteristic of high-strength steels. (V.F. Zackay, E.R. Parker, D. Fahr, and R. Busch, *Trans. Am. Soc. Met.*, 60:252 (1967), by permission.)

IX-7. BRITTLE FRACTURE

Any metal that exhibits strong temperature and strain-rate dependence of the yield strength is susceptible to brittle fracture. As discussed in Chap. I, cross slip of screw dislocations in iron becomes more difficult as temperature is reduced or as strain rate is increased. The relief of high local stresses by fracture rather than by slip then becomes more likely. Rosenfield et al.[1] contrast slip in Nb and in W. Cross slip of dislocations in Nb is profuse whereas those in W scarcely cross-slip at all. The difference is reflected in the ductile–brittle transition temperature, which is about −120°C for Nb and about 300°C for W.

The brittle fractures that concern us in bcc Fe are transcrystalline cleavage fractures on {100}. The brittle grain boundary fractures caused by solute segregation, as in temper embrittlement, will not be considered.

Brittle fracture in metals is nucleated by plastic deformation, and plastic deformation always accompanies the propagation of brittle cracks. Because of the energy required for this plastic deformation, the Griffith theory[2] in its original form cannot

[1] A.R. Rosenfield, E. Votava, and G.T. Hahn, in *Ductility*, Am. Soc. Met., Metals Park, Ohio, 1968, p. 63.

[2] A.A. Griffith, *Philos. Trans. Roy. Soc. (London)*, 221A:163 (1920); also, *Trans. Am. Soc. Met.*, 61:861 (1968).

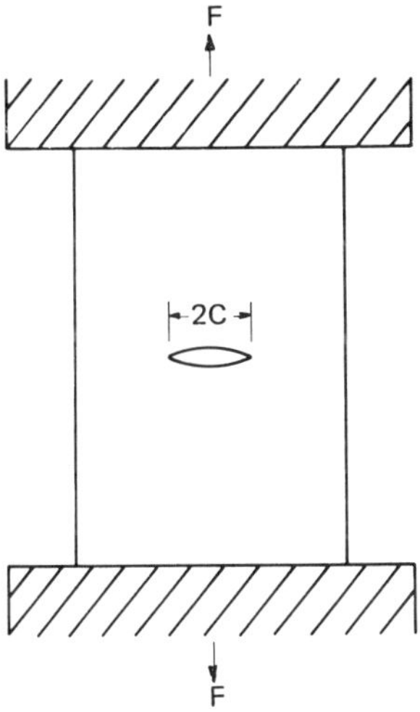

Figure IX-5. Brittle specimen loaded between fixed grips.

be strictly applied to semibrittle metals. Nevertheless, Griffith's concept that cracks would propagate when the decrease in elastic strain energy was equal to or greater than the energy required to create the new fracture surfaces need only be modified to account for plastic deformation.

From Griffith, the stress required to propagate a preexisting crack of radius C in a brittle material is

$$\sigma = \left(\frac{2E\gamma}{\pi C}\right)^{1/2} \tag{IX-1}$$

where E = Young's modulus
γ = surface energy

Because $(2/\pi)^{1/2}$ is close to unity, this can be written as

$$\sigma \cong \left(\frac{E\gamma}{C}\right)^{1/2} \tag{IX-2}$$

After Cottrell,[1] the elastic energy of a crack in a brittle solid can be estimated by considering a crack of length $2C$ in a body stressed in tension between fixed grips, as indicated in Fig. IX-5. In regions near the crack surface the tensile stress is zero, so the material within a radius C contains no elastic strain energy. The elastic strain energy per unit volume in a body can be obtained by considering a cube of edge l_0 strained in uniaxial tension by a force increasing from zero to F. The loaded faces of the cube move apart a distance δl. The work done by the applied force is $\frac{1}{2}F\,\delta l$, the factor $\frac{1}{2}$ entering because δl increases in proportion to F. The tensile stress is given by

[1] A.H. Cottrell, *The Mechanical Properties of Matter,* J. Wiley & Sons, New York., 1964, p. 345.

$$\sigma = \frac{F}{l_0^2} \tag{IX-3}$$

and the strain by

$$\epsilon = \frac{\delta l}{l_0} \tag{IX-4}$$

The work done per unit volume (the density of strain energy in the body) is given by

$$\tfrac{1}{2}\sigma\epsilon = \frac{\sigma^2}{2E} \tag{IX-5}$$

The strain energy relieved by the crack is then roughly $(\sigma^2/2E)\pi C^2$ per unit thickness of the body and exactly $(\sigma^2/E)\pi C^2$.

When the crack is about to grow under the stress σ, the increase in surface energy is just balanced by the decrease in strain energy, or

$$\frac{d}{dc}\left(\frac{\pi C^2 \sigma^2}{E}\right) = \frac{d}{dc}\,(4\gamma C) \tag{IX-6}$$

The factor 4 enters because two surfaces are growing at each end of the crack. Therefore,

$$\sigma = \left(\frac{2\gamma E}{\pi C}\right)^{1/2} \cong \left(\frac{\gamma E}{C}\right)^{1/2} \tag{IX-7}$$

as in Eqs IX-1 and 2. For metals, the γ term must be increased by several orders of magnitude to allow for plastic deformation ahead of the crack.

Effective cracks can exist in metals, these being nonmetallic inclusions that cannot support tensile stresses normal to their surface. Fortunately, these are usually below the critical size at the service stress. Cleavage cracks can be nucleated at the intersection of slip bands on two different slip systems, at the intersection of twins when the plane of intersection is {100}, at the intersection of a twin with a grain boundary, or by the intersection of a slip band with a grain boundary. However, we are still lacking experimental evidence of dislocation interactions producing a crack. Cottrell's proposal[1] that dislocations on two {110} slip bands can meet along a (010) line of intersection to form a new dislocation

$$\frac{a}{2}[\bar{1}\bar{1}1] + \frac{a}{2}[111] \rightarrow a[001]$$

has been mentioned in Chap. I. The $a[100]$ dislocation is of pure edge character. The Burgers vector of this dislocation can be increased by coalescence of more slip dislocations until a "crack" is formed.

Although several theoretical models of cleavage have been proposed, none is amenable to testing and all are deficient in ignoring rate effects, departures from linear elasticity in dislocation cores, and plastic relaxation.[1]

[1] A.H. Cottrell, *Trans.-AIME,* 212:192 (1958).

In steels, as contrasted to iron, cleavage fracture in ferrite can be initiated by cracks in cementite films.[1] These cracks form in the carbide because of stress imposed by the plastic deformation of the surrounding ferrite. Cementite particles, when less than about 1000 nm thick, behave like whiskers and can accommodate considerable elastic strain. The interface between ferrite and cementite seems to be quite strong and therefore is seldom a source of cracks.

During plastic deformation, the ferrite serves to concentrate stresses in the carbide, which can react by deforming elastically, plastically, or by breaking. If the cementite breaks, the crack will quickly arrive at the ferrite-cementite interface, at which point the ferrite can crack or deform plastically. If we regard the ferrite as a brittle material, the Griffith equation (IX-2) can be applied:

$$\sigma = \left(\frac{E\gamma}{C}\right)^{1/2}$$

Assuming the following values,

$\sigma = 300$ MPa (lower yield stress)
$E = 20.8 \times 10^4$ MPa
$\gamma = 1\ \mathrm{J/m^2}$

the critical Griffith crack size, $2C$, is about 4 μm. When a carbide of this size cracks, the crack can propagate into the ferrite. Whether the ferrite will crack or deform plastically depends upon orientation of the $\{100\}_\alpha$ cleavage planes with respect to the crack in the cementite.

When carbides are smaller than required to support a Griffith crack, their presence in low concentrations, as in Fe-0.013%C quenched from 720°C and aged at 200-480°C, serves to improve ductility (uniform elongation) at −195°C.[2] The particles serve as dislocation sources at low temperatures, thereby distributing slip over systems that would otherwise not operate, and the particles inhibit twinning, thus eliminating cracks that can form at twin intersections.

IX-8. DUCTILE FRACTURE

Ductility is an elusive quality and difficult to define. Rogers' definition[3] is ". . . the ability of a material to undergo substantial amounts of general deformation." This ability is limited by the onset of ductile fracture. Considering such fracture in uniaxial tension, a ductile metal loaded beyond its yield strength (and beyond any Lüders strain) elongates uniformly. This process is terminated by the formation of internal voids at inclusions or second-phase particles. These voids can occur by fracture of the particles

[1] C.J. McMahon Jr. and M. Cohen, *Acta Metall.*, 13:991 (1962).
[2] W.C. Leslie, D.S. Hutton, and G.B. Williams, *J. Iron Steel Inst.*, 198:21 (1961).
[3] H.C. Rogers, in *Ductility*, Am. Soc. Met., Metals Park, Ohio, 1968, p. 31.

or by separation at particle–matrix interfaces. Voids form most readily at the largest particles, then at progressively smaller particles as plastic strain increases. Void formation is usually favored by low temperatures. The creation of voids can lead to a local reduction in the load-bearing section, and thereafter deformation is localized in the neck. Formation of a neck produces triaxial tensile stresses which lead to further growth of voids and their coalescence into cracks.

In tensile tests, this mechanism leads to the familiar "cup and cone" fracture. A central "crack" formed by coalescence of voids concentrates deformation at its tips into bands of high shear strain at an angle of about 60° from a plane normal to the tensile axis. Within these bands, sheets of voids are nucleated. The voids within these sheets grow and coalesce. The crack progresses in zig-zag fashion by such void-sheet formation until it reaches the surface and fracture is complete.

Voids are formed at preexisting defects or particles; there is no evidence that they can be formed by dislocation interactions. Voids have been seen to nucleate at particles as small as 5 nm. The creation of voids leads to "dimples" on the fracture, and a particle is often seen by scanning electron microscopy at the center of each dimple. In very clean, high-strength alloys, voids can be generated at grain corners.

IX-9. METALLURGICAL FACTORS AFFECTING FRACTURE

For steels made to yield strengths of 700 to 1750 MPa, notch toughness is of paramount importance. Metallurgical factors can alter notch toughness in at least three ways[1]:

1. By changing the energy absorbed in ductile fracture
2. By shifting the ductile–brittle transition temperature
3. By changing embrittlement caused by adsorption phenomena

Item 3 (temper embrittlement) has been discussed in Chap. VII. The metallurgical factors that can be controlled are:

1. Austenite and ferrite grain sizes
2. Elements in solid solution in ferrite
3. Dislocation density and distribution
4. Second phases and their composition, amount, shape, and number density
5. Segregation of impurities of limited solubility

Let us consider ductile fracture first. As just described, this occurs by the formation, growth, and joining of voids. In steels, the sources of such voids are particles of oxides, sulfides, nitrides, and carbides. Therefore, to improve toughness in the ductile range it is necessary to eliminate these particles to the greatest extent possible. The remaining factor is then the toughness of the matrix itself. Toughness in the ductile fracture region is roughly inversely proportional to yield strength, at any constant inclusion content. As the matrix becomes stronger, it is less able to resist void growth.

[1] L.F. Porter, *W.H. Eisenman Conf. Program,* Am. Soc. Met., Philadelphia, 1970.

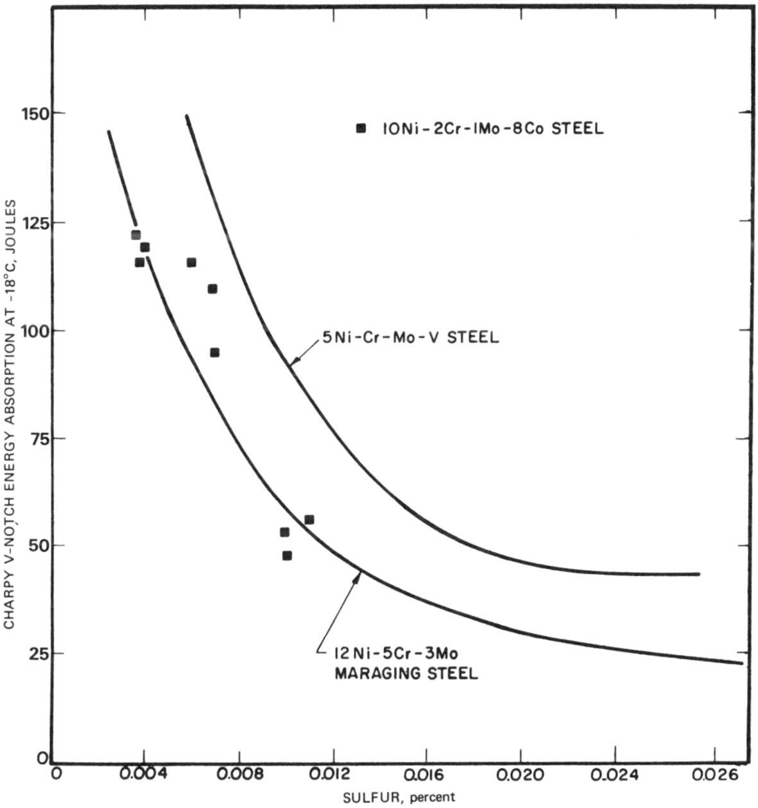

Figure IX-6. Effect of sulfur content and sulfide shape on shelf energy of high-strength steels. (L.F. Porter, *W.H. Eisenman Conf. Program*, Am. Soc. Met., Philadelphia, 1970, by permission.)

Less energy is dissipated in plastic deformation around the growing void. Also, the density of sites at which voids are created increases because the critical stress for void formation is reached at sites of lower stress concentration. Any procedure that improves steel cleanliness will improve toughness in the ductile range. The limit to this improvement is set by the nucleation of voids at grain corners.

The effect of reduction of S content on toughness in the ductile region is particularly striking, as seen in Fig. IX-6. When S contents exceed 0.015%, a common occurrence, the toughness is insensitive to S content; but at lower S contents the toughness improves dramatically. It is not always necessary to reduce the sulfur content to very low levels; an addition of rare-earth metals to change the sulfide shape from stringers to equiaxed particles can be even more effective. Despite having the highest yield strength of the three steels in Fig. IX-6, the 10%Ni steel had the greatest toughness because of such sulfide shape control. The relation shown in Fig. IX-6 holds for steels of various compositions and microstructures.

When S exceeds 0.015%, large variations in O_2 content make little difference in toughness; but when S content is less than this, reducing the O_2 content increases toughness. Unfortunately, when the yield strength of a steel approaches 1700 to 2100 MPa, the beneficial effects of purification no longer operate.

To obtain a matrix of highest toughness Mn and Si in solid solution must be minimized. Manganese reduces toughness even when present in amounts as small as 0.12%. Although Ni is generally known to increase toughness, this is brought about by lowering the transition temperature rather than by raising the energy absorbed in ductile fracture (the shelf energy).

The maximum toughness at a given strength is not obtained by tempering at a low temperature, but by adjusting the composition of the steel so that secondary hardening occurs. The carbides formed during low-temperature tempering tend to form on crystallographic planes in the martensite, or on martensite lath boundaries. In contrast, the very fine alloy carbides formed during secondary hardening are uniformly dispersed. Care must be taken to balance the composition so that all of the C is combined as fine alloy carbides, with no massive or grain-boundary carbides.

The prevention of brittle fracture—that is, the lowering of the ductile-brittle transition temperature—must be handled in a different manner. The most effective technique for lowering the transition temperature is reduction of grain size, as is strikingly shown in Fig. IX-7. Note that the fine-grained steel has a markedly higher yield strength, in addition to a lower transition temperature.

The effect of Ni in lowering the transition temperature was discussed in Chap. III

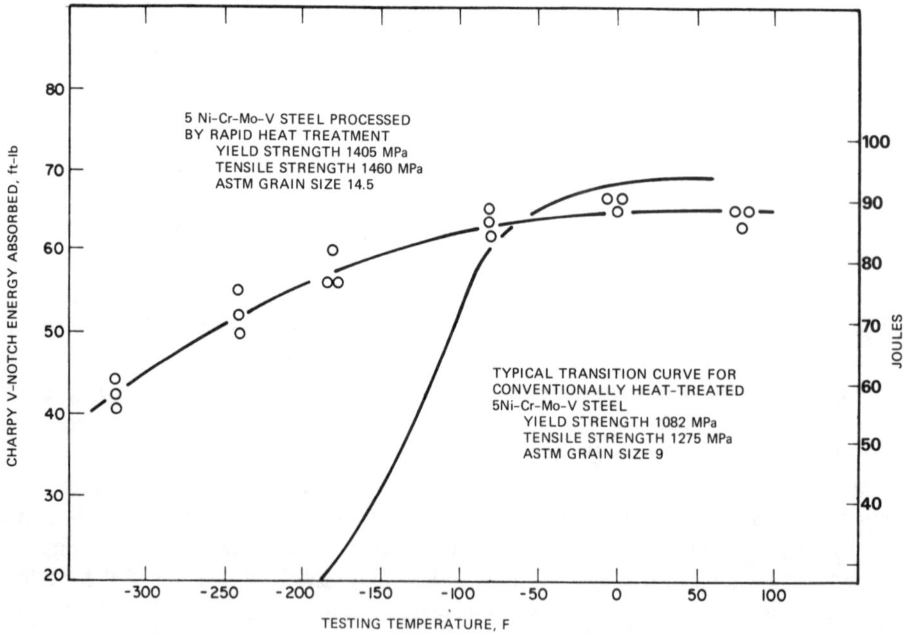

Figure IX-7. Effect of grain size on transition curves for rapidly heat-treated and conventionally heat-treated steels. (L.F. Porter, S.J. Manganello, D.S. Dabkowski, and J.H. Gross, *Met. Eng. Q.*, 6:17 (Aug., 1966), by permission.)

and it was concluded that it was due to the promotion of cross slip of screw dislocations. This characteristic of Ni, unique among commercial alloying elements, accounts for its use in high-strength steels in concentrations from 1 to 10%. At a constant grain size and Ni content, the transition temperature increases with increasing yield strength. Therefore, after possibilities of grain refinement have been exhausted, the only way to maintain a low transition temperature is to increase the Ni content as the yield strength increases.

Void-nucleating particles have no effect on the ductile to brittle transition. Their effect is to lower the energy absorbed in ductile fracture.

The embrittling effects of adsorption phenomena have been discussed previously.

In summary, to produce a high-strength steel of maximum toughness with maximum energy absorption in ductile fracture and a low fracture-mode transition,

1. Void-nucleating sites must be minimized; that is, the steel must be clean.
2. The matrix must have the maximum ductility, which is obtained by (a) minimizing Mn and Si, (b) avoiding massive carbides, and (c) providing the proper size, shape, and distribution of strengthening particles.
3. The grain size must be reduced to the maximum extent possible.
4. Ni must be added to reduce the fracture transition to the desired temperature.

Tough, strong steels must be melted to the lowest obtainable S, N, and O contents. This can be done by

1. Electric furnace melting
2. Vacuum C deoxidation
3. Vacuum arc remelting or electroslag remelting

Strong deoxidizers such as Al or Ti should be avoided.

CHAPTER

TEN

SURFACE-HARDENING PROCESSES

X-1. INTRODUCTION

Surface-hardening processes for steel are of two general types, those which involve changes in composition of the steel at or near the surface (carburizing and nitriding) and those which obtain hardness by phase transformation through rapid heating and cooling of the surface layers (flame hardening, induction hardening, and laser hardening). The objectives of all of these are the same: to obtain a hard, wear-resistant surface on a tougher core and to impart a favorable pattern of residual stresses, with compressive stresses at the surface, to increase life under fatigue loading.

X-2. CARBURIZING

Carburizing is the most widely used surface-hardening technique[1]; it has remained nearly unchanged since the introduction of Al-killed steels in the early 1930s allowed the use of reasonably high carburizing temperatures without the danger of excessive grain growth. Carburizing is used in the production of bearing races, gears, bushings, rock-drilling bits, etc., wherever a high-carbon hard case and a low-carbon tough core are desirable. The steels commonly used in carburizing contain from 0.10 to 0.25%C. Typical grades are AISI 1018, 1117, 4023, 4118, 5015, and 8620.

Carburizing is usually done in the temperature range 900–930°C with gas having

[1] *Carburizing and Carbonitriding,* Am. Soc. Met., Metals Park, Ohio, 1977.

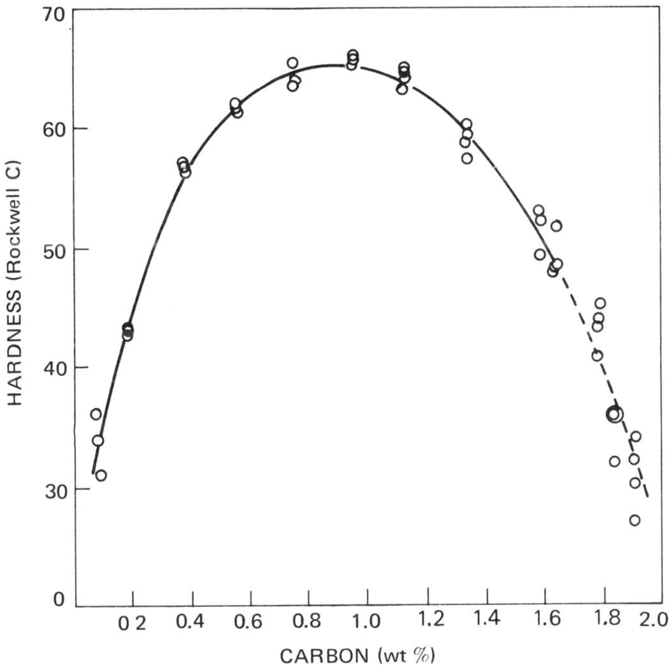

Figure X-1. Hardness of Fe–C alloys, brine-quenched from austenite. (A. Litwinchuk, F.X. Kayser, H.H. Baker, and A. Henkin, *J. Mater. Sci.*, 11:1200 (1976), copyright Chapman and Hall, Ltd., London.)

a carbon potential of about 0.9%C in austenite. The principal carburizing reaction is

$$CO + H_2 \rightleftarrows \underline{C} + H_2O$$

where $\underline{C}$ is carbon in solid solution in austenite. Most commercial carburizing processes are not done under conditions approaching equilibrium, but in two parts, a carburizing period and a diffusion period. In the first, the gas has a higher carbon potential than is desired in the finished part. In the diffusion period the atmosphere is held at the desired chemical potential while the C diffuses into the part.

In carbon steels, the maximum hardness of the case after quenching is reached at a C content of 0.8%. If this C content is exceeded, the hardness of the as-quenched case decreases because of the retention of austenite (Fig. X-1). The carbon content at which maximum hardness is attained can vary with alloy content because of the tendency for more austenite to be retained (Fig. X-2). The difficulties with current carburizing practices are low hardness in the outer case because of retained austenite (Fig. X-3) and the time required.

The carburizing cycles currently in use are summarized in Fig. X-4. In the large majority of applications, cycle no. 1 is used, often without a tempering operation. In whatever cycle is used, good practice requires that the highest C content should reside at the surface. During quenching, the transformation from austenite to martensite

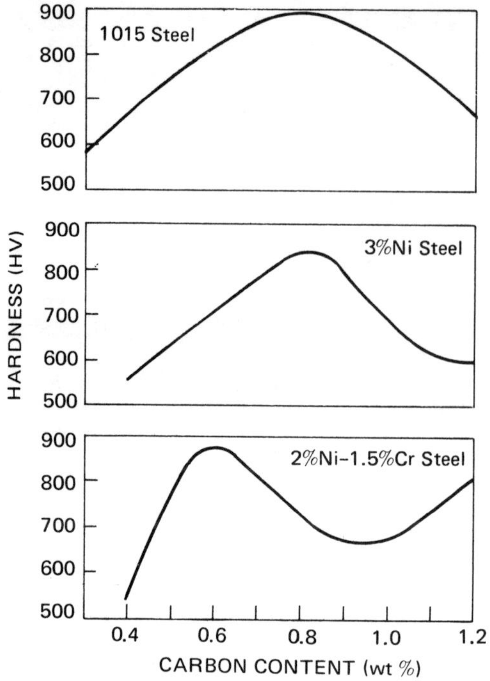

Figure X-2. Variation of maximum case hardness with surface carbon content for three carburizing grades of steel. (C. Dawes and D.F. Tranter, *Met. Technol.*, 1:397 (1974), by permission.)

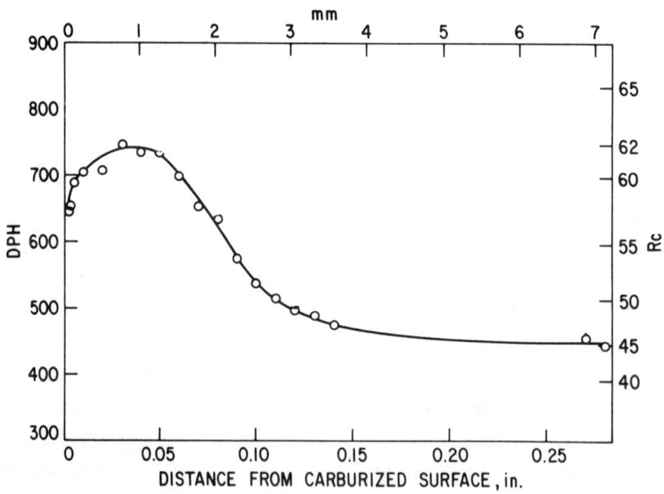

Figure X-3. Microhardness traverse through the carburized case of 8620 steel pack carburized 16 h at 925°C, then tempered 1 h at 205°C. (R.A. Grange, *SAE Trans.*, 82:556 (1973), by permission.)

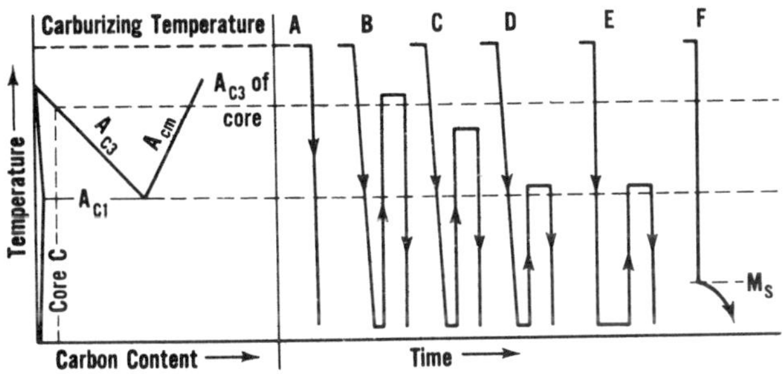

Treatment	Case	Core
A	Unrefined. Solution of excess carbide favored. Austenite retained. Distortion low to medium.	Unrefined but hardened.
B	Refined. Solution of excess carbide favored. Austenite retention promoted in highly alloyed steels.	Refined. Maximum core strength and hardness.
C	Refined. Some solution of excess carbide.	Partially refined. Stronger and tougher than Treatment D.
D	Refined. Excess carbide not dissolved.	Unrefined. Soft and machinable.
E	Refined. Some solution of excess carbide. Austenite retention minimized.	Low hardness. High toughness. Machinable.
F (Interrupted quench; marquenching)	Unrefined. Solution of excess carbide favored. Austenite retained. Distortion minimized.	Fully hardened.

Figure X-4. Schematic representation of treatments subsequent to carburizing and the case and core characteristics obtained. Adequate hardenability and fine-grained steel are assumed. (*Carburized Nickel Alloy Steels*, INCO Ltd, New York, 1975, by permission.)

begins at the boundary between the case and the core and moves outward toward the surface of the piece. Because of the constraints imposed by the core, the volume expansion accompanying the transformation leaves the surface layers in compression. However, if the carbon gradient takes the form shown in Fig. X-5, the outer layers can be in tension because they are not the last to transform.

The presence of retained austenite at the surface can be harmful in bearing surfaces. The unstable austenite can transform under stress, and the transformation causes a slight upheaval of the surface. This roughening of a precisely ground surface can lead to early bearing failure. Low-carbon steels high in Ni, which are desirable for tough carburized parts, are especially susceptible to retention of austenite in the case.

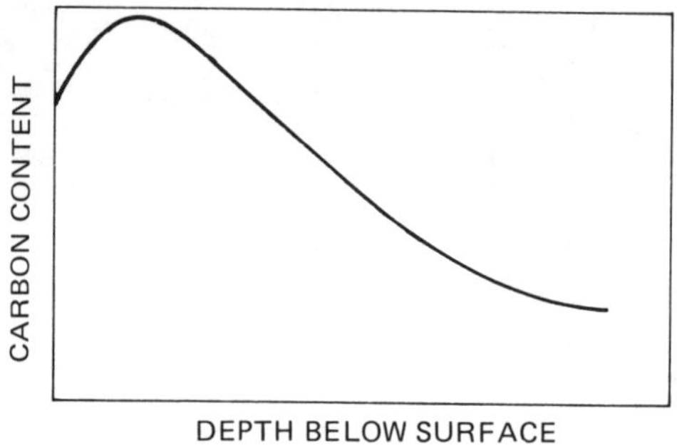

Figure X-5. Carbon profile of an incorrectly carburized steel.

Grange[1] has devised a procedure to circumvent the difficulties of low case hardness and long carburizing time. It is summarized by Fig. X-6. Carburizing is done at about 1035°C, about 115°C above the normal range, and for a shorter period; then the part is oil quenched. The second step in the four-step process is a temper at about 370°C to convert retained austenite to bainite. The third step is rapid austenitizing to just above the A_3 temperature of the steel, followed immediately by an oil quench. This step would probably require induction heating. The fourth and final step is a temper in the range 175–225°C. The result is a case of maximum hardness, between 800 and 900 DPH, as shown in Fig. X-7, combined with very small grains in both case and core. The process is applicable to less expensive Si-killed steels, and the toughening provided by grain refinement allows use of Ni-free steels.

The so-called vacuum carburizing process involves heating parts in an evacuated chamber to a temperature between 950 and 1050°C, which is somewhat higher than is used in conventional carburizing. When the parts have reached the desired temperature, natural gas, methane, or propane is admitted to a pressure between 7 and 55 kPa. The gas flow can be continuous, or repeated filling and evacuation can be used. The carbon potential is controlled by the partial pressure of the gas, which dissociates at the steel surface:

$$Fe + CH_4 \rightleftarrows Fe\underline{C} + 2H_2$$

The carburizing stage is followed by a diffusion period, during which the furnace is evacuated. Because of the high temperature used for carburizing, it is desirable to cool the parts by admitting nitrogen, evacuating again, reheating to a low temperature in the austenite range for grain refinement, then oil quenching. The advantages of the process are clean parts, elimination of the atmosphere generator, and short cycles. The disadvantages are high cost of equipment and nonuniform case depth when a part is unfavorably positioned relative to the gas flow.

Grube and Gay[2] report on glow-discharge carburizing, another technique for shortening the time spent in carburizing. This process takes place in a reduced pressure (0.1 to 2.7 kPa) of CH_4. A potential of a few hundred volts is applied between the part

[1] R.A. Grange, *SAE Trans.*, 82:556 (1973).
[2] W.L. Grube and J.G. Gay, *Met. Trans. A*, 9A:1421 (1978).

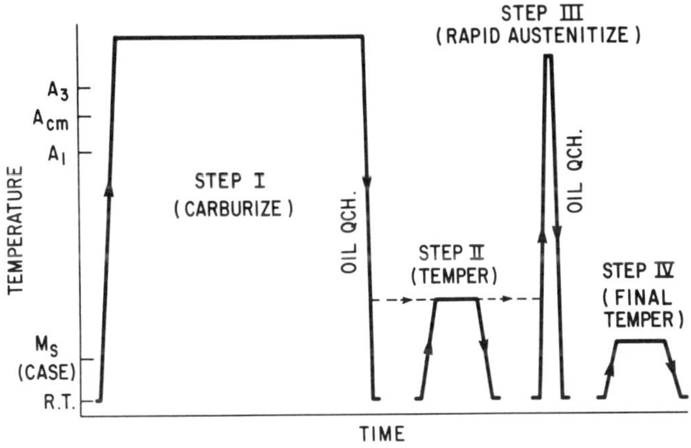

Figure X-6. Schematic representation of procedure to obtain maximum case hardness and very fine grain size. (R.A. Grange, *SAE Trans.*, 82:556 (1973), by permission.)

to be carburized (cathode) and an anode. This creates a thin envelope of plasma that closely follows the contours of even the most irregular part, a plasma being a gas consisting of ions, electrons, and both charged and neutral atoms and molecules. It is a good conductor of electricity. Figure X-8 shows results obtained by this means at 1050°C, as compared to vacuum carburizing at the same temperature and to conventional atmosphere carburizing at 900°C. The C potential of the plasma is very high, so the short carburizing stage (~10 min) must be followed by a longer diffusing period (~30

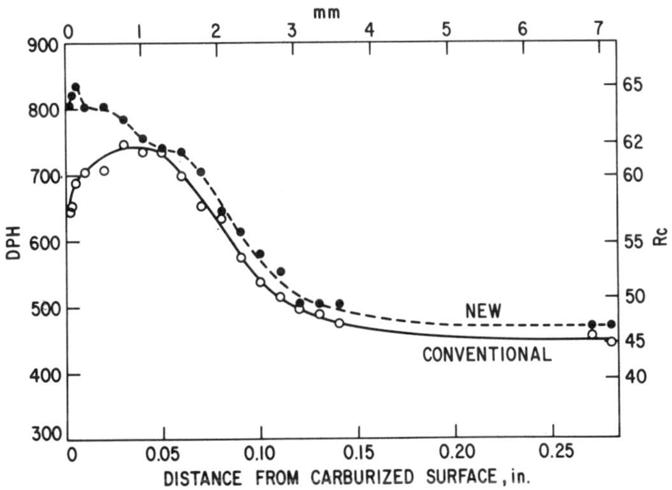

Figure X-7. Microhardness traverse through the carburized case of 8620 steel, pack-carburized at 925°C for 16 h, then tempered at 205°C for 1 h. The comparison is between the procedure outlined in Fix. X-6 and conventional oil quench and temper. (R.A. Grange, *SAE Trans.*, 82:556 (1973), by permission.)

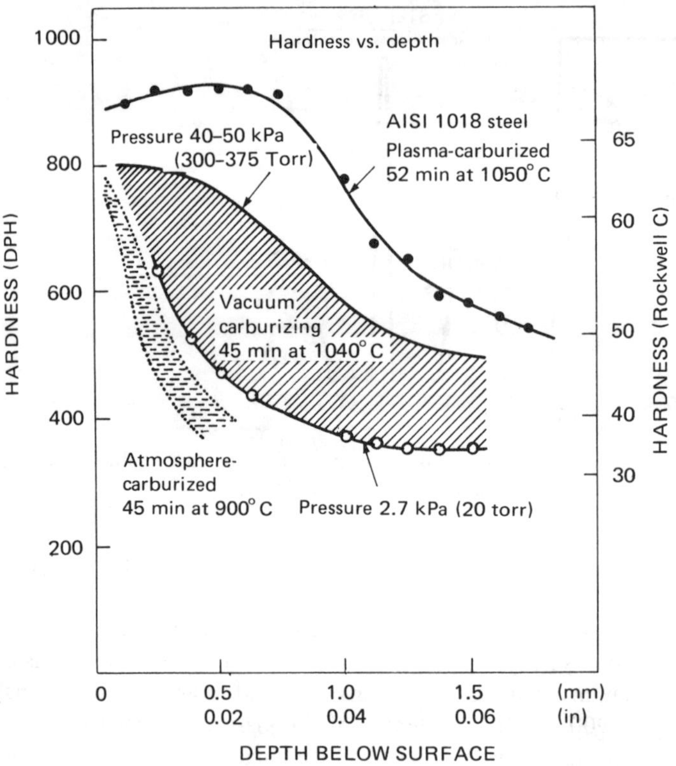

Figure X-8. Hardness profiles through cases on AISI 1018 steel, comparing atmosphere carburizing at 900°C, vacuum carburizing at 1040°C, and plasma carburizing at 1050°C. (W.L. Grube and J.G. Gay, *Met. Trans. A*, 9A:1421 (1978), by permission.)

min). Hardness profiles after this combined treatment are shown in Fig. X-9. The high rate of carburizing arises from a high rate of supply of C atoms to the surface of the austenite. There is no change in rate of diffusion of C in austenite. One major advantage of glow-discharge carburizing is the exceptional uniformity of case, even on reentrant surfaces and even on surfaces shielded from the anode. The consumption of methane is very small as compared to conventional atmosphere carburizing.

Pack carburizing, the oldest method, employing a solid carburizing medium, is still used in industry for short runs or where only a portion of a piece need be carburized. It is also a convenient laboratory tool when atmosphere carburizing furnaces are not available.

X-3. CARBONITRIDING

Despite its somewhat misleading name, carbonitriding is simply a variation of carburizing in which some N is absorbed in austenite from ammonia added to the carburizing gas.[1] It is used when a hard, thin case, not more than 0.75 mm thick, is desired. The N content of the case increases hardenability, thereby allowing oil quenching of C steels and minimizing distortion. The N content also increases resistance to tempering. The surface of the case can contain about 0.8%C and 0.5%N. Usually, steels to be carbonitrided contain less than 0.25%C. The temperature employed is about 815°C and the time, which will vary with the desired case depth, can be about 30 min.

The problem of retained austenite is greater for carbonitrided than for carburized steels, because of the stabilization of austenite by N. The austenite content can be reduced by cooling to −40 to −100°C, or even by removing surface layers by grinding. The impact resistance of carbonitrided cases is less than that of carburized cases, so use of carbonitrided parts in applications involving impact should be avoided.

X-4. NITRIDING

There are two general types of nitriding processes in common use. One is based upon the development, on unalloyed low-carbon steel, of a thin hard layer of ϵ iron nitride

[1] *Carburizing and Carbonitriding,* Am. Soc. Met., Metals Park, Ohio, 1977.

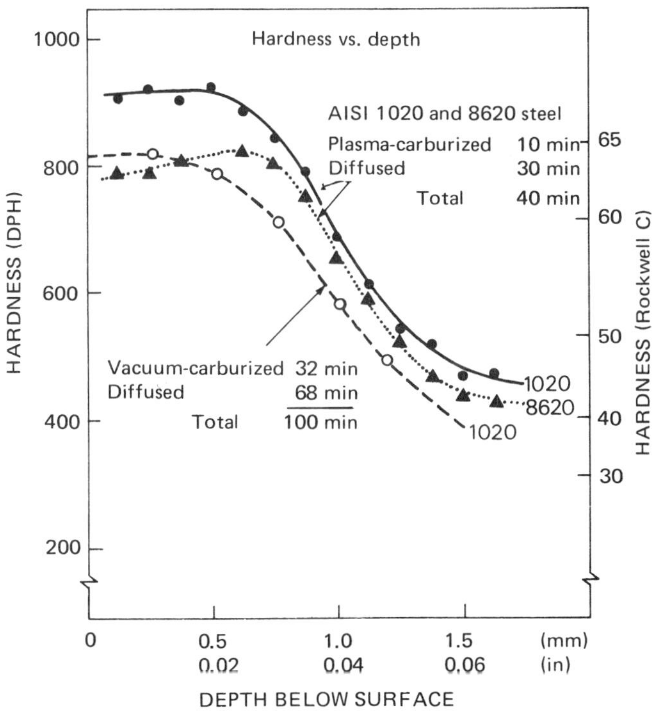

Figure X-9. Hardness profiles through carburized cases on AISI 1020 and 8620 steels after plasma carburizing for 10 min and diffusing for 30 min. (W.L. Grube and J.G. Gay, *Met. Trans. A*, 9A:1421 (1978), by permission.)

(Fe_3N, Fig. X-10) or carbonitride. The other depends for hardening upon interactions between N and solute atoms, principally Al, Cr, and Mo, in solution in ferrite. Both processes require that the steel be ferritic, so the nitriding temperatures are restricted to the range 500 to 575°C.

Surface hardening of C steels by formation of an ϵ nitride or carbonitride compound layer can be accomplished by heating in[1]

1. NH_3
2. NH_3-N_2 mixtures
3. NH_3-endothermic gas mixture (endothermic gas is approximately 40%N_2, 20%CO, and 40%H_2, with traces of CO_2)
4 NH_3-N_2-CO_2 mixtures

The mechanical properties and corrosion resistance of the compound layer are not affected by C content, but the presence of CO or CO_2 in the nitriding atmosphere accelerates the nitriding reaction. Any CO_2 present in the gas initially, or formed by the reaction $2CO \rightarrow \underline{C} + CO_2$, reacts with H_2 created by $2NH_3 \rightarrow 2\underline{N} + 3H_2$ to increase the rate of dissociation of NH_3 at the iron surface.

For maximum resistance to scuffing and galling at lowest cost, the compound layer on C steel should be about 0.018 mm thick. This can be obtained in about 2 h at 570°C. The hardness of the ϵ "Fe_3N" layer is about 1100 on the Knoop scale, which corresponds to a slightly larger number on the Vickers scale.

In addition to providing resistance to wear, the compound layer also provides

[1] C. Dawes, D.F. Tranter, and C.G. Smith, *Met. Technol.*, 6:345 (1979).

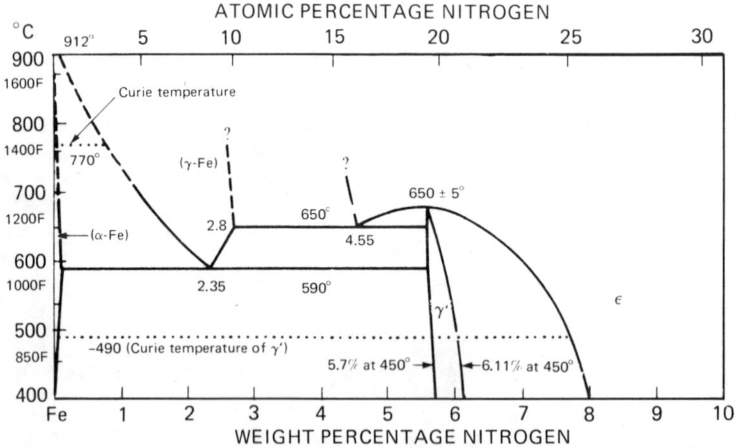

Figure X-10. The iron–nitrogen system. (*Metals Handbook*, 8th ed., vol. 8, Am. Soc. Met., Metals Park, Ohio, 1973, by permission.)

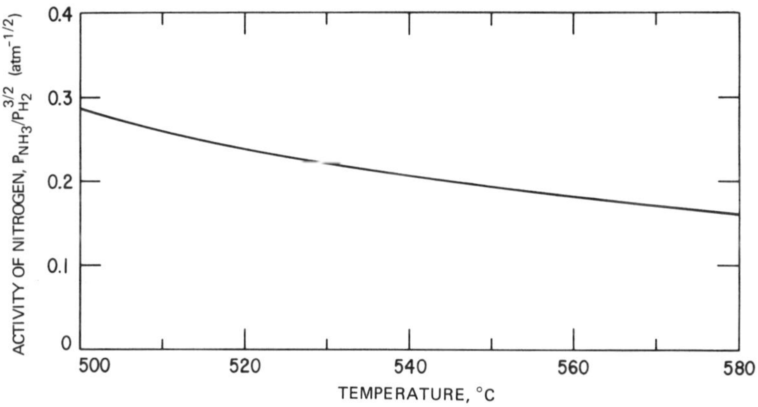

Figure X-11. Activity of nitrogen in equilibrium with Fe_4N. (H.E. Knechtel and H.H. Podgurski, *U.S. Patent 3,339,085*, Aug. 27, 1968.)

corrosion resistance in moist atmospheres, or in NaCl solutions, when the solutions are not acidic.

The improved resistance to fatigue obtained by this nitriding process is not obtained from the compound layer, but from the diffusion of N into the underlying ferrite. Residual stresses can be produced by gradients of plastic strain, temperature, or chemical concentration. By diffusing N into α Fe it is possible to achieve a residual compressive stress at the surface of a part of about 170 MPa.[1]

The more familiar type of nitriding in NH_3 or NH_3–H_2 mixtures is also restricted to the temperature range from 500 to 575°C, in the ferrite region. It is commonly done with steels having a composition near

C	Mn	Si	Cr	Mo	Al
0.40	0.60	0.26	1.65	0.30	1.0

Because of the low temperature, diffusion of nitrogen is slow, so the process often requires 30 h or more. The hardness obtained can be as high as 1100 DPH, giving parts that can be used when very high wear resistance is needed.

Aside from the time required, the greatest problem encountered in nitriding is the formation of "white layer" at the surface. This layer is mostly composed of Fe_4N, and it tends to crack. It must be removed by grinding or pickling before the part can be put into service. The formation of this layer can be prevented by using a mixture of NH_3 and H_2 with an activity of N less than that required to form Fe_4N at the nitriding temperature (Fig. X-11). Nitriding can begin with a higher activity than indicated in Fig. X-11 in order to minimize the nitriding time; then the NH_3/H_2 ratio is reduced before Fe_4N is nucleated at the surface. The relation between N activity and the NH_3/H_2 ratio is shown in Fig. X-12. Although calling for a higher degree of control

[1] O. Richmond, W.C. Leslie, and H.A. Wriedt, *Trans. Am. Soc. Met.*, 57:294 (1964).

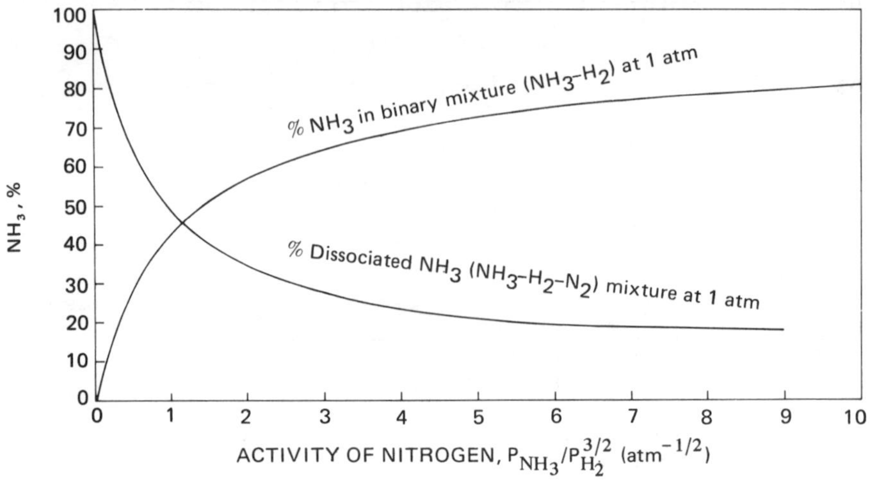

Figure X-12. Nitriding gas compositions corresponding to nitrogen activities. (H.E. Knechtel and H.H. Podgurski, *U.S. Patent 3,399,085*, Aug. 27, 1968.)

than is usually exercised in commercial practice, this procedure will give any desired depth of case, high surface hardness, and no surface damage.

Research into the mechanisms involved in the nitriding of alloy steels[1–5] has improved greatly our understanding of the process. The attainment of very high hardness requires that an alloy steel for nitriding contain Al but if one is looking at mechanisms an alloy steel is excessively complicated. It is better to examine the process of nitriding of an Fe–Al binary alloy. Figure X-13 shows the results obtained when an Fe-2%Al alloy, previously cold rolled to 0.25 mm (50% reduction in thickness), was nitrogenated at 500°C in an 11%NH_3-89%H_2 atmosphere. The nitrogen potential was below that required to form Fe_4N. If the alloy had been in the annealed condition, difficulties in nucleation would have prevented the formation of AIN.[4] Nitrogen sorption isotherms were obtained by equilibration with NH_3-H_2 atmospheres and reduction by H_2. The point Y in Fig. X-13 represents the concentration of N combined as AlN. The uncertainty shown about this point arises from the uncertainty in determination of the Al content of the alloy. The point P is the limiting N content obtained on reduction with H_2; that is, the amount of N represented by the distance Y-P could not readily be removed by H and was considered to be present at Fe–AlN interfaces. Point X is the intercept at zero activity of N of the sorption points at 400,

[1] D.L. Speirs, W. Roberts, P. Grieveson, and K.H. Jack, *Proc. 2nd ICSMA*, 2:601 (1970).
[2] H.H. Podgurski and R.A. Oriani, *Physical Chemistry in Metallurgy* (Darken Conf.), United States Steel Corporation, Pittsburgh, 1976, p. 421.
[3] H.H. Podgurski, R.A. Oriani, and F.N. Davis, *Trans. AIME*, 245:1602 (1969).
[4] H.H. Podgurski and H.E. Knechtel, *Trans. AIME*, 245:1595 (1969).
[5] J.C.M. Li and Y.T. Chou, *Trans. AIME*, 245:1606 (1969).

450, and 500°C. Nitrogen in excess of point X is held in the strain fields of dislocations, and that concentration of N represented by the distance P–X is presumed to be relatively weakly absorbed at Fe–AlN interfaces.

The conclusion reached from these studies was that N was strongly bound in a chemisorbed N monolayer at the ferrite-AlN interface, and less strongly bound at a variety of sites in the stress fields of the AlN particles and of dislocations. These inhomogenous stress fields enhance the lattice solubility of N and cause the enthalpy of solution of N to vary.[1]

One of the most striking features of nitriding is a kind of "bootstrap" process. Dislocations seem to be required for the nucleation of AlN particles. At temperatures

[1] J.C.M. Li and Y.T. Chou, *Trans. AIME,* 245:1606 (1969).

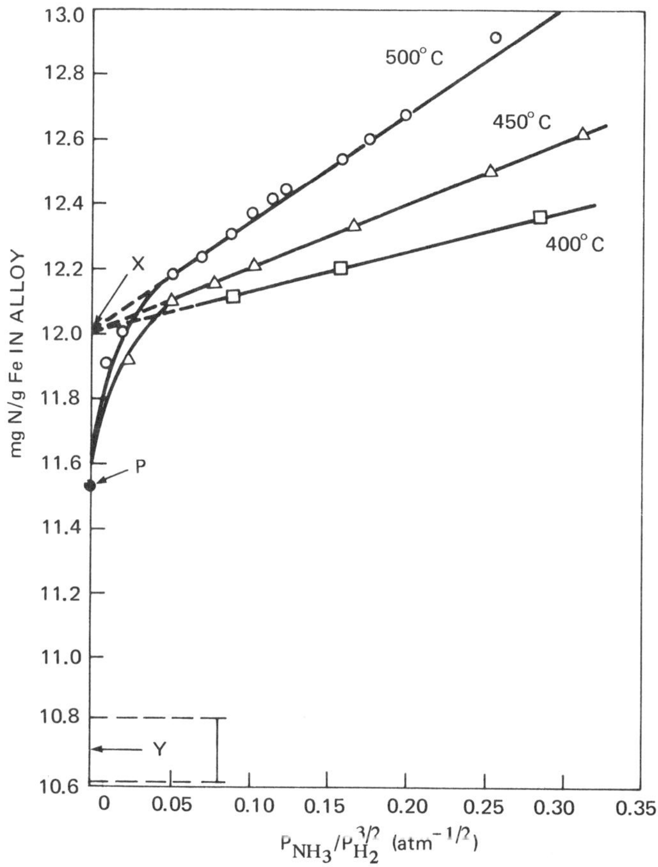

Figure X-13. Nitrogen sorption isotherms for nitrided Fe–2%Al alloy; details in the text. (H.H. Podgurski, R.A. Oriani, and F.N. Davis, *Trans. AIMI,* 245:1602 (1969), by permission.)

near 500°C, the growth of AlN particles strains the ferrite lattice to such an extent that dislocations are created, thereby providing more nucleation sites. The great hardening obtained by commercial nitriding is due to a combination of small (5 to 15 nm) closely spaced particles, a very high density of dislocations (at least 10^{10} cm^{-2}), and a high concentration of N associated with dislocations and with Fe–AlN interfaces.

Although it is not a surface-hardening phenomenon, it seems appropriate to add here a brief discussion of the drastic strengthening that can be obtained in dilute alloys of Ti or Nb in α Fe.[1,2] "Interstitial-free" alloys can be prepared containing 0.2%Ti or 0.3%Nb. As cold-rolled and annealed sheets, these have excellent formability. After forming into complex shapes they can be nitrogenated (internally nitrided) in NH_3–H_2 mixtures at temperatures below 590°C to convert all the Ti or Nb to the nitride. To remove excess N, specimens can be treated in dry H_2 at 575 to 600°C. With a volume fraction of precipitate of only 0.002, the yield strength of a 0.19%Ti–0.39%Mn alloy increases from 83 to 630 MPa. The total elongation in 25 mm decreases from 30% to 15%. The nitrogenation presumably produces TiN molecules or clusters of a few molecules, for no precipitates can be seen by electron microscopy until the specimens have been aged at 850°C for the Ti alloy or 600°C for the Nb alloy. The strengthening is consistent with a model based on interactions of screw dislocations with the strain fields of coherent particles,[3]

$$\Delta\sigma = 2G\,|\epsilon|^{3/2}\,(R/b)^{1/2}\,f_v^{1/2} \tag{X-1}$$

where G = shear modulus of the matrix
ϵ = misfit of the particle in the matrix
b = magnitude of Burgers vector of the dislocation
f_v = volume fraction of precipitate
R = particle radius

Whether this process can be employed in the production of very high-strength parts formed from sheet metal will depend upon costs.

X-5. SURFACE-HARDENING PROCESSES THAT DO NOT CHANGE THE STEEL COMPOSITION

There are at least three processes which harden steel surfaces by rapid heating and quenching: induction hardening, flame hardening, and laser hardening. The first two are described in detail in the *Metals Handbook.*[4] Case depths from 0.25 to 6.0 mm can be obtained by induction heating, with the depth being regulated by the current frequency and duration of heating. Flame hardening can be used to produce cases

[1] D.H. Jack, *Acta Metall.*, 24:137 (1976).
[2] L.J. Cuddy and H.H. Podgurski, *Met. Trans. A.* 8A:245 (1977).
[3] V. Gerold and H. Haberkorn, *Phys. Status Solidi,* 16:675 (1966).
[4] *Metals Handbook,* 8th ed., vol. 2, Am. Soc. Met., Metals Park, Ohio, 1964, p. 167.

from 0.75 to 6 mm in depth, or to through-harden parts as large as 75 mm in diameter. Both processes can be automated, with spray-quenching after heating.

Laser heating differs considerably from the other two in rate of heat input and the lack of a quenching step, the quench being supplied by the mass of unheated metal. In laser heat-treating, melting of the steel surface usually must be avoided to minimize distortion and to prevent cracking. When a laser is used for surface heating a lens is used to produce a defocused spot on the steel surface. The surface must be coated with a material such as manganese phosphate to aid in the absorption of energy from the beam, for at the wavelength of radiation from a CO_2 laser (10.6 μm), 80 to 90% of the incident radiation can be reflected.[1] The spot size for heat treating with a 1-kW laser can vary from about 0.5 to 25 mm.

If melting does not occur, the structure of a laser-heated steel resembles the heat-affected zone of a weld. For an AISI 1045 steel with an annealed initial structure, the microstructure varies from bainite + ferrite at the surface of the heated spot to pearlite + ferrite at the unaltered metal. Because of the very short time of heating, the structure is highly dependent on the starting structure. The depth of hardening is not great, about 0.25 mm, if melting is avoided. This depth can be increased by increasing the laser power and spot size while maintaining the energy and power density at levels that will not produce melting.

For depth of hardening of a heat-treatable steel, Steen and Courtney[2] propose the relation

$$Y = -0.11 + 3.02\, P/\sqrt{D_b V} \qquad \text{(X-2)}$$

where Y = depth of hardening (mm)
P = laser power (W)
D_b = incident beam diameter (mm)
V = traverse speed (mm/s)

but with considerable scatter of experimental data. At a constant value of $P/\sqrt{D_b V}$, the depth of hardening can vary by a factor of 2.

Laser heat-treating has the advantages of precise control over the areas to be hardened, an ability to harden reentrant surfaces, very high speed of hardening, no separate quenching operation, little or no distortion, and adaptibility to automation, coupled with the disadvantages of very shallow hardening and variation in depth of hardening.

[1] G.H. Harth, W.C. Leslie, V.G. Gregson, and B.A. Sanders, *J. Met. Sci,*. 28:5 (Apr. 1976).
[2] W.M. Steen and C. Courtney, *Met. Technol.*, 6:456 (1979).

CHAPTER

ELEVEN

STEELS FOR MAGNETIC AND ELECTRICAL APPLICATIONS

Steels for magnetic and electrical applications are of three types; in order of increasing cost and complexity of manufacture, these are

1. Low-carbon steels
2. Nonoriented Si steels
3. Oriented Si steels

They are sold on the basis of core loss in the Epstein test.[1] For this test, at least 12 strips, 3 cm wide and 28 cm long, are cut from the sheets to be tested, two sets in the rolling direction and two sets in the transverse direction. The total mass of specimens is usually 2 kg. The strips are arranged as the four sides of a square, with their ends overlapping and clamped at the corners. Four solenoids surround the four sides of the square magnetic circuit. A wattmeter measures the power consumed in the test specimen and in the secondary circuit. The specific core loss is reported in watts per unit mass, from the equation

$$\frac{\mathrm{W}}{\mathrm{kg}} = \left(\mathrm{W} - \frac{E^2}{R}\right)\frac{l}{0.0235m} \tag{XI-1}$$

where m = total specimen mass in kg
l = actual strip length
W = watts read on a wattmeter
E = rms volts
R = parallel resistance of wattmeter potential circuit

[1] ASTM A343-69.

Table XI-1 Magnetic properties, 1.5 T (15 kG), cold-rolled sheet, magnetic lamination steel–type 25[1]

Actual micrometer thickness		Maximum core loss 60 Hz		Typical ac peak Permeability
in	mm	W/lb	W/kg	
0.016	0.41	3.3	7.3	1900
0.018	0.46	3.6	7.9	2000
0.021	0.53	4.1	9.0	2000
0.025	0.64	4.9	10.8	2000
0.028	0.71	5.6	12.3	2000

[1] ASTM standard A706-76.

The core loss is the amount of electrical energy dissipated as heat when the steel is magnetized by an alternating current. The test is done at fixed induction and frequency, as specified. These can be 1.5 or 1.7 tesla (15 or 17 kilogauss) and 50 or 60 Hz.

Carbon steel sheets for magnetic applications are made to low C content, less than 0.08%, and normal Mn content, 0.25 to 0.75%. Sometimes P is added to the steel to improve the punching properties of the sheet. These steels can be sold without guarantee of magnetic properties, or according to the requirements listed in Table XI-1.

Silicon is added to steels for magnetic or electrical applications because it is the most effective element for increasing the electrical resistivity of iron, thereby reducing eddy current losses in alternating current magnetization. The nonoriented grades of Si steel are low-carbon iron alloys containing from 0.80 to 3.5% (Si + Al) (Table XI-2).

Table XI-2 Core-loss types and maximum core loss, as sheared Epstein, at 1.5 T (15 kG)[1]

Approx alloy content (Si + Al), %	0.014 in (0.36 mm) Thick			0.0185 in (0.47 mm) Thick			0.025 in (0.64 mm) Thick		
		Max core loss			Max core loss			Max core loss	
	Core loss type	W/lb, 60Hz	W/kg, 50Hz	Core loss type	W/lb, 60Hz	W/kg, 50Hz	Core loss type	W/lb, 60Hz	W/kg, 50Hz
3.5	36F145	1.45	2.53	47F168	1.68	2.93	–	–	–
3.3	36F158	1.58	2.75	47F174	1.74	3.03	64F208	2.08	3.62
3.2	36F168	1.68	2.93	47F185	1.85	3.22	64F218	2.18	3.80
2.8	36F180	1.80	3.13	47F190	1.90	3.31	64F225	2.25	3.92
2.65	36F190	1.90	3.31	47F205	2.05	3.57	64F240	2.40	4.18
2.35	36F190	–	–	47F230	2.30	4.01	64F270	2.70	4.70
1.85	–	–	–	47F305	3.05	5.31	64F360	3.60	6.27
1.05	–	–	–	47F400	4.00	6.96	64F490	4.90	8.53
0.80	–	–	–	47F475	4.75	8.27	64F610	6.10	10.62

[1] ASTM standard A677-77.

Table XI-3 Core loss types and thicknesses for transformer cores[1]

ASTM core loss type	Thickness class		Maximum specific core loss[2] at 1.7 T (17 kG)			
			W/lb		W/kg	
	mm	in	60 Hz	50 Hz	60 Hz	50 Hz
27HO76	0.27	0.0106	0.76	0.58	1.68	1.27
30H083	0.30	0.0118	0.83	0.63	1.83	1.39
35H094	0.35	0.0138	0.94	0.71	2.07	1.57

[1] ASTM standard A725-75.
[2] Based on parallel-grain Epstein specimens, stress-relief annealed after shearing.

The Al addition, if any, is usually less than 0.50%. Nonoriented grades containing 0.80 to 1.85%Si are used in small motors for household appliances. Those containing 2.35 to 3.5%Si are used in refrigerator compressor motors, industrial motors, and television and radio transformers. The C content is less than 0.025%, the S is less than 0.025, and the Mn is 0.10 to 0.35%. Again, P can be added to improve punching qualities.

Oriented grades are made at only one Si content, 3.0 to 3.3%. This limit is set by the necessity for severe cold reductions in the manufacturing process. Higher Si contents would lead to higher scrap losses because of cracking during cold rolling. These materials are the most sophisticated mass-produced products of the steel industry. To refer to them as Si steels is a misnomer, for they are high-purity alloys of Fe and Si. They are made for two applications, transformer cores[1] and rotating electrical machinery.[2] They are sold in only three thicknesses, 0.27, 0.30, and 0.35 mm, with maximum core losses listed in Tables XI-3 and XI-4. These products are unique in that quality can be measured and expressed as a single number, the core loss per

[1] ASTM A725-75.
[2] ASTM A665-75.

Table XI-4 Core loss types and thicknesses for rotating machinery[1]

ASTM core loss type	Thickness class		Maximum specific core loss[2] 15 kG (1.5 T)			
			W/lb		W/kg	
	mm	in	60 Hz	50 Hz	60 Hz	50 Hz
27G053	0.27	0.0106	0.53	0.40	1.17	0.89
30G058	0.30	0.0118	0.58	0.44	1.28	0.97
35G066	0.35	0.0138	0.66	0.50	1.46	1.11

[1] ASTM standard A665-75.
[2] Based on parallel-grain Epstein specimens, stress-relief annealed after shearing.

unit mass. Therefore, competition in quality is severe, with sales going to those who can consistently produce sheets with the lowest core loss. The efficiency of electric power generation and transmission can be related directly to core loss in generator laminations and transformer cores; the importance of this property is self-evident.

The processing of grain-oriented Si steels is the most complicated for any product in the industry. Starting with the slab, the process is:

1. Reheat slab to 1315–1370°C
2. Hot roll, coil
3. Pickle
4. Cold roll about 40%, or → cold roll about 85%
5. Continuous anneal (to recrystallize) — omit
6. Cold roll about 40% — omit
7. Continuous anneal to decarburize
8. Coat with separating medium
9. Anneal at about 1150°C in dry H_2 in coils
10. Continuous anneal to remove coil set
11. Apply insulating coating
12. Epstein tests to classify

The high slab heating temperature is very important. Figure XI-1 shows the closed gamma loop in the Fe–Si system. The presence of C greatly enlarges the $\alpha + \gamma$ field, with the amount of austenite sharply increasing with C content (Fig. XI-2) at a

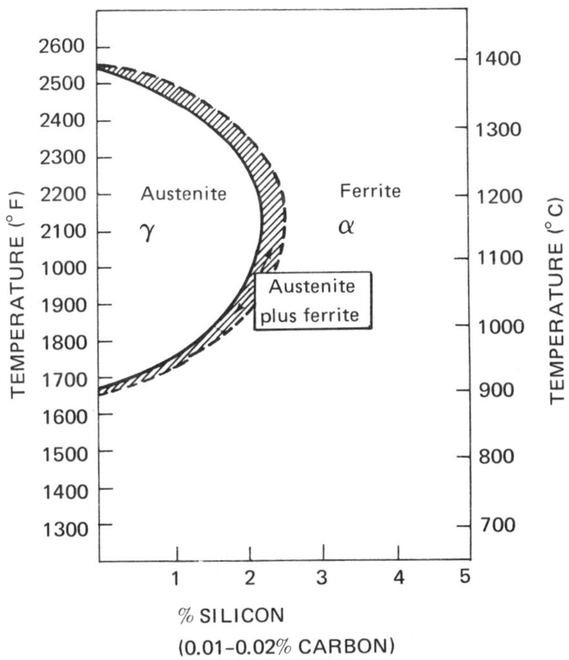

Figure XI-1. Section of the ternary Fe–Si–C diagram at 0.01–0.02%C. (*Making, Shaping and Treating of Steel*, 9th ed., 1971, U.S. Steel Corporation, by permission.)

constant Si concentration of 3.25%. The amount of austenite varies from zero at 0.02%C to 20% at about 0.05%C, at 1150°C. It is important that the hot rolling be done in the ferrite range, therefore the slab must be heated to a temperature at which all austenite disappears. Fortunately, the formation of austenite during cooling is quite sluggish, so that hot rolling can be completed in the single-phase region. If any transformation products of austenite are present in the steel after hot rolling, the texture and the final magnetic properties suffer.

The second reason for the high slab heating temperature is to ensure sufficient solution of second-phase particles so that these can precipitate later during the processing and serve as grain-boundary pinning agents. This pinning process leads to the secondary recrystallization that develops the desired {110}⟨001⟩ texture. There are numerous compounds that can serve this purpose, among them AlN, MnS, BN, TiC, Si_3N_4, VN, and SiO_2. The solubility products of the first three in 3.25%Si iron (bcc) have been determined:

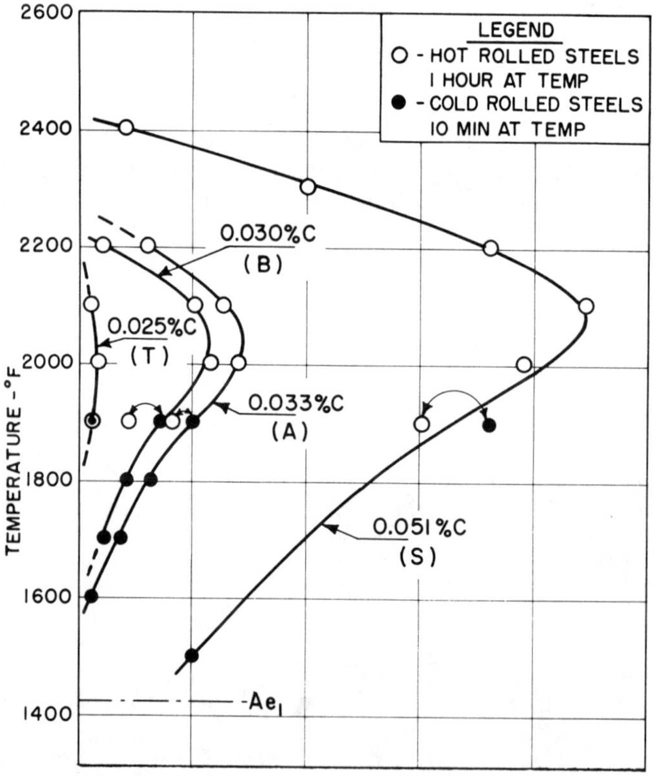

Figure XI-2. Amount of austenite in four 3.25%Si steels at high temperatures. (W.C. Leslie, R.L. Rickett, C.P. Stroble, and G. Konoval, *Trans. Am. Soc. Met.*, 53:715 (1961).)

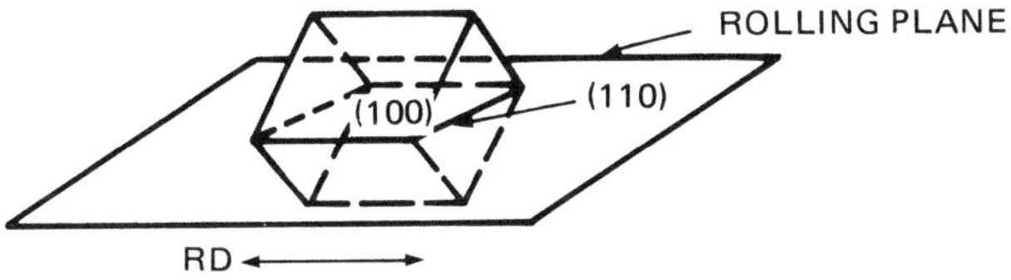

Figure XI-3. Illustration of cube-on-edge texture, {110}⟨001⟩.

$$\text{AlN}^1: \quad \log\,[\%\text{Al} \times \%\text{N}] = \frac{-11{,}900}{T} + 3.56 \qquad \text{(XI-2)}$$

$$\text{MnS}^2: \quad \log\,[\%\text{Mn} \times \%\text{S}] = \frac{-10{,}590}{T} + 4.092 \qquad \text{(XI-3)}$$

$$\text{BN}^3: \quad \log\,[\text{ppm B} \times \text{ppm N}] = \frac{-19{,}560}{T} + 15.75 \qquad \text{(XI-4)}$$

The control of variables in the two cold-rolling and two continuous-annealing steps is proprietary information. The separating medium is put on the sheet before coiling for the high-temperature anneal, to prevent sticking of the wraps of the coil.

The high-temperature anneal in dry H_2 has a two-fold purpose: (1) to develop a strong cube-on-edge {110}⟨001⟩ texture, and (2) to improve magnetic properties by removing C, N, S, O, and B.

The development of texture in this alloy is the single most important application of secondary recrystallization. During the high-temperature anneal, secondary recrystallization begins at about 815°C and is complete at about 870°C. Obviously, the process does not depend upon the solution of second-phase particles at these temperatures, but upon the ability of some few grain boundaries to break away from the pinning points. The anneal in H_2 is continued to high temperatures because the texture continues to improve with time at higher temperatures and the high temperature is required for efficient removal of impurities.

When grain-oriented Si steel is properly made, the grains are large, with each grain extending through the thickness of the sheet. The texture is very sharp, with at least 90% of the grains falling within ±10% of the ideal orientation (Fig. XI-3). As shown in Fig. XI-4, the easy direction of magnetization in bcc metals is [100]; [110] is intermediate; and [111] is the most difficult to magnetize. With grain-oriented Si steels, magnetic cores are made so that the main flux path is in the rolling direction [100]. In the development of this texture certain grains of the final (110) [001] orientation are able to grow at the expense of their neighbors.

[1] H.A. Wriedt, in *Physics of Materials (W. Boas Festschrift)*, Univ. of Melbourne, Australia, 1979, p. 325.

[2] H.A. Wriedt and H. Hu, *Met. Trans. A*, 7A:711 (1976).

[3] H.C. Fiedler, *Met. Trans. A*, 9A:1489 (1978).

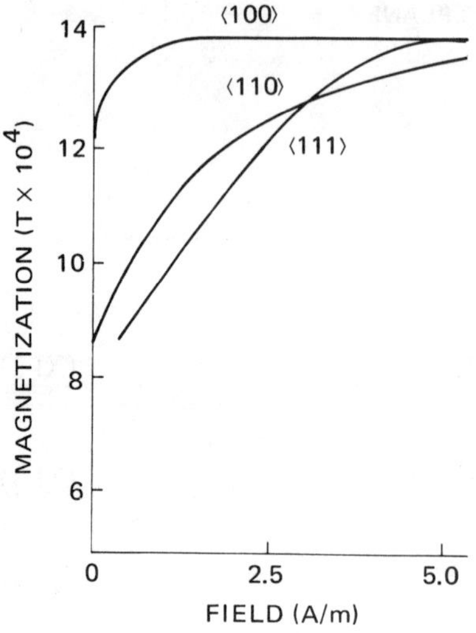

Figure XI-4. Magnetization field curves for iron, with the field applied parallel to ⟨100⟩ and to ⟨111⟩. (K. Honda, H. Masumoto, and S. Kaya, *Sci. Rep. Tohoku Imp. Univ.*, Ser. 1, 17:111 (1928).)

The texture of Si iron after primary recrystallization is weak, but from this weak annealed texture comes a very sharp cube-on-edge secondary recrystallization texture. The primary annealing texture contains a minor component in the (110) [001] orientation. During secondary recrystallization, growth rates decrease as the orientation of the secondary grains deviates from (110) [001]. In a very general sense, the cube-on-edge texture develops because grains in this orientation have a minimum free energy and therefore a maximum driving force for grain growth. The free energy of the grains consists of three parts:

1. Strain energy associated with dislocations
2. Surface energy of grain boundaries
3. Surface energy of gas–metal interfaces

Without understanding the detailed mechanism of growth selectivity preference for (110) [001] grains, we can say that there seems to be a minimum in the total of these three factors when grains are in the cube-on-edge orientation.

Grain-oriented Si steels were developed by Goss in 1934[1]—the {110} [001] texture is often called the Goss texture—and there has been a steady improvement in their properties since that time. The steel is made and processed in the best manner known, always aiming for the lowest possible core loss; then the fully-processed steel is graded for core loss by the Epstein test and sold on that basis. Neither the initial development nor the subsequent improvement of quality has been the result of basic research. The reduction of core loss has been the result of painstaking and expensive study of every phase of manufacture, from melting to the third continuous anneal. The latest improve-

[1] N.P. Goss, *U.S. Patent 1,965,559,* July 3, 1934.

ment[1,2] has come as a result of studies of the effect of residual stresses on core loss and the generation of desirable tensile stresses by proper selection of an insulating medium. Magnetic properties currently available at 0.3 mm thickness are:

Permeability at 800 A/m	1920
Induction at 800 A/m	1.92 T
Core loss at 60 Hz	
$B = 1.5$ T	0.81 W/kg
$B = 1.7$ T	1.45 W/kg

Magnetic aging was discussed in Chap. II. The practical effect of such aging is a gradual increase of core loss with time. This increase is reduced by the presence of Si, for, as described in Chap. IV, Si inhibits precipitation of Fe_3C. The metastable carbide (probably ϵ) is less effective in increasing the core loss. Magnetic aging is seldom of any importance in grain-oriented steels because of their very low carbon contents in the fully processed condition.

Much research and publicity has been devoted to cube-oriented $\{100\}\langle 001\rangle$ Si steels, which would have the advantage of two directions of easy magnetization, the rolling direction and the transverse direction, in the plane of the sheet. Such orientation can be developed by a variety of processes—directional solidification, cross rolling, and surface energy control—but unfortunately the advantage gained has not been sufficient to offset the cost of the additional processing.

[1] S. Taguchi, T. Yamamoto, and A. Sakakura, *IEEE Trans. Magn.*, MAG-10:123 (1974).
[2] G.W. Chen, *J. Magn. Mag. Mater.*, 7:308 (1978).

CHAPTER

TWELVE

STAINLESS STEELS

XII-1. INTRODUCTION

There is a wide variety of stainless steels, but they all have a common characteristic—they contain more than 11.5%Cr. Obviously, the existence of this important class of alloys depends upon the availability of Cr at reasonable cost. Stainless steels are used because of their corrosion resistance, oxidation resistance, and pleasing appearance, all of which are derived from their Cr content. They are customarily divided into five types:

1. Ferritic, containing 12-30%Cr and low C
2. Martensitic, with 12-17%Cr and 0.1 to 1.0%C
3. Austenitic, containing 17-25%Cr and 8-20%Ni
4. Duplex alloys, with 23-30%Cr, 2.5-7%Ni, and additions of Ti or Mo
5. Precipitation hardening alloys, which can have an austenitic or martensitic base, with additions of Cu, Ti, Al, Mo, Nb, or N

Streicher[1] has provided a summary of the history of these alloys, their current applications, and future prospects.

The physical metallurgy of these commercial steels is best approached through the three binary phase diagrams, Fe-Cr, Fe-Ni, and Ni-Cr, and the ternary diagram Cr-Fe-Ni.

[1] M.A. Streicher, *Stainless Steel 77*, Climax Molybdenum Co., Greenwich, Connecticut, 1977, p. 1.

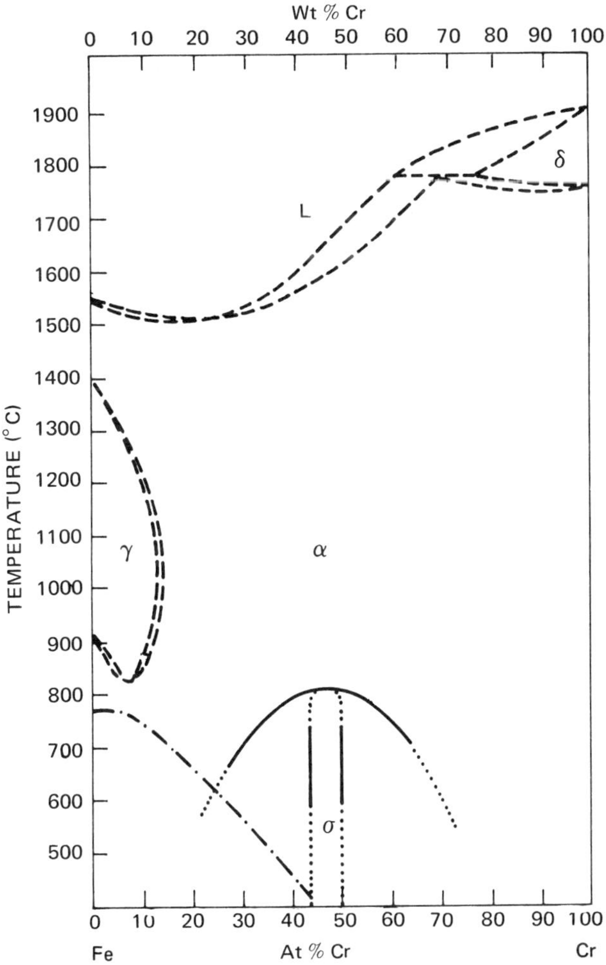

Figure XII-1. The iron–chromium system. (N. Yukawa, N. Hida, T. Imura, M. Kawamura, and Y. Mizuno, *Met. Trans.*, 3:887 (1972), by permission.)

XII-2. PHASE DIAGRAMS

The Fe–Cr system (Fig. XII-1) is the basis for all stainless steels, so we will begin there. Chromium stabilizes ferrite, leading to a closed gamma loop, with a maximum solubility of about 12%Cr in austenite at about 1000°C. Below about 830 and above about 1390°C austenite does not exist. Yukawa et al.[1] report the presence of a Cr-rich, high-temperature phase which they call δ, which should not be confused with "delta ferrite." This has a bcc structure, WO_3 type.

Addition of C or N, which are strong austenite stabilizers, expands the γ loop, thereby increasing the solubility of Cr in austenite. The $\alpha + \gamma$ region is especially expanded, 0.2%(C + N) extending it from 13 to 27%Cr.

[1] N. Yukawa, N. Hida, T. Imura, M. Kawamura, and Y. Mizuno, *Met. Trans.*, 3:887 (1972).

The lower portion of the Fe–Cr system is shown in Fig. XII-2. The σ phase, FeCr, is tetragonal, with 30 atoms per unit cell. This phase forms very slowly in the temperature range between 800 and 600°C. Its composition range in the binary alloy is small, about 46 to 53 at % Cr. At temperatures below about 520°C it decomposes to $\alpha + \alpha'$ by a eutectoidal transformation, but this reaction is also very slow. Sigma phase can be prevented from forming by rapid cooling, and when this is done the α' phase forms at temperatures below about 520°C by spinodal decomposition. The α' contains about 90%Cr (Fig. XII-2). Its precipitation leads to the so-called 475°C embrittlement. The α and α' phases, Fe-rich and Cr-rich ferrite, differ in lattice parameter by only about 0.2%.

The Fe-Ni system (Fig. XII-3) at temperatures above about 450°C is now generally accepted. The solubility of Ni in α Fe reaches a maximum between 400 and 500°C (Fig. XII-4). Nickel stabilizes γ so that the field extends at 30%Ni from about 500 to about 1450°C. The transformation of γ on cooling is very sluggish, so a high degree of undercooling can be obtained, which provides the driving force for the highly useful martensitic transformation in Fe-Ni alloys. The Ms temperature drops very sharply with increasing Ni content, from about 200°C at 20%, to 0°C at 30%, to about −220°C at 34%.

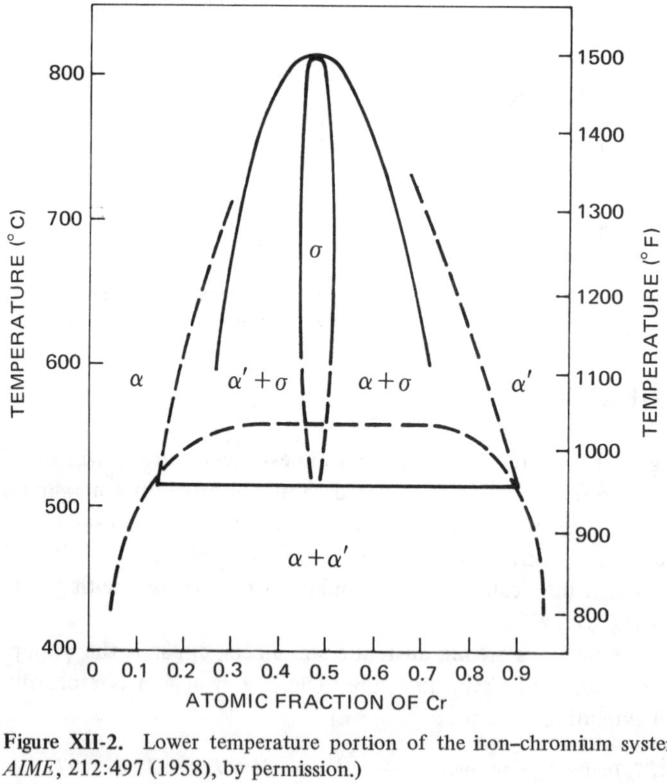

Figure XII-2. Lower temperature portion of the iron–chromium system. (R.O. Williams, *Trans. AIME*, 212:497 (1958), by permission.)

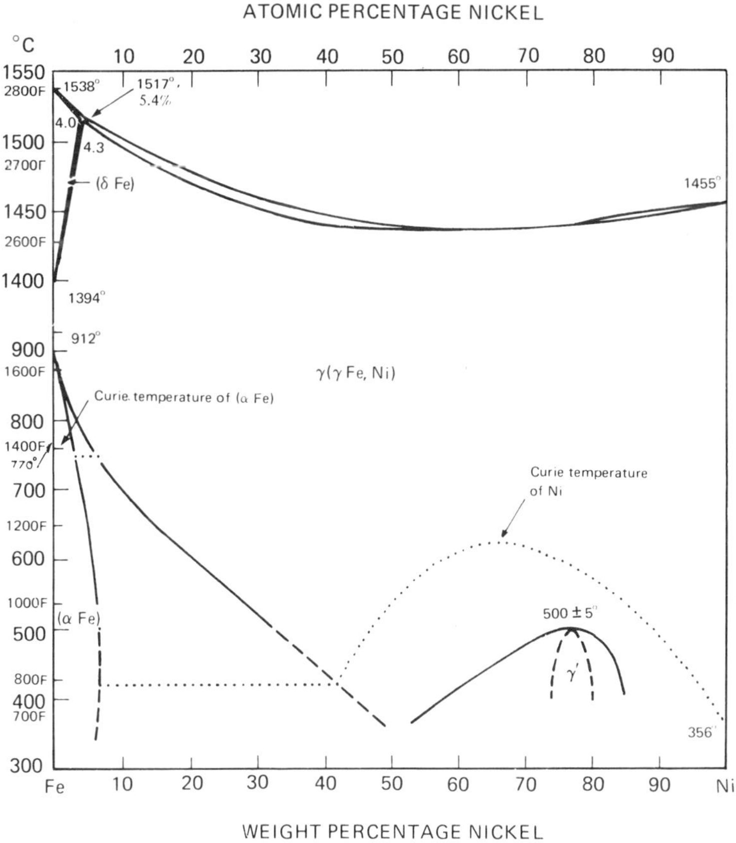

Figure XII-3. Iron–nickel phase diagram. (*Metals Handbook*, 8th ed., vol. 8, 1973, Am. Soc. Met., Metals Park, Ohio, by permission.)

However, the features of the Fe–Ni phase diagram below about 450°C are still very much in dispute. Evidence is accumulating[1] that the ordered face-centered tetragonal (fct) phase FeNi can form at temperatures below about 400°C; that is, the equilibrium phases in Fe-rich alloys at these temperatures are α and FeNi. The equilibria on the Ni-rich side are also in dispute. Heumann and Karsten[2] claim that the face-centered cubic (fcc) Fe–Ni solid solution decomposes eutectoidally at 345°C and 52%Ni to a mixture of $FeNi_3$ and α Fe. Much work remains to be done on this important system.

The Cr–Ni system shown in Fig. XII-5 is also due to Yukawa et al.[3] As in Fig. XII-1,

[1] J.T. Albertsen, G.B. Jensen, and J.M. Knudsen, *Nature,* 273:453 (1978).

[2] T. Heumann and G. Karsten, *Arch. Eisenhuettenwes,* 34:781 (1963).

[3] N. Yukawa, N. Hida, T. Imura, M. Kawamura, and Y. Mizuno, *Met. Trans.,* 3:887 (1972).

they show an allotropic transformation in pure Cr. Both the σ and the δ phases can be preserved to room temperature by quenching from 1250°C. As concerns stainless steels, the principle feature of this diagram is the large extent of the γ field; up to 50 at % Cr can be dissolved in fcc Ni at 1350°C.

The uncertainties in the three binary systems are carried over into the Fe–Cr–Ni phase diagram. Also, because of the sluggishness of reactions, the equilibrium phase diagram cannot always be used to predict the structures that will be present in commercial alloys. The commonly cited instances are Type 302 stainless steel (17–19%Cr, 8-10%Ni), which, from a section of the ternary diagram at 18%Cr, should be austenitic only when quenched from temperatures in the range 800–1200°C (Fig. XII-6). However, this ignores the very strong effects of C and N in stabilizing austenite, which result in the alloy being entirely austenitic at ambient and higher temperatures. Plastic deformation or cooling well below ambient temperature is required to move this metastable austenite toward the equilibrium ferrite, and then the transformation is martensitic.

Although σ phase is a major constituent of the Fe–Cr–Ni phase diagram (Fig. XII-7), its rate of formation at temperatures below 500°C is so slow that it is not an important factor in determining the properties of stainless steels used at these temperatures.

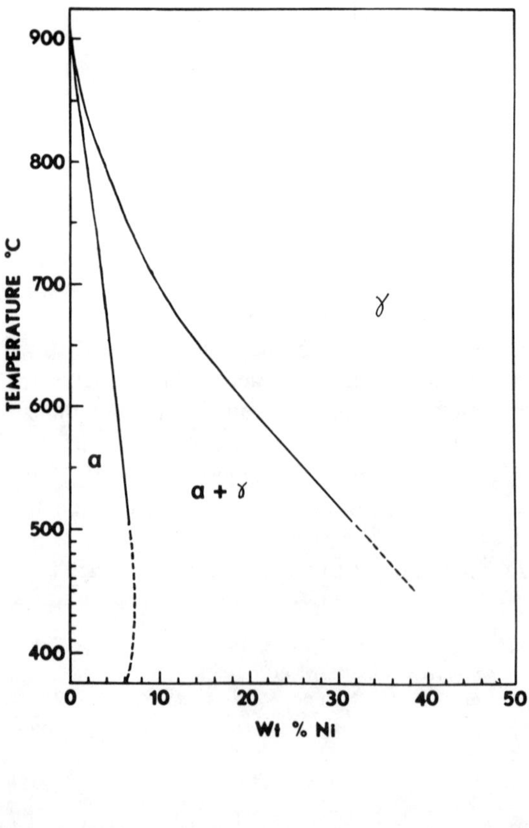

Figure XII-4. Iron-rich side of the iron–nickel diagram. (J.I. Goldstein and R.E. Ogilvie, *Trans. AIME*, 233:2083 (1965), by permission.)

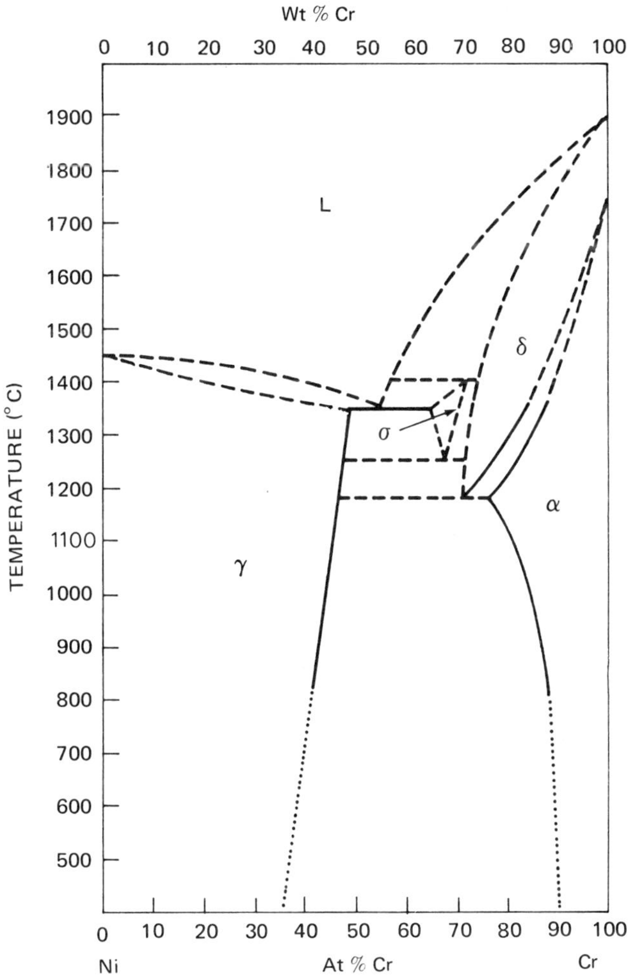

Figure XII-5. The nickel-chromium system. (N. Yukawa, N. Hida, T. Imura, M. Kawamura, and Y. Mizuna, *Met. Trans.*, 3:887 (1972), by permission.)

XII-3. EFFECTS OF SINGLE ADDITIONS TO Fe-Cr-Ni ALLOYS

Carbon is only slightly soluble in austenitic Fe-Cr-Ni alloys, but nevertheless it is a potent stabilizer of the austenite. Gordon[1] suggests the following limits to C solubility in 18%Cr-8 to 10%Ni: 0.04% at 800, 0.01% at 700, and about 0.003% at 600°C. This produces the plot of Fig. XII-8. Rosenberg and Irish[2] reported the

[1] G.M. Gordon, in *Proc. Firminy Conf.*, NACE, Houston, Texas, 1977, p. 893.
[2] S.J. Rosenberg and G.R. Irish, *J. Res. Nat. Bur. Standards,* 48:40 (1951).

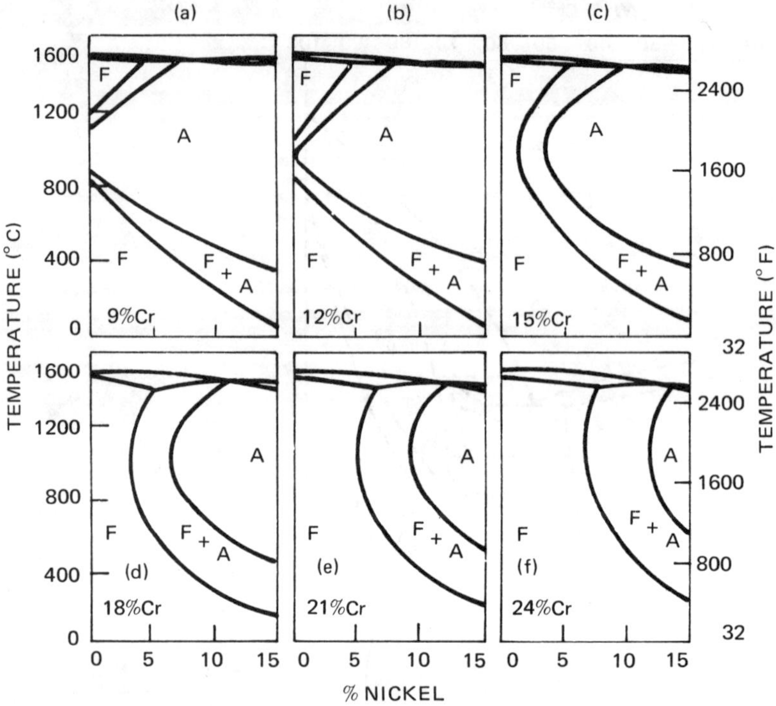

Figure XII-6. Sections through the chromium–iron–nickel system at constant chromium contents. (E.C. Bain and R.H. Aborn, *Metals Handbook*, 1939 ed., Am. Soc. Met., Metals Park, Ohio, by permission.)

solubility at 700°C to be 0.007%. Carbon precipitated as $Cr_{23}C_6$ in austenite grain boundaries renders the steel susceptible to intergranular corrosion by depleting Cr in solid solution.[1] Such precipitation is likely to occur during cooling after welding. Keeping the C content below about 0.03% prevents this deterioration. Although this C concentration may exceed the solubility of $Cr_{23}C_6$ at temperatures of 700°C and below, the carbides form slowly at such temperatures and the cooling rates after welding are high enough to avoid precipitation.

Nitrogen is more soluble than C in Fe-Cr-Ni austenite (Table XII-1); it is a strong stabilizer of austenite; and it does not produce intergranular corrosion. Nitrogen can be an economical substitute for a portion of the Ni in austenitic stainless steel.

The effects of other elements are often reported as "Ni equivalents" if they stabilize austenite and as "Cr equivalents" if they stabilize ferrite, as follows, in wt %[2]:

[1] R.H. Aborn and E.C. Bain, *Trans. Amer. Soc. Steel Treating,* 18:837 (1930).

[2] F.B. Pickering, *Physical Metallurgy and the Design of Steels,* Appl. Science Pub., London, 1978, p. 228.

$$\text{Cr equivalent} = \text{Cr} + 2(\text{Si}) + 1.5(\text{Mo}) + 5(\text{V}) + 5.5(\text{Al}) + 1.75(\text{Nb}) + 1.5(\text{Ti}) + 0.75(\text{W}) \quad \text{(XII-1)}$$

$$\text{Ni equivalent} = \text{Ni} + \text{Co} + 0.5(\text{Mn}) + 0.3\text{Cu} + 30(\text{C}) + 25(\text{N}) \quad \text{(XII-2)}$$

Similar equations have been applied in the form of Schaeffler diagrams, as in Fig. XII-9.[1,2] Although originally intended for use with air-cooled weld deposits, these diagrams can be employed to estimate the phases that will be present in stainless steels when compositions are changed.

Martensitic transformations are used in Fe-Cr-C stainless steels to produce structures with high hardness and good corrosion resistance. They occur also in metastable austenitic stainless steels when cooled below room temperature or when deformed at room temperature. In the former instance, the Ms temperature varies with composition in the manner shown[3]:

[1] A.L. Schaeffler, *Met. Prog.*, 56:620 (1949).

[2] C.J. Long and W.T. DeLong, *Weld. J.*, 52:281s (1973).

[3] F.B. Pickering, *Physical Metallurgy and the Design of Steels*, Appl. Science Publ., London, 1978, p. 229.

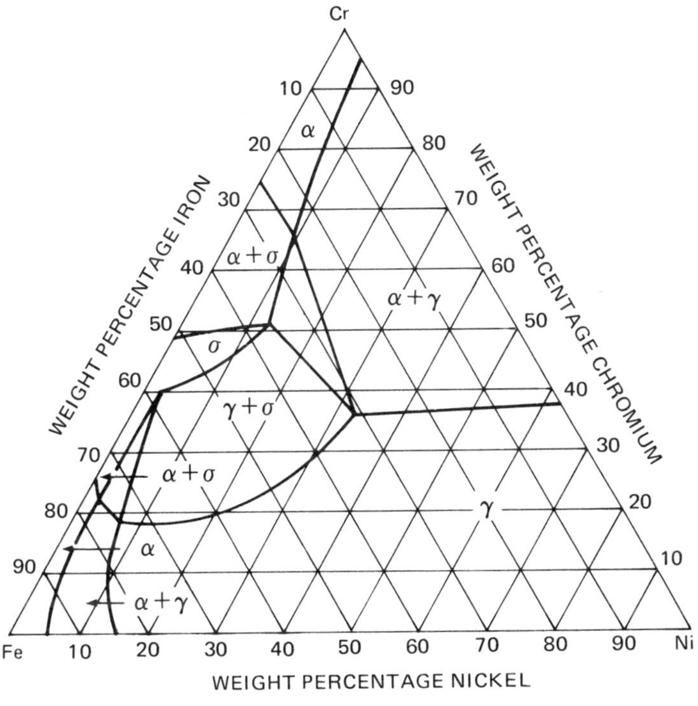

Figure XII-7. The isothermal section at 650°C through the chromium–iron–nickel system. (G.R. Speich, *Metals Handbook*, 8th ed., vol. 8, 1973, Am. Soc. Met., Metals Park, Ohio, by permission.)

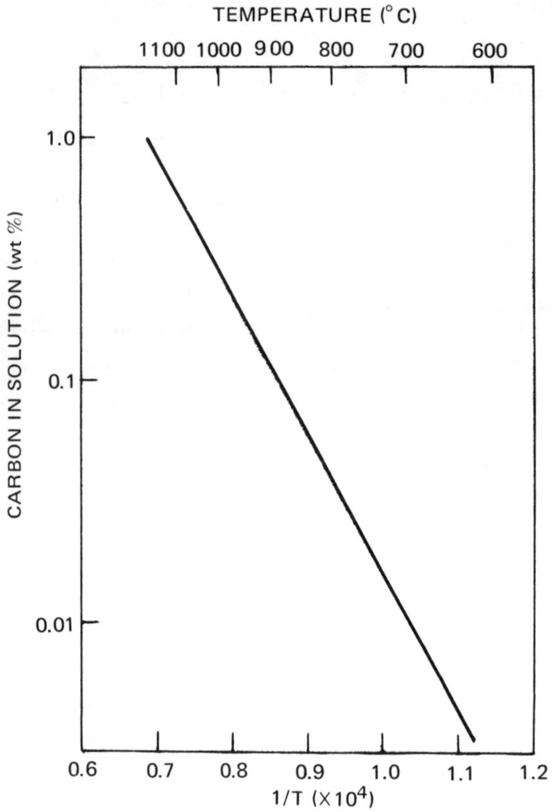

Figure XII-8. Approximate solubility of carbon in 18%Cr, 8–10%Ni, Fe alloys.

Table XII-1 Temperature dependence of nitrogen solubility in types 304 and 308 stainless steel[1]

	Solid solubility (wt %)	
Temperature (°C)	Type 304	Type 308
838	0.125	0.115
866	0.145	0.132
893	0.177	0.165
927	0.190	—
954	0.258	0.254
982	0.281	0.273
1010	–	0.297

[1] J.B. Cox and J.T. Eckel, *J. Mater. Sci.*, 4:282 (1969).

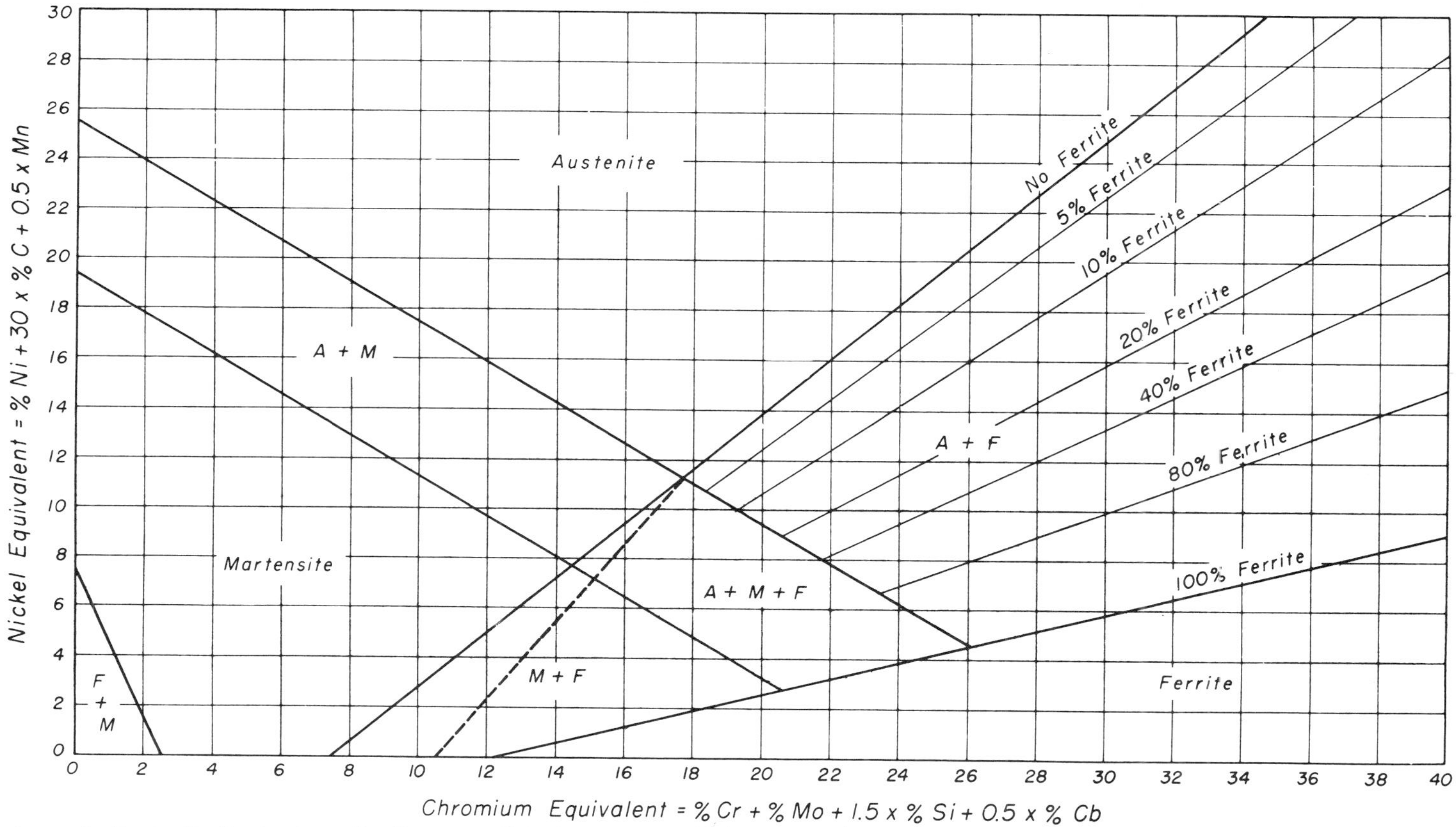

For austenite-plus-ferrite structures, the diagram predicts the percentage ferrite within 4% for the following stainless steels: 308, 309, 309 Cb, 310, 312, 316, 317, 318 (316 Cb), and 347. Dashed line is the martensite/M + F boundary modification by Eberhard Leinhos, "Mechanische Eigenschaften und Gefugeausbildung von mit Chrom und Nickel legiertem Schweissgut," VEB Deutscher Verlag fur Grundstoffindustrie, Leipzig, 1966.

Figure XII-9. Constitution diagram (Schaeffler diagram) for stainless steel weld metal. (*Metal Progress Data Book*, vol. 112, 1977, Am. Soc. Met., Metals Park, Ohio, by permission.)

$$\text{Ms}\ (^\circ\text{C}) = 502 - 810(\text{C}) - 1230(\text{N}) - 13(\text{Mn}) - 30(\text{Ni}) - 12(\text{Cr}) - 54(\text{Cu}) - 46(\text{Mo}) \qquad \text{(XII-3)}$$

where ( ) indicates wt % of the alloying element. This is based on the assumptions

1. That the effect of an element is proportional to its concentration
2. There are no interactions between elements

It is also desirable to know the relation between M_d and composition. M_d is the upper limit of temperature at which plastic deformation can initiate a martensitic transformation. Tensile strains are more effective in initiating transformation than are compressive strains. In metastable austenitic stainless steels the quantity (Md − Ms) can vary between 30 and 330°C.

A knowledge of the effect of composition on M_d is desirable, but the M_d is difficult to determine. Angel[1] avoided this difficulty by measuring M_{d30}, the temperature at which a true strain of 0.30 transforms 50% of the specimen to martensite. The equation he proposed was

$$M_{d30}\ (^\circ\text{C}) = 413 - 462(\text{C}+\text{N}) - 9.2(\text{Si}) - 8.1(\text{Mn}) - 13.7(\text{Cr}) - 9.5(\text{Ni}) - 18.5(\text{Mo}) \qquad \text{(XII-4)}$$

Metastable austenite in stainless steels can transform martensitically to ϵ phase (hexagonal close-packed) or to α' (bcc). The latter can have a microstructure of lath martensite or of plate martensite, depending upon the Ms temperature. The formation of ϵ is favored by a low stacking fault energy, because the transformation can be accomplished by propagation of intrinsic stacking faults on every other $(111)_\gamma$ plane—that is, by glide of $\frac{1}{6}\langle 112\rangle$ partial dislocations on every other $(111)_\gamma$. In γ and ϵ, the close-packed planes and directions coincide.

$$(111)_\gamma \parallel (0001)_\epsilon$$
$$[1\bar{1}0]_\gamma \parallel [1\bar{2}10]_\epsilon$$

Wirth and Bickerstaffe,[2] considering steels containing 8%Ni and either 12, 14, or 16%Cr, found that the type of martensite and its substructure depended upon the structure of the austenite before transformation, which in turn depended upon the stacking fault energy. Increasing the Cr content or increasing the cooling rate increased the amount of internal twins within the lath martensite. These internally twinned laths were formed by a sequence of

$$\gamma \rightarrow \epsilon\ (\text{or faulted}\ \gamma) \rightarrow \alpha'$$

The correspondence between the three phases was

$$(111)_\gamma \parallel (0001)_\epsilon \parallel (011)_{\alpha'}$$

[1] T. Angel, *J. Iron Steel Inst.*, 177:165 (1954).
[2] A. Wirth and J. Bickerstaffe, *Met. Trans.*, 5:799 (1974).

$$[1\bar{1}0]_\gamma \parallel [1\bar{2}10]_\epsilon \parallel [1\bar{1}1]_{\alpha'}$$

which is the Kurdjumov-Sachs relation.

In most stainless steels, the martensite is of the lath form, with a habit plane near $(225)_\gamma$. Plate martensite occurs when the Ms temperature is low, as in steels with a high Ni content.

XII-4. EMBRITTLEMENT OF Fe–Cr ALLOYS

As indicated in Fig. XII-2, there is a miscibility gap in the Fe–Cr system; the solid solution of Cr in Fe divides into two parts, Fe-rich α and Cr-rich α', at temperatures between 350 and 500°C. When α' forms it causes an embrittlement of the alloy.[1,2] This embrittlement can be removed by reheating to temperatures above 550°C.

At Cr contents of 14 to 18% the formation of α' occurs by normal nucleation and growth. At higher Cr contents reaction is by spinodal decomposition (Fig. XII-10). Within the spinodal—the area enclosed by dashed lines in the figure—the solid solution decomposes by an increase in amplitude of compositional differences, in competition with nucleation and growth. Outside the spinodal, only the latter is possible.

For a given binary system, the free energy may vary with composition at a constant temperature T_1 in the manner shown in Fig. XII-10*a*. Between the points A and B, both α and α' phases can coexist; that is, the boundary of the two-phase region in the phase diagram (Fig. XII-10*b*) is the locus of points at which $dF(x)/dx = 0$. The points marked C and D are inflection points on the free energy curve, where $d^2F(x)/dx^2 = 0$. The locus of such points, the dashed line in Fig. XII-10*b*, defines the spinodal region. A calculated free-energy diagram for the Fe–Cr system at 482°C is shown in Fig. XII-11, from which it appears that spinodal decomposition at this temperature is possible between about 0.25 and 0.60 atom fraction of Cr.

The embrittlement caused by precipitation of α' can be extreme, the energy absorbed at room temperature in breaking a half-width V-notch Charpy specimen of an 18%Cr alloy dropping from 100 ft · lb to 2 ft · lb after 250 h at 482°C.[1] The rate of the embrittling process varies with composition and melting practice, the time to reach an energy absorption of 10 ft · lb in a half-width V-notch Charpy specimen ranging from 20 to 2000 h at 480°C, the temperature at which the reaction is most rapid. Lower Cr contents and vacuum melting delayed the onset of embrittlement. The α' precipitates formed first at dislocations. They remain small, in the range 2.5 to 25 nm, and coherent with the matrix.

Unlike temper embrittlement, the 475°C embrittlement is not an internal surface effect; single crystals can be embrittled. The fracture is transgranular. The embrittlement in Fe–Cr alloys can be enhanced by the presence of Al, Mo, Ti, Si, Nb, or

[1] R.M. Fisher, E.J. Dulis, and K.G. Carroll, *Trans. AIME*, 197:690 (1953).

[2] P.J. Grobner, *Met. Trans.*, 4:251 (1973).

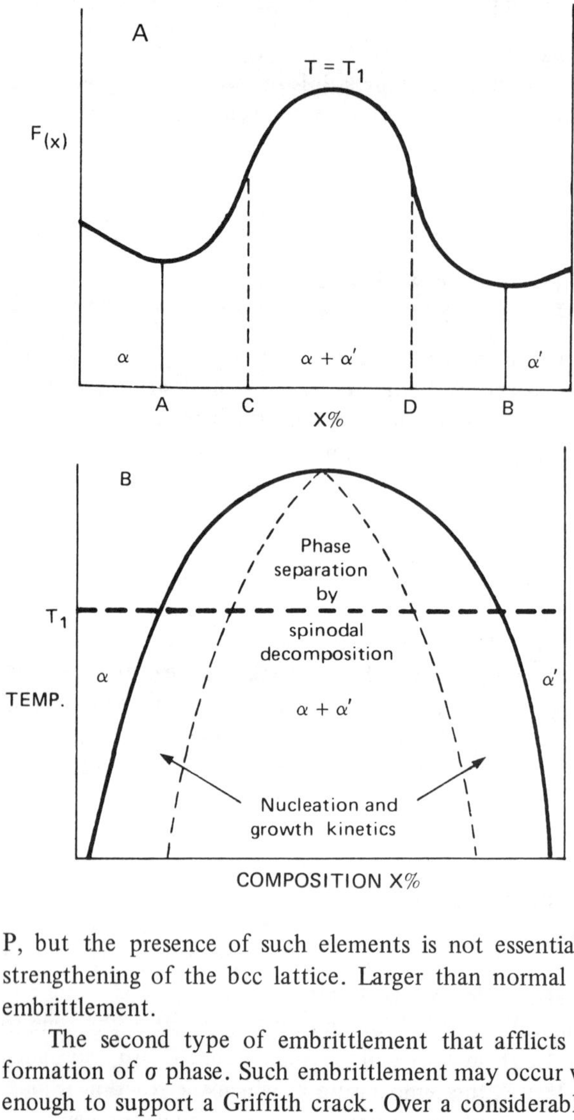

Figure XII-10. Schematic free energy-composition diagram and associated miscibility gap phase diagram.

P, but the presence of such elements is not essential. They may act only through strengthening of the bcc lattice. Larger than normal amounts of Ni can reduce the embrittlement.

The second type of embrittlement that afflicts Fe–Cr alloys is caused by the formation of σ phase. Such embrittlement may occur when the particles of σ are large enough to support a Griffith crack. Over a considerable range of temperature, α' and σ can coexist. In 12 to 16%Cr ferritic steels, both can exist at 480°C.[1] After very long holding at this temperature, α' particles can redissolve, with the Cr atoms diffusing to σ particles. The formation of σ is promoted by cold work, presumably because dislocations can assist nucleation. The ferrite-stabilizing elements Mo, Si, Ti, and P increase the rate of formation of σ, as does Mn. As shown in Fig. XII-12, the rate of formation and the range of temperatures in which σ forms are both increased by increased Cr content.

[1] D.C. Ludwigson and H.S. Link, in *Advances in the Technology of Stainless Steels and Related Alloys,* ASTM STP 369 (1965), p. 299.

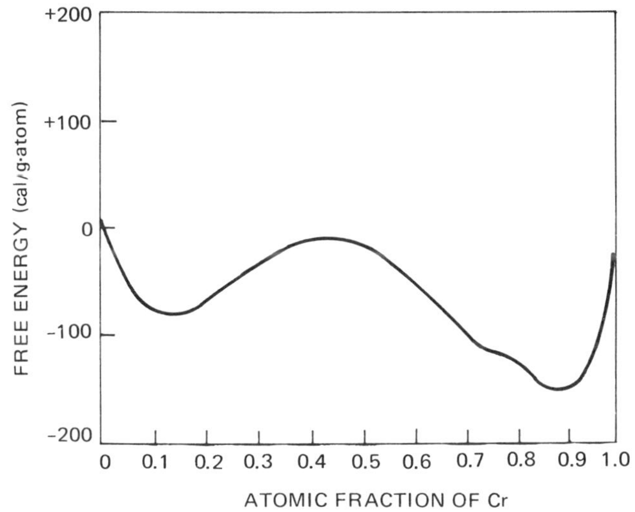

Figure XII-11. Free energy diagram of the chromium–iron system at 482°C. (P.J. Grobner, *Met. Trans.*, 4:251 (1973), by permission.)

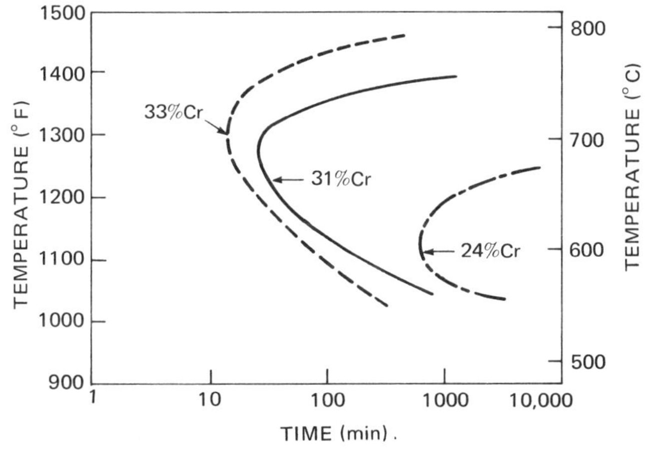

Figure XII-12. Approximate isothermal transformation curves for the start of sigma phase formation in cold-worked Fe–24, 31, and 33%Cr alloys containing 0.06 to 0.08%C. (P. Duwez and H. Martens, in *Ductile Chromium*, Am. Soc. Met., Metals Park, Ohio, 1957, p. 322, by permission.)

Although σ can form in the AISI 300 series of austenitic stainless steels in periods as brief as 24 h at 650°C,[1] embrittlement is not a major problem except in the 25%Cr Type 310 and in the 17%Cr-2%Mo Type 316.

In addition to α' and σ, another intermetallic compound, called the χ phase, can form in ferritic stainless steels that contain Mo. The existence of this phase was established by Andrews[2] and by McMullin et al[3]; its effects on mechanical properties and corrosion behavior are discussed by Steigerwald and coworkers.[4] Chi phase has the nominal composition Fe_2CrMo. It can form in the temperature range 550–925°C, often in conjunction with σ. Like σ, it reduces the toughness of the steels and can promote susceptibility to crevice corrosion because of removal of Cr and Mo from the ferrite matrix.

XII-5. STACKING FAULT ENERGIES OF AUSTENITIC STAINLESS STEELS

The stacking fault energy (SFE) of a fcc metal is always reduced by alloying with one other element, but addition of a third element can either decrease or increase the stacking fault energy. In metals with a high stacking fault energy, dislocations are not dissociated and cross slip tends to be easy. This leads to formation of dislocation cells. With a low SFE, dislocations tend to dissociate readily and cross slip is difficult. Dislocations are largely restricted to {111} slip planes, forming planar arrays. The stacking fault, bounded by two partial dislocations, amounts to four atom layers in the hcp configuration, and its width is proportional to the SFE. The SFE controls the rate of work hardening of austenitic steels, the frequency of annealing twins, the ease of transformation to ϵ martensite, the deformation and annealing textures, the fine structure of α martensite, the kinetics of recovery and recrystallization, and perhaps the susceptibility to stress-corrosion cracking. The stacking fault energies of a number of fcc alloys are listed in Table XII-2. The composition corresponding to commercial 18–8 stainless steel has the lowest SFE, in agreement with its tendency to form ϵ martensite during straining.

The cold-rolling texture of austenite with a low stacking fault energy is principally $\{110\}\langle 112\rangle$ (brass type), whereas with a high SFE the texture is $\{123\}\langle 412\rangle$ and $\{146\}\langle 211\rangle$ (copper type). As discussed in Chap. VIII, transformation of these textured austenites to martensite can produce a texture in the martensite, through the Kurdjumov-Sachs relation:

$$(111)_\gamma \parallel (011)_{\alpha'}$$

$$[1\bar{1}0]_\gamma \parallel [1\bar{1}1]_{\alpha'}$$

[1] M.E. Nicholson, C.H. Samans, and J.F. Shortsleeve, *Trans. Am. Soc. Met.*, 44:601 (1952).

[2] K.W. Andrews, *Nature*, 164:1015 (1949).

[3] J.G. McMullin, S.F. Reiter, and D.G. Ebeling, *Trans. Am. Soc. Met.*, 46:799 (1954).

[4] R.F. Steigerwald, H.J. Dundas, J.R. Redmond, and R.M. Davison, *Stainless Steel 77,* Climax Molybdenum Co., Greenwich, Connecticut, 1977, p. 57.

Table XII-2 Stacking fault energies of Fe–Ni–Cr alloys at 25°C

Alloy (%)	SFE (mJ/m²)
Ni	30–40
90Ni–10Cr	13
72Ni–28Cr	5
20Ni–20Cr–0.015C	2.5
16Ni–19Cr–0.005C	3.0
14Ni–18Cr	1.8
12Ni–18Cr–0.003C	3.0
9Ni–19Cr–0.07C	1.5

XII-6. FERRITIC STAINLESS STEELS

The ferritic stainless steels are alloys of Fe and Cr, containing from 11.5 to 27%Cr, with additions of Mn and Si and occasionally Ni, Al, Mo, or Ti (see Table XII-3). The C content is kept as low as economically feasible to improve toughness and to minimize sensitization, which occurs through precipitation of Cr carbides. Because of this low C content, these are not high-strength steels; their yield strengths in the annealed condition are usually in the range 275-415 MPa. They are used because of their good resistance to general liquid corrosion and high-temperature oxidation, good cold-formability, resistance to pitting and stress-corrosion cracking, and generally lower cost than the austenitic grades.

The decade of the 1970s saw great advances in the techniques of removing C from stainless steel melts. Such removal must be done by oxidation, but without major loss of Cr. This is accomplished by reducing the partial pressure of O (and of CO) as the C content is decreased. The most commonly used procedure is argon-oxygen decarburization (AOD). The steel is melted in an electric furnace, then transferred to a second vessel in which it is blown with a mixture of argon and oxygen. The partial pressure of CO is reduced by gradually increasing the argon content of the mixture, thus driving down the concentration of C in the melt. The use of this second vessel allows the introduction of desulfurizing slags. A similar technique, called Creusot-Loire-Uddenholm (CLU), employs a mixture of oxygen, steam, and nitrogen or argon. However, attaining the very low C contents required for good performance from high-Cr ferritic stainless steels requires refining under vacuum. One variant, vacuum oxygen decarburization (VOD), calls for melting in an electric arc furnace, initial decarburization in a basic oxygen converter, and final decarburization in a vacuum vessel, with introduction of oxygen and argon. Other processes involving vacuum are vacuum induction melting (VIM) and electron beam refining (EBR). In the latter, purification is achieved by continuous hearth refining under high vacuum.

This brief discussion of these melting and refining practices is introduced because they are required for the manufacture of the new types of ferritic steels listed in Table

Table XII-3 Standard ferritic stainless steels[1]

Type number	UNS number	Chemical composition (%) (maximum unless otherwise shown)								
		C	Mn	P	S	Si	Cr	Ni	Mo	Other elements
405	(S40500)	0.08	1.00	0.040	0.030	1.00	11.50/14.50			Al 0.10/0.30
409	(S40900)	0.08	1.00	0.045	0.045	1.00	10.50/11.75			Ti 6 × C min./0.75 max.
429	(S42900)	0.12	1.00	0.040	0.030	1.00	14.00/16.00			
430	(S43000)	0.12	1.00	0.040	0.030	1.00	16.00/18.00			
430F	(S43020)	0.12	1.25	0.060	0.15 (min.)	1.00	16.00/18.00		0.60	
430FSe	(S43023)	0.12	1.25	0.060	0.060	1.00	16.00/18.00			Se 0.15 min.
434	(S43400)	0.12	1.00	0.040	0.030	1.00	16.00/18.00		0.75/1.25	
436	(S43600)	0.12	1.00	0.040	0.030	1.00	16.00/18.00		0.75/1.25	Cb + Ta 5 × C min./0.70 max.
442	(S44200)	0.20	1.00	0.040	0.030	1.00	18.00/23.00			
446	(S44600)	0.20	1.50	0.040	0.030	1.00	23.00/27.00			N 0.25

[1] AISI Steel Products Manual 13, *Stainless and Heat Resisting Steels*, 1974.

Table XII-4 New ferritic stainless steels

Alloy	C and N Limits (wt %)		Stabilizing additions (wt %)
Fe–18% Cr–2% Mo–Ti	C	0.025 max	Ti + Nb, 0.20 + 4(C + N) min, 0.80 max
	N	0.025 max	
Fe–25% Cr–1% Mo	C	0.005 max	Nb, 13 to 29(N)
	N	0.015 max	
Fe–26% Cr–1% Mo–Ti	C	0.040 max	Ti, 0.2 to 1.0
	N	0.040 max	
Fe–28% Cr–2% Mo–4% Ni	C	0.015 max	Nb $\geqslant$ 12(C + N) + 0.2
	N	0.035 max	
Fe–29% Cr–4% Mo	C	0.010 max	
	N	0.020 max	
	C + N	0.025 max	
Fe–29% Cr–4% Mo–2% Ni	same		

XII-4. These steels were developed after the realization that austenitic stainless steels had an Achilles heel—stress-corrosion cracking—especially in the presence of chloride ions. Mo is introduced to improve resistance to pitting and crevice corrosion. Ni is added to increase resistance to attack by reducing acids. Nb and/or Ti are added to getter C and N.

Because ferritic stainless steels have a bcc structure they suffer from a ductile-to-brittle fracture transition, which has been a serious obstacle to their employment in load-bearing structures. However, if $C + N < 0.015\%$, the transition temperature can be kept well below ambient, even with Cr contents of 29%[1] Of course, even steels made to this extra-low interstitial content are still susceptible to embrittlement by formation of α' or σ, so if subsequent room-temperature toughness is important, the ferritic steels should not be used in extended service above about 325°C.

The low-interstitial ferritic steels are less susceptible than the standard grades to intergranular corrosion caused by Cr depletion through the precipitation of chromium carbides in grain boundaries, as can occur in the heat-affected zone of a weld. This sensitization occurs after heating to temperatures in excess of 900°C, where the solubility of C in Cr ferrite is sufficient to allow subsequent precipitation. Because of the much higher rates of diffusion in the bcc lattice as compared to those in the fcc lattice at the same temperature, sensitization is much more rapid in ferritic than austenitic steels. The sensitization process consists of the precipitation of sheets of $(Cr, Fe)_7C_3$ or $(Cr, Fe)_{23}C_6$ in grain boundaries.[2] In a ferritic steel containing 20%Cr, the minimum Cr content in the depleted zone adjacent to the grain boundary can lie in the range from zero to 5%. The width of this depleted zone, where corrosion occurs, can vary from 0.05 to 0.7 μm. Sensitization can be removed by annealing in the temperature range 650–850°C to allow Cr atoms to diffuse into the Cr-depleted zones around the

[1] H. Abo, T. Nakazawa, S. Takemura, M. Onoyama, H. Ogawa, and H. Okada, *Stainless Steel 77*, Climax Molybdenum Co., Greenwich, Connecticut, 1977, p. 35.

[2] R.P. Frankenthal and H.W. Pickering, *J. Electrochem. Soc.*, 120:23 (1973).

grain boundary carbides. Alloys also can be stabilized by the addition of Ti or Nb which form stable carbides throughout the matrix, removing C from solid solution and thereby preventing formation of $Cr_{23}C_6$ or Cr_7C_6.

About half of the stainless steels produced are rolled to sheet which is subsequently cold-drawn into articles such as cooking utensils, sinks, ornamental automotive trim, etc. The r_m values of the low-cost grades of ferritic stainless steels, such as Type 430 (17%Cr), are generally close to 1, but some of the newer, low-carbon grades have been produced with r_m values as high as 2, indicating superior drawability.[1] This also indicates that there is need for research in processing to control texture, to ensure uniform high r_m values. In contrast, austenitic stainless steels, such as Types 302, 304, and 316, have r_m values near 1, with little hope of improvement.

Ferritic stainless steels have relatively poor stretchability (low uniform elongation, low n values), also with little hope of improvement. The austenitic steels, and especially those that transform to martensite during plastic straining, have excellent stretchability with n values as high as 0.45.

The grain size in steels, and especially in Type 430 (17%Cr), can be controlled during hot rolling if some austenite forms; in the two-phase structure the growth of both phases is hampered.

A problem peculiar to ferritic stainless steel sheet is *ridging* or *roping.*[2,3,4] it mars the appearance of parts formed from sheet. When the sheet is cold drawn, markings form parallel to the original rolling direction, differing in this respect from stretcher strains. They also differ from stretcher strains in that they do not outline regions that are thinned locally, but rather a depression on one side of the sheet matches an elevation on the other side; the thickness is constant. The cause is a preferred orientation in the sheet, with bands of grains of (001)[011] orientation within the matrix, which has a texture consisting of (111)[011] and (111)[112] components. These strings of (001)[011] orientation are produced during hot rolling. The explanation that seems to be in the best accord with observations is that offered by Weight.[4] When plastically strained in the rolling direction, the strings of (001)[110] grains tend to contract in the short transverse (through-thickness) direction. The surrounding grains of $(111)[11\bar{2}]$ orientation contract in the transverse direction, with the result that the bands of (001)[110] orientation buckle plastically, leading to a corrugated sheet of substantially constant thickness, as shown in Fig. XII-13.

XII-7. MARTENSITIC STAINLESS STEELS

The martensitic stainless steels differ from the ferritic types in their higher C content (Table XII-5). They contain from 12 to 18%Cr and 0.12 to 1.20%C. As can be seen

[1] K.-J. Blom, *Stainless Steel 77,* Climax Molybdenum Co., Greenwich, Connecticut, 1977, p. 213.

[2] H.C. Chao, *Trans. Am. Soc. Met.,* 60:37 (1967).

[3] J.D. Defilippi and H.C. Chao, *Met. Trans.,* 2:3209 (1971).

[4] R.N. Wright, *Met. Trans.,* 3:83 (1972).

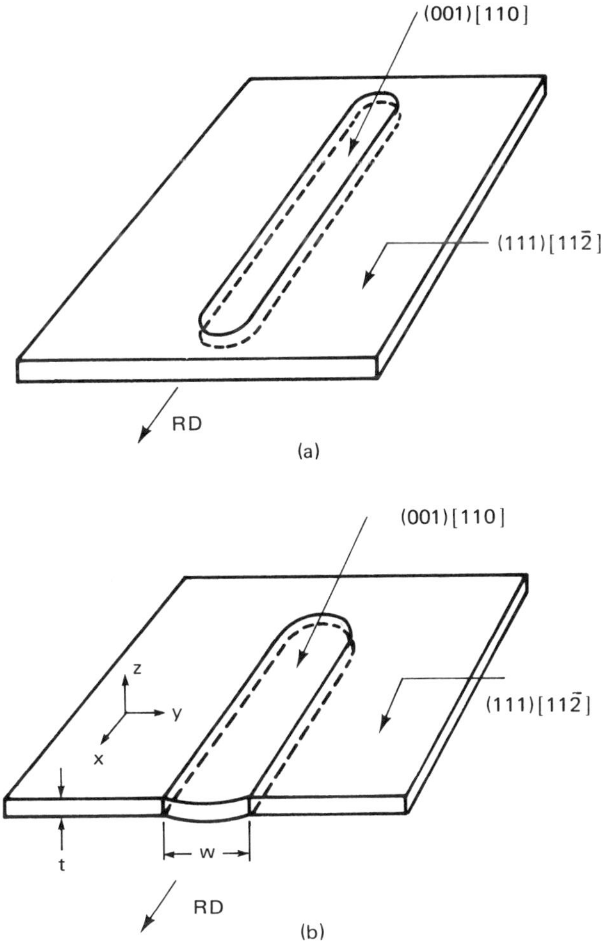

Figure XII-13. Schematic representation of plastic buckling mechanism for "roping." (*a*) Before rolling direction elongation. (*b*) After rolling direction elongation. (R.N. Wright, *Met. Trans.*, 3:83 (1972), by permission.)

from Fig. XII-14, such steels can be heated to form austenite or austenite plus carbide, then cooled to form martensite. For the steels to fall within the γ loop the relation %Cr − 17(%C) ⩽ 12.5 should hold. They can be hardened and tempered to yield strengths in the range 550 to 1860 MPa. The Cr content gives these steels such high hardenability that they can be air-hardened even in large sections. If they are to be heat-treated for maximum strength the δ ferrite content should be minimized.

The most widely used of the martensitic stainless steels is Type 410, which contains 11.5 to 13.5%Cr and 0.15%(max)C. An isothermal transformation diagram for a typical 12%Cr steel is shown in Fig. XII-15, taken from Rickett et al.[1] This reference is an excellent starting point for anyone interested in these steels and their heat treatment.

[1] R.L. Rickett, W.F. White, C.S. Walton, and J.C. Butler, *Trans. Am. Soc. Met.*, 44:138 (1952).

Table XII-5 Standard martensitic stainless steels[1]

Type number	UNS number	Chemical composition (%) (maximum unless otherwise shown)								
		C	Mn	P	S	Si	Cr	Ni	Mo	Other
403	(S40300)	0.15	1.00	0.040	0.030	0.50	11.50/13.00			
410	(S41000)	0.15	1.00	0.040	0.030	1.00	11.50/13.50			
414	(S41400)	0.15	1.00	0.040	0.030	1.00	11.50/13.50	1.25/2.50		
416	(S41600)	0.15	1.25	0.060	0.15 (min.)	1.00	12.00/14.00		0.60[2]	
416Se	(S41623)	0.15	1.25	0.060	0.060	1.00	12.00/14.00			Se 0.15 min.
420	(S42000)	Over 0.15	1.00	0.040	0.030	1.00	12.00/14.00			
420F	(S42020)	Over 0.15	1.25	0.060	0.15 (min.)	1.00	12.00/14.00		0.60[2]	
422	(S42200)	0.20/0.25	1.00	0.025	0.025	0.75	11.00/13.00	0.50/1.00	0.75/1.25	V 0.15/0.30 W 0.75/1.25
431	(S43100)	0.20	1.00	0.040	0.030	1.00	15.00/17.00	1.25/2.50		
440A	(S44002)	0.60/0.75	1.00	0.040	0.030	1.00	16.00/18.00		0.75	
440B	(S44003)	0.75/0.95	1.00	0.040	0.030	1.00	16.00/18.00		0.75	
440C	(S44004)	0.95/1.20	1.00	0.040	0.030	1.00	16.00/18.00		0.75	

[1] AISI Steel Products Manual 13, *Stainless and Heat-Resisting Steels*, 1974.
[2] May be added at manufacturer's option.

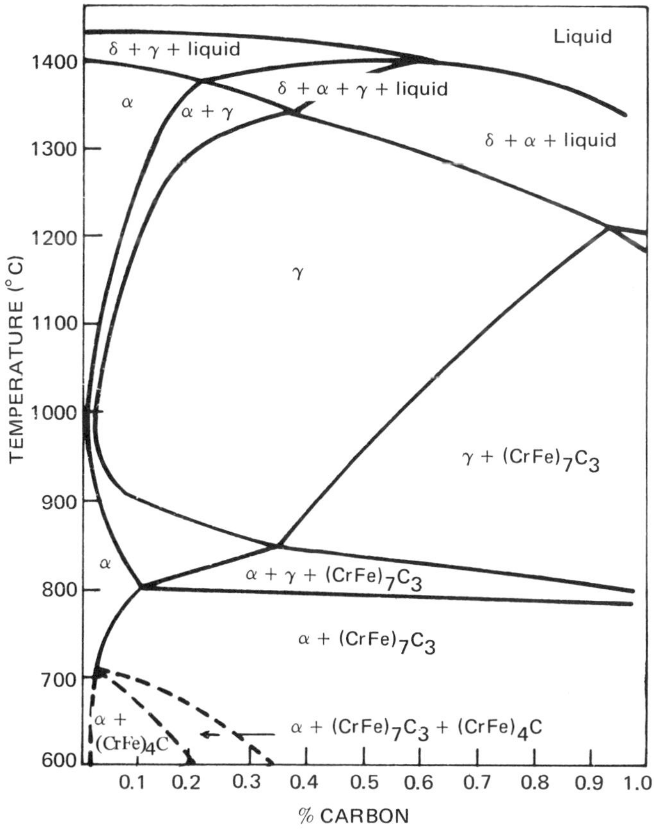

Figure XII-14. Effect of carbon on equilibrium relationships in 12%Cr steels. (W. Tofaute, A. Sponheuer, and H. Bennek, *Arch. Eisenhuettenwes.*, 8:499 (1935), by permission.)

In tempering the low-carbon martensitic stainless steels, the temperature range from about 440 to 540°C should be avoided because of a sharp reduction in notch-impact resistance that accompanies a small secondary hardening peak. This kind of temper embrittlement must be attributed to carbide morphology rather than to any segregation phenomenon.[1] The temper embrittlement can be reduced by the addition of a small quantity of Nb, which appears to affect the nucleation of $Cr_{23}C_6$. It also produces grain refinement.[2] Temper embrittlement in 12%Cr steels also increases stress-corrosion cracking.

[1] B.R. Banerjee, J.J. Hauser, and J.M. Capenos, in *Advances in the Technology of Stainless Steels and Related Alloys,* ASTM, STP 369, 1969, p. 291.

[2] H. Tanczyn, in *Advances in the Technology of Stainless Steels and Related Alloys,* ASTM, STP 369, 1969, p. 82.

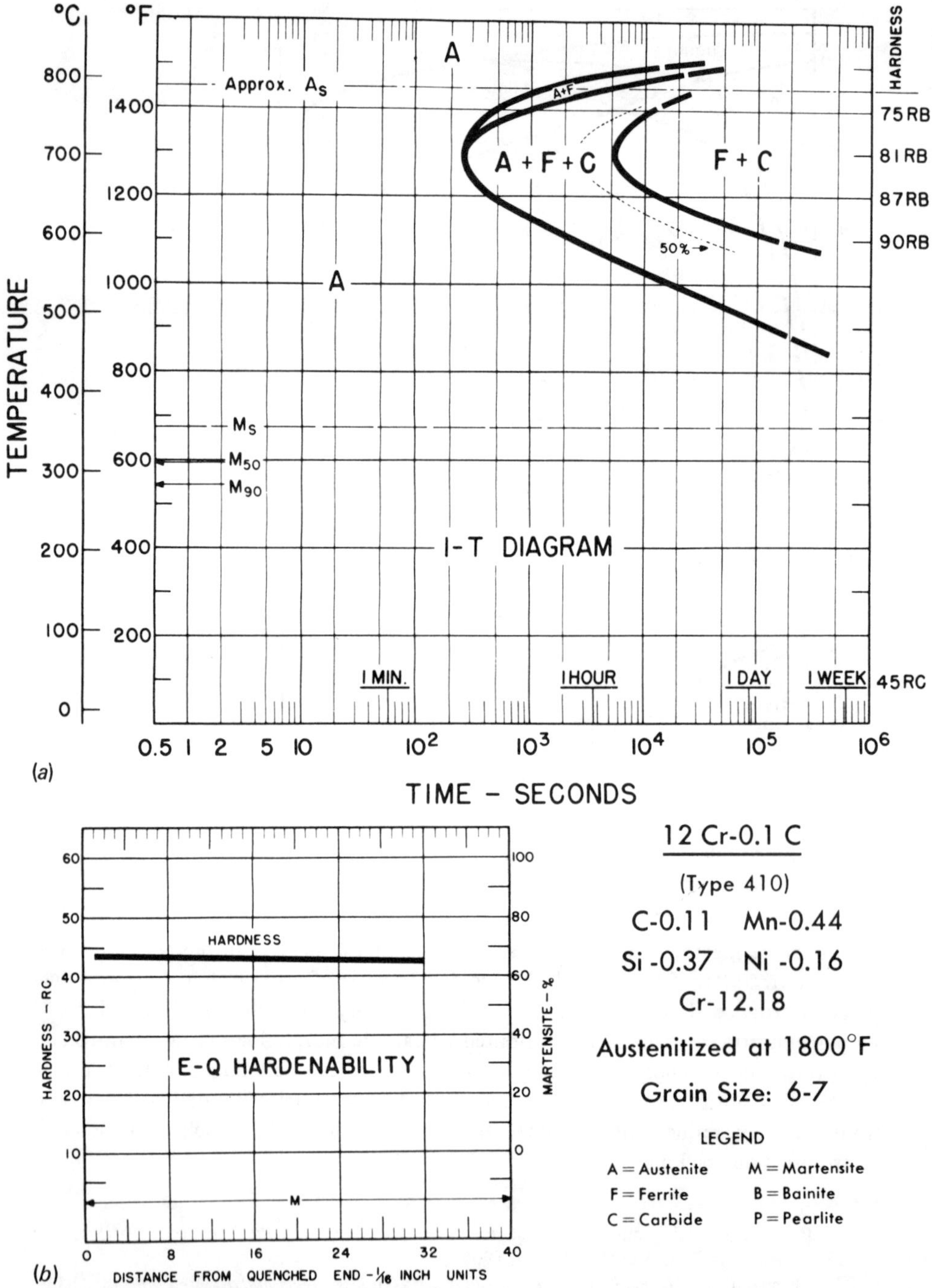

Figure XII-15. (*a*) Isothermal transformation and (*b*) end-quench diagrams for a Type 410 stainless steel. (R.L. Rickett, W.F. White, C.S. Walton, and J.C. Butler, *Trans. Am. Soc. Met.*, 44:138 (1952), by permission.)

Nitrogen can be added to low-carbon 12%Cr steel to the maximum attainable without ingot porosity, about 0.3%. This, combined with an addition of about 1.6%Ni, produces Type 404 stainless steel. The function of the N is to increase the as-quenched hardenss to about 56 R_C.

At the expense of weldability and toughness, the strength of Cr steels can be raised by increasing the C content to 0.20% max (type 431), 0.60–0.75% (440A), 0.75–0.95% (440B), or to 0.95–1.20% (type 440C). In the first instance, the Cr must be increased to 16% to compensate for the Cr removed from the matrix as carbides. After hardening and tempering, these steels must not be heated to temperatures in excess of about 500°C because of the possibility of copious precipitation of $Cr_{23}C_6$ at prior austenite grain boundaries. Type 440C is used for the best cutting implements, for it can be tempered to a hardness of about 600 DPH, with good corrosion resistance.

XII-8. DUPLEX FERRITE–AUSTENITE STEELS

Steels with a microstructure consisting of ferrite grains in austenite, or vice versa, have one principal advantage over austenitic stainless steels: their yield strength can be twice as great. If the grain size is sufficiently fine, 3 to 10 μm, such alloys can be superplastic at temperatures near 950°C,[1,2] with elongations in tension of about 500%. Other advantages are corrosion resistance about the same as austenitic steels, improved resistance to stress-corrosion cracking, good weldability, and lack of grain-boundary corrosive attack because of Cr depletion. The disadvantages are susceptibility to embrittlement by α' and σ phase, some difficulties in hot-working, and limited sources of supply.

Although listed under austenitic stainless steels (Table XII-6), type 329 is a ferrite-base duplex alloy containing about 26%Cr, 4%Ni, and 1.5%Mo. An austenite-base alloy, Uranus 50, contains 21.5%Cr, 7.5%Ni, 2.5%Mo, and 1.5%Cu. Type 329 has a minimum yield strength at room temperature of 435 MPa compared with 214 MPa for type 304. The yield strength increases with increasing ferrite content. A law of mixtures does not apply to the strength, probably because of strain-induced transformation of austenite grains. The ultimate tensile strength at room temperature rises to a maximum at about 70 to 80% ferrite, then decreases as the ferrite goes to 100%. This behavior is attributed to the greater rate of work-hardening of austenite, which has a lower yield strength than the ferrite but a higher tensile strength.[3]

If the alloys are predominantly ferrite, they have the ductile–brittle impact transition typical of ferritic steels,[4] with the transition temperature rising sharply as the ferrite content increases over about 60%. The failures occur by cleavage of the ferrite grains and by ductile fracture of the austenite grains.

[1] H.W. Hayden, R.C. Gibson, H.F. Merrick, and J.H. Brophy, *Trans. Am. Soc. Met.*, 60:3 (1967).

[2] R.C. Gibson, H.W. Hayden, and J.H. Brophy, *Trans. Am. Soc. Met.*, 61:85 (1968).

[3] H.W. Hayden and S. Floreen, *Met. Trans.*, 4:561 (1973).

[4] S. Floreen and H.W. Hayden, *Trans. Am. Soc. Met.*, 61:474 (1968).

Table XII-6 Standard austenitic stainless steels[1]

Type number	UNS number	Chemical composition (%) (maximum unless otherwise shown)								
		C	Mn	P	S	Si	Cr	Ni	Mo	Other elements
201	(S20100)	0.15	5.50/7.50	0.060	0.030	1.00	16.00/18.00	3.50/5.50		N 0.25
202	(S20200)	0.15	7.50/10.00	0.060	0.030	1.00	17.00/19.00	4.00/6.00		N 0.25
205	(S20500)	0.12/0.25	14.00/15.50	0.060	0.030	1.00	16.50/18.00	1.00/1.75		N 0.32/0.40
301	(S30100)	0.15	2.00	0.045	0.030	1.00	16.00/18.00	6.00/8.00		
302	(S30200)	0.15	2.00	0.045	0.030	1.00	17.00/19.00	8.00/10.00		
302B	(S30215)	0.15	2.00	0.045	0.030	2.00/3.00	17.00/19.00	8.00/10.00		
303	(S30300)	0.15	2.00	0.20	0.15 (min.)	1.00	17.00/19.00	8.00/10.00	0.60[2]	
303Se	(S30323)	0.15	2.00	0.20	0.060	1.00	17.00/19.00	8.00/10.00		Se 0.15 min.
304	(S30400)	0.08	2.00	0.045	0.030	1.00	18.00/20.00	8.00/10.50		
304L	(S30403)	0.030	2.00	0.045	0.030	1.00	18.00/20.00	8.00/12.00		
	(S30430)	0.08	2.00	0.045	0.030	1.00	17.00/19.00	8.00/10.00		Cu 3.00/4.00
304N	(S30451)	0.08	2.00	0.045	0.030	1.00	18.00/20.00	8.00/10.50		N 0.10/0.16
305	(S30500)	0.12	2.00	0.045	0.030	1.00	17.00/19.00	10.50/13.00		
308	(S30800)	0.08	2.00	0.045	0.030	1.00	19.00/21.00	10.00/12.00		
309	(S30900)	0.20	2.00	0.045	0.030	1.00	22.00/24.00	12.00/15.00		

309S	(S30908)	0.08	2.00	0.045	0.030	1.00	22.00/24.00	12.00/15.00		
310	(S31000)	0.25	2.00	0.045	0.030	1.50	24.00/26.00	19.00/22.00		
310S	(S31008)	0.08	2.00	0.045	0.030	1.50	24.00/26.00	19.00/22.00		
314	(S31400)	0.25	2.00	0.045	0.030	1.50/3.00	23.00/26.00	19.00/22.00		
316	(S31600)	0.08	2.00	0.045	0.030	1.00	16.00/18.00	10.00/14.00	2.00/3.00	
316F	(S31620)	0.08	2.00	0.20	0.10 (min.)	1.00	16.00/18.00	10.00/14.00	1.75/2.50	
316L	(S31603)	0.030	2.00	0.045	0.030	1.00	16.00/18.00	10.00/14.00	2.00/3.00	
316N	(S31651)	0.08	2.00	0.045	0.030	1.00	16.00/18.00	10.00/14.00	2.00/3.00	N 0.10/0.16
317	(S31700)	0.08	2.00	0.045	0.030	1.00	18.00/20.00	11.00/15.00	3.00/4.00	
317L	(S31703)	0.030	2.00	0.045	0.030	1.00	18.00/20.00	11.00/15.00	3.00/4.00	
321	(S32100)	0.08	2.00	0.045	0.030	1.00	17.00/19.00	9.00/12.00		Ti 5 × C min.
329	(S32900)	0.10	2.00	0.040	0.030	1.00	25.00/30.00	3.00/6.00	1.00/2.00	
330	(N08330)	0.08	2.00	0.040	0.030	0.75/1.50	17.00/20.00	34.00/37.00		
347	(S34700)	0.08	2.00	0.045	0.030	1.00	17.00/19.00	9.00/13.00		Cb + Ta 10 × C min.
348	(S34800)	0.08	2.00	0.045	0.030	1.00	17.00/19.00	9.00/13.00		Cb + Ta 10 × C min. Ta 0.10 max, Co 0.20 max.
384	(S38400)	0.08	2.00	0.045	0.030	1.00	15.00/17.00	17.00/19.00		

[1] AISI Steel Products Manual 13, *Stainless and Heat-Resisting Steels*, 1974.
[2] May be added at manufacturer's option.

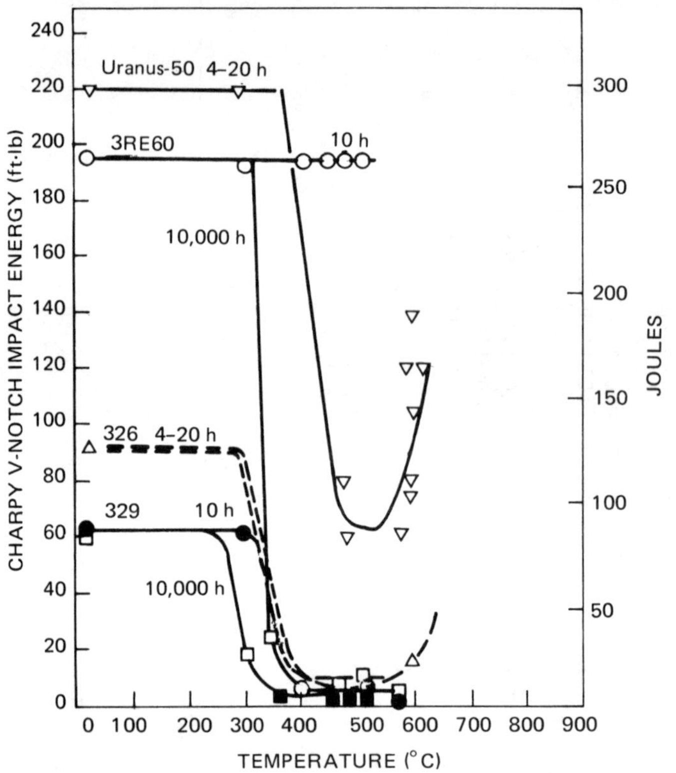

Figure XII-16. Isothermal embrittlement of duplex stainless steels. (G.M. Gordon, in *Stress Corrosion Cracking and Hydrogen Embrittlement of Iron-Base Alloys*, NACE, Houston, Texas, 1977, p. 914, by permission.)

Because of the high Cr content of the duplex alloys and the segregation of Cr in the ferrite grains, the duplex alloys can be embrittled by holding in or slow cooling through the temperature range from 300 to 550°C (Fig. XII-16). This embrittlement is due to the formation of α'. Aging at temperatures between about 550 and 950°C can lead to severe embrittlement by σ phase, with the maximum rate of formation occurring at 700°C. In the usual manner, cold work accelerates the precipitation of sigma.

Ferrite-base duplex stainless steels should be hot-rolled in the all-ferrite range during ingot breakdown and in the two-phase region during subsequent hot-rolling, the latter to prevent ferrite grain growth.[1]

The excellent resistance of duplex stainless steels to intergranular corrosive attack can be attributed to one or both of two mechanisms. The first is that γ–γ boundaries are not continuous. They frequently intersect α–γ boundaries. At tempera-

[1] H.F. Merrick, H.W. Hayden and R.C. Gibson, *Met. Trans.*, 4:827 (1973).

tures at which Cr carbides precipitate, the higher concentration and diffusivity of Cr in ferrite leads to rapid repair of Cr depletion . The second mechanism involves the motion of γ–α grains into α if Cr carbides form along the boundary. The decreased Cr content of the α allows it to transform to γ, thereby isolating the carbides.

More important than freedom from intergranular attack is freedom from transgranular stress-corrosion cracking, and the presence of the two phases also minimizes this problem.

XII-9. AUSTENITIC STAINLESS STEELS

The austenitic stainless steels contain from 16 to 25%Cr, plus sufficient Ni, Mn, or N to make them austenitic at room temperature (Table XII-6). Because of their fcc structure they are nonmagnetic and tough at low temperatures. Because of their single-phase microstructure they are readily weldable, but also because of this single-phase structure their yield strength is low and they can be strengthened only by cold work or by elements in solid solution. Their disadvantages are high cost and susceptibility to stress-corrosion cracking.

Type 304 stainless steel in the annealed condition has a yield strength (0.2% offset) at room temperature of about 240 MPa and a tensile strength of about 585 MPa, so work-hardening is an obvious means of strengthening. Considering only compositions that are stable with respect to strain-induced transformation to martensite, the yield strength can be raised to 1035 MPa or more (Fig. XII-17). Cold-work strengthening is effective only for sheet or wire products which will be used at temperatures below about 500°C. Also, such products cannot be welded without a loss of strength. When there is substantial transformation to martensite during cold working, the yield strength can exceed 1500 MPa.

Some of the problems associated with cold-rolled austenitic steels can be alleviated by warm-working at temperatures between 700 and 900°C.[1] This process can be made more effective by adding a small amount of Nb to the steel to inhibit recrystallization through the precipitation of NbC or NbN, and by reheating to about 850°C after rolling. The result is a structure that has improved resistance to softening at elevated temperatures and one that can be attained with less severe restrictions on section size than apply to cold-worked steels.

The austenitic steels can be strengthened by elements in solid solution. Interstitial solutes are by far the most effective, but C additions are out of the question because of the deleterious effects of the precipitation of Cr carbides. This leaves N as the most effective element for solid solution strengthening. To achieve a reasonable solubility of N in austenite, it is desirable to add also some substitutional solute that will lower the activity of N in austenite without leading to precipitation. It is also desirable that the substitutional solute should stabilize austenite. Manganese is the

[1] D.T. Llewellyn and V.J. McNeely, *Sheet Met. Ind.*, 49(1):17 (1972).

Table XII-7 Precipitation hardening stainless steels[1]

UNS number	Chemical composition (%) (maximum unless otherwise shown)								
	C	Mn	P	S	Si	Cr	Ni	Mo	Other
(S13800)	0.05	0.10	0.01	0.008	0.10	12.25/13.25	7.50/8.50	2.00/2.50	Al 0.90/1.35 N 0.010
(S15500)	0.07	1.00	0.040	0.030	1.00	14.00/15.50	3.50/5.50		Cu 2.50/4.50 Cb + Ta 0.15/0.45
(S17400)	0.07	1.00	0.040	0.030	1.00	15.50/17.50	3.00/5.00		Cu 3.00/5.00 Cb + Ta 0.15/0.45
(S17700)	0.09	1.00	0.040	0.040	1.00	16.00/18.00	6.50/7.75		Al 0.75/1.50

[1] AISI Steel Products Manual 13, *Stainless and Heat-Resisting Steels*, 1974.

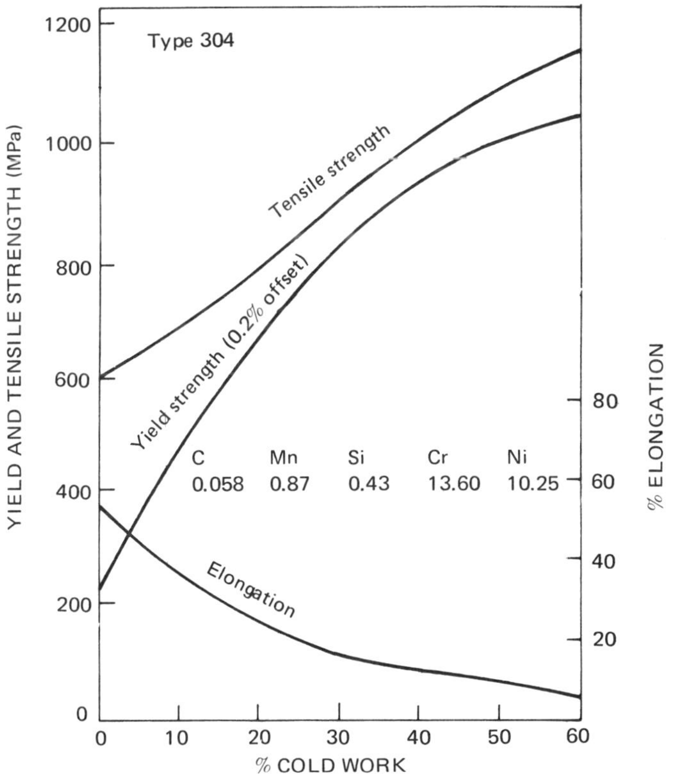

Figure XII-17. Tensile properties of cold-worked Type 304 stainless steel. (Courtesy Allegheny-Ludlum Steel Corp.)

only element that fills both requirements. This had led to the development of the AISI 200 series of Cr-Mn-Ni-N steels.

The comparatively minor effects of substitutional solutes on yield and tensile strengths of austenitic stainless steels are shown by the equations quoted by Pickering,[1] which also include the effects of grain size, δ ferrite, and twin spacing:

$$\begin{aligned}\text{Yield strength (0.2\% offset, MPa)} = 15.4[&4.4 + 23(\text{C}) + 1.3(\text{Si}) + 0.24(\text{Cr}) \\ &+ 0.94(\text{Mo}) + 1.2(\text{V}) + 0.29(\text{W}) \\ &+ 2.6(\text{Nb}) + 1.7(\text{Ti}) + 0.82(\text{Al}) \\ &+ 32(\text{N}) + 0.16(\delta\ \text{ferrite}) + 0.46d^{-1/2}]\end{aligned}$$

$$\begin{aligned}\text{Tensile strength (MPa)} = 15.4[&29 + 35(\text{C}) + 55(\text{N}) + 2.4(\text{Si}) + 0.11(\text{Ni}) \\ &+ 1.2(\text{Mo}) + 5.0(\text{Nb}) + 3.0(\text{Ti}) + 1.2(\text{Al}) \\ &+ 0.14(\delta\ \text{ferrite}) + 0.82t^{-1/2}]\end{aligned}$$

[1] F.B. Pickering, *Physical Metallurgy and the Design of Steels,* Appl. Science Publ., London, 1978, p. 231.

where $(\delta \text{ ferrite})$ = volume % δ ferrite
d = mean linear intercept grain diameter (mm)
t = twin spacing (mm)

The coefficients for Nb, Ti, and Al must be regarded with some reservation, for they may be due to a combination of solid solution and precipitation (or preprecipitation) effects.

The low solubility of C in Fe–Ni–Cr austenite has been shown previously (Fig. XII-8). The usual C content of commercial alloys is in excess of the limits indicated there, so that if they are heated to temperatures between about 450 and 850°C, Cr carbides will precipitate in austenite grain boundaries, resulting in depletion of Cr and acceleration of corrosive attack at the boundaries. This process can be prevented by (a) lowering the C content to 0.03%, or less, as previously discussed, or by (b) adding Ti or Nb to the steel, with Nb amount equal to 10 × C, Ti to 5 × C.

The stabilized steels, types 321 (Ti addition), 347, and 348 (Nb additions), require a minimum Ni content of 9% to compensate for the loss of the austenite-stabilizing effects of C and N removed from solution and for the ferrite-stabilizing effects of Ti and Nb in solid solution. These grades may crack in the heat-affected zone after welding, probably because the precipitation of finely dispersed Ti or Nb carbonitrides causes local strengthening.

The most persistent problem plaguing the austenitic stainless steels is chloride stress-corrosion cracking.[1] This phenomenon can occur in the absence of general corrosion, in the presence of stress (which can be applied or residual) and chloride ions, even in trace amounts. Fracture occurs by transgranular cracking with little or no plastic deformation and often with catastrophic effect. This type of cracking affects steels containing Ni, with the greatest damage occurring in the most common range of Ni content, as indicated in Fig. XII-18. The remedies are to reduce stress and eliminate chloride ions from solutions in contact with the steel, but application of these measures can be very difficult.

XII-10. PRECIPITATION-HARDENED STAINLESS STEELS

Precipitation-hardened stainless steels (Table XII-7) can have a matrix structure of either austenite or martensite, with the former being less frequently used. The austenitic precipitation-hardened steels have the advantage for some applications of being nonmagnetic. The austenite contains at least 10%Ni or Ni + Mn in order to remain austenitic even when some Ni or Mn is removed from solution by formation of an intermetallic compound. The strengthening elements can be P, Mo, Cu, Nb, or Ti. The alloys are solution treated at about 1200°C, quenched, then aged at 700 to 800°C to yield strengths up to about 700 MPa. Because of this high aging temperature they can serve at higher temperatures than the martenisitic grades.

[1] *Stress Corrosion Cracking and Hydrogen Embrittlement of Iron-Base Alloys,* National Association of Corrosion Engineers, NACE-5 Houston, Texas, 1977.

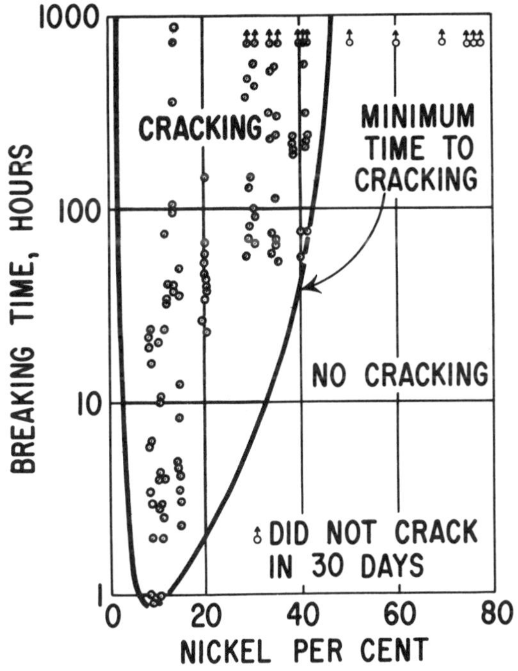

Figure XII-18. Effect of nickel content on stress-corrosion cracking of iron–chromium–nickel wires in boiling 42% magnesium chloride. (H.R. Copson, *Research Seminar on Constructional Alloy Steels* (Pittsburgh), INCO, New York, 1959, p. 16-5, by permission.)

The martensitic precipitation-hardened alloys are analogous to maraging steels, except for their Cr content, better corrosion resistance, and lower strength. In both instances a lath martensite structure is used to provide some strength, but the primary function of this structure is to supply an abundance of nucleation sites for the precipitation of intermetallic compounds. As in maraging steels, C is kept low, $<0.05\%$, to improve toughness.

The composition of the austenite is balanced to give an Ms temperature between about 125 to 250°C, by lowering the Ni content to between 4 and 7%. Elements added to form precipitates are Cu, Mo, Al, Ti, Nb, and N. The alloys are solution treated, quenched, then aged at temperatures between 400 and 500°C. The precipitation process is quite complex, but apparently some of the hardening agents are Cu (with Ni in solution), Ni_3Ti, Ni_3Al, and NiAl.

To allow forming at room temperature while the alloy is in the soft austenitic condition, higher Ni contents can be used to make the so-called semiaustenitic precipitation-hardened steels. After forming, these can be chilled to form martensite, then aged, or given a "trigger anneal" at about 750°C. This removes Cr and C from the austenite, thereby raising the Ms temperature. The alloy is aged after cooling to below room temperature.

CHAPTER

THIRTEEN

TOOL STEELS

XIII-1. CATEGORIES, COMPOSITIONS, AND USES

Most tools that the average person handles, such as hammers, saws, wrenches, pliers, pruning shears, screwdrivers, etc., are not made of tool steels, but of mass-produced carbon or low-alloy steels heat-treated or case-hardened to give the desired combination of properties. The products called tool steels are intended to be used for the shaping of other metals by cutting, shearing, battering, drawing, extruding, die casting, or rolling, or for the shaping and cutting of wood, paper, rock, or concrete. The increasing employment of tool steels in structural applications was mentioned in Chap. IX. All applications for tool steels are not equally demanding, so there is a variety of compositions available, ranging from comparatively simple and inexpensive high-carbon steels to the complicated and costly high-speed steels (Table XIII-1). The seven categories of tool steels listed in Table XIII-1 are each produced in about equal volume, but of course the value of each product is not the same.

The choice of a tool steel should depend upon the requirements of the application, and the steel chosen should be no more expensive than is needed to meet these requirements. Files, woodworking tools, axes, taps, reamers, and most drill bits can be made satisfactorily from water-hardening tool steels. Shearing tools must be resistant to wear and tough enough not to chip. Battering tools, such as chisels, railway track tools, punches, and rivet sets must have toughness as their first requirement. Steels for molds for plastics must be soft to facilitate machining and capable of being carburized for wear resistance: types P2–P6. Types P20 and P21 are prehardened, then machined into die-casting dies, which must be resistant to thermal shock. Cold-work tool steels, as noted in Table XIII-1, can have high carbon and high chromium contents, are hardened by cooling in air, and therefore have a high degree of dimensional stability. They are used for cold-forming dies and thread rolling dies. High wear resistance goes with high carbon and vanadium content, with the reverse being true for toughness.

Hot-work tool steels are used for forging dies, piercing tools, extrusion dies, and die-casting dies, where resistance to softening at elevated temperatures is of paramount importance. They combine high hardenability and high toughness, by virtue of their high alloy and low carbon contents. The most complex and important of the tool steels are the high-speed types, which combine high resistance to abrasion with high resistance to softening at elevated temperatures. These are used as lathe tools, milling cutters, end mills, and other cutting tools that must remove metal at a high rate and withstand the high temperatures that result. The relative abrasion resistance and toughness of some of the more popular of these steels are compared in Fig. XIII-1.

Only a brief overview of tool steels can be given in this text. Those interested in more detailed treatments should refer to the books by Roberts, Hamaker, and Johnson[1] and by Payson.[2] There is also a wealth of literature distributed by the makers of tool steels. Because of the diverse uses of tool steels, the variety of compositions that can meet the requirements of such usage, and the difficulty of devising tests to measure relative performance, tool steels were the last category of steels to be rationalized and standardized. The number of compositions offered for sale has been considerably reduced in recent years, and under the harsh dictates of economics this winnowing is likely to continue.

XIII-2. HEAT TREATMENT; AUSTENITIZING AND SOLUTION OF CARBIDES

Tool steels are customarily sold in the annealed condition; the buyer is responsible for the heat treatment, which consists of austenitizing, quenching, and tempering. A glance at Table XIII-2 indicates that to harden tool steels properly requires heating to temperatures ranging from 770 to 1300°C. The austenitizing temperature used for a particular composition of steel is determined by the temperature needed to dissolve all carbides and to produce a reasonably homogenous austenite, or to dissolve a sufficient fraction of carbides so that subsequent secondary hardening can occur during tempering. The majority of tool steels are hypereutectoid. In vanadium-containing high-speed steels such as T15, as much as 40% of the structure in the annealed condition can be carbide. Figure XIII-2 shows that in high-speed steels austenitized for hardening, undissolved carbides constitute about 10% of the volume of the steel. These comparatively large particles serve to enhance the wear resistance of the steels after quenching and tempering and also have the merit of inhibiting austenite grain growth. It should be noted in Fig. XIII-2 that all $M_{23}C_6$ carbide is dissolved at the hardening temperature, and some MC and M_6C remain. The amount of MC depends strongly on the vanadium content of the steel. The amount of undissolved carbide is nearly constant after 10 min at the hardening temperature,[3] but often the time at temperature is less than that.

[1] G.A. Roberts, J.C. Hamaker, and A.R. Johnson, *Tool Steels,* 3d ed., Am. Soc. Met., Metals Park, Ohio, 1962.

[2] P. Payson, *The Metallurgy of Tool Steels,* John Wiley and Sons, New York, 1962.

[3] J. Malkiewicz, Z. Bojarski, and J. Foryst, *J. Iron Steel Inst.*, 193:25 (1959).

Table XIII-1 Tool steel types and nominal compositions[1]

Type	UNS	Identifying elements (%)									
		C	Mn	Si	W	Mo	Cr	V	Co	Ni	Al
1. Water-Hardening Tool Steels (Symbol W)											
W1[2]	(T72301)	.60/1.40[4]	–	–	–	–	–	–	–	–	–
W2[2]	(T72302)	.60/1.40[4]	–	–	–	–	–	.25	–	–	–
W5	(T72305)	1.10	–	–	–	–	.50	–	–	–	–
2. Special-Purpose Tool Steels (Symbol L)											
L2	(T61202)	.50/1.10[4]	–	–	–	–	1.00	.20	–	–	–
L6	(T61206)	.70	–	–	–	.25[3]	.75	–	–	1.50	–
3. Shock-Resisting Tool Steels (Symbol S)											
S1[2]	(T41901)	.50	–	–	2.50	–	1.50	–	–	–	–
S2	(T41902)	.50	–	1.00	–	.50	–	–	–	–	–
S5[2]	(T41905)	.55	.80	2.00	–	.40	–	–	–	–	–
S6	(T41906)	.45	1.40	2.25	–	.40	1.50	–	–	–	–
S7	(T41907)	.50	–	–	–	1.40	3.25	–	–	–	–
4. Mold Steels (Symbol P)											
P2	(T51602)	.07	–	–	–	.20	2.00	–	–	.50	–
P3	(T51603)	.10	–	–	–	–	.60	–	–	1.25	–
P4	(T51604)	.07	–	–	–	.75	5.00	–	–	–	–
P5	(T51605)	.10	–	–	–	–	2.25	–	–	–	–
P6	(T51606)	.10	–	–	–	–	1.50	–	–	3.50	–
P20	(T51620)	.35	–	–	–	.40	1.70	–	–	–	–
P21	(T51621)	.20	–	–	–	–	–	–	–	4.00	1.20

5. Cold-Work Tool Steels (Symbol D, high-carbon high-chromium types)[5]

D2[2]	(T30402)	1.50	–	–	–	1.00	12.00	1.00	–	–
D3[2]	(T30403)	2.25	–	–	–	–	12.00	–	–	–
D4[2]	(T30404)	2.25	–	–		1.00	12.00	–	–	–
D5	(T30405)	1.50	–	–	–	1.00	12.00	–	3.00	–
D7	(T30407)	2.35	–	–	–	1.00	12.00	4.00	–	–
		(Symbol O, oil-hardening types)								
O1	(T31501)	.90	1.00	–	.50	–	.50	–	–	–
O2	(T31502)	.90	1.60	–	–	–	–	–	–	–
O6[6]	(T31506)	1.45	.80	1.00	–	.25	–	–	–	–
O7	(T31507)	1.20	–	–	1.75	–	.75	–	–	–
		(Symbol A, medium-alloy air-hardening types)[5]								
A2[2]	(T30102)	1.00	–	–	–	1.00	5.00	–	–	–
A3	(T30103)	1.25	–	–	–	1.00	5.00	1.00	–	–
A4	(T30104)	1.00	2.00	–	–	1.00	1.00	–	–	–
A6	(T30106)	.70	2.00	–	–	1.25	1.00	–	–	–
A7	(T30107)	2.25	–	–	1.00[3]	1.00	5.25	4.75	–	–
A8	(T30108)	.55	–	–	1.25	1.25	5.00	–	–	–
A9	(T30109)	.50	–	–	–	1.40	5.00	1.00	–	1.50
A10[6]	(T30110)	1.35	1.80	1.25	–	1.50	–	–	–	1.80

[1] AISI Steel Products Manual, *Tool Steels,* 1976.

[2] Commonly available grades.

[3] Optional.

[4] Various carbon contents are available.

[5] Some of the types can be produced with a sulfur addition to improve machinability.

[6] Contains free graphite in the microstructure to improve machinability.

Table XIII-1 Tool steel types and nominal compositions *(Continued)*

		Identifying elements (%)							
Type	UNS	C	Mn	Si	W	Mo	Cr	V	Co
6. Hot-Work Tool Steels (Symbol H, chromium types)[1]									
H10	(T20810)	.40	–	–	–	2.50	3.25	.40	–
H11[2]	(T20811)	.35	–	–	–	1.50	5.00	.40	–
H12[2]	(T20812)	.35	–	–	1.50	1.50	5.00	.40	–
H13[2]	(T20813)	.35	–	–	–	1.50	5.00	1.00	–
H14	(T20814)	.40	–	–	5.00	–	5.00	–	–
H19	(T20819)	.40	–	–	4.25	–	4.25	2.00	4.25
(Tungsten types)									
H21[2]	(T20821)	.35	–	–	9.00	–	3.50	–	–
H22	(T20822)	.35	–	–	11.00	–	2.00	–	–
H23	(T20823)	.30	–	–	12.00	–	12.00	–	–
H24	(T20824)	.45	–	–	15.00	–	3.00	–	–
H25	(T20825)	.25	–	–	15.00	–	4.00	–	–
H26	(T20826)	.50	–	–	18.00	–	4.00	1.00	–
(Molybdenum types)									
H42	(T20842)	.60	–	–	6.00	5.00	4.00	2.00	–

7. High-Speed Tool Steels (Symbol M, molybdenum types)[1]									
M1	(T11301)	.85[3]	–	–	1.50	8.50	4.00	1.00	–
M2[2]	(T11302)	.85; 1.00[3]	–	–	6.00	5.00	4.00	2.00	–
M3 Class 1[2]	(T11313)	1.05	–	–	6.00	5.00	4.00	2.40	–
M3 Class 2[2]	(T11323)	1.20	–	–	6.00	5.00	4.00	3.00	–
M4	(T11304)	1.30	–	–	5.50	4.50	4.00	4.00	–
M6	(T11306)	.80	–	–	4.00	5.00	4.00	1.50	12.00
M7	(T11307)	1.00	–	–	1.75	8.75	4.00	2.00	–
M10[2]	(T11310)	.85; 1.00[3]	–	–	–	8.00	4.00	2.00	–
M30	(T11330)	.80	–	–	2.00	8.00	4.00	1.25	5.00
M33	(T11333)	.90	–	–	1.50	9.50	4.00	1.15	8.00
M34	(T11334)	.90	–	–	2.00	8.00	4.00	2.00	8.00
M36	(T11336)	.80	–	–	6.00	5.00	4.00	2.00	8.00
M41	(T11341)	1.10	–	–	6.75	3.75	4.25	2.00	5.00
M42	(T11342)	1.10	–	–	1.50	9.50	3.75	1.15	8.00
M43	(T11343)	1.20	–	–	2.75	8.00	3.75	1.60	8.25
M44	(T11344)	1.15	–	–	5.25	6.25	4.25	2.00	12.00
M46	(T11346)	1.25	–	–	2.00	8.25	4.00	3.20	8.25
M47	(T11347)	1.10	–	–	1.50	9.50	3.75	1.25	5.00
(Symbol T, tungsten types)[1]									
T1[2]	(T12001)	.75[3]	–	–	18.00	–	4.00	1.00	–
T2[2]	(T12002)	.80	–	–	18.00	–	4.00	2.00	–
T4	(T12004)	.75	–	–	18.00	–	4.00	1.00	5.00
T5	(T12005)	.80	–	–	18.00	–	4.00	2.00	8.00
T6	(T12006)	.80	–	–	20.00	–	4.50	1.50	12.00
T8	(T12008)	.75	–	–	14.00	–	4.00	2.00	5.00
T15	(T12015)	1.50	–	–	12.00	–	4.00	5.00	5.00

[1] Some of the types can be produced with a sulfur addition to improve machinability.
[2] Commonly available grades.
[3] Other carbon contents may be available.

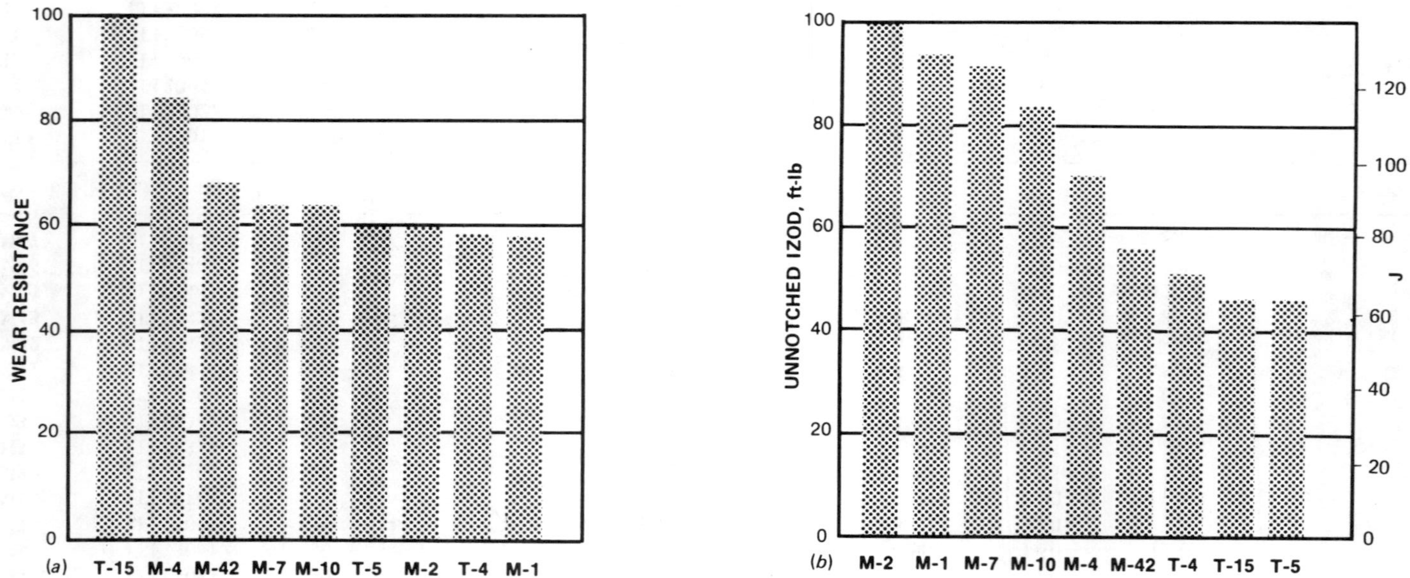

Figure XIII-1. Relative wear resistance (*a*) and toughness (*b*) of high-speed tool steels. (R.J. Henry and T.A. Myers, Teledyne Vasco, 1977, by permission.)

Table XIII-2 Heat-treating conditions for tool steels

Type	Quenchant	Hardening temperature (°C)	Hardness after quenching (R_C)	Tempering temperature (°C)	Hardness after tempering (R_C)
W1	water	770–790	65–67	175–275	65–56
W2	water	770–815	65–67	175–275	65–56
L6	oil	815–870	62–64	205–425	62–48
S1	oil	900–980	57–59	150–535	57–45
S2	water	845–855	60–62	150–260	57–45
	oil	870–885	58–60	150–260	58–54
S5	water	845–870	60–62	150–345	60–54
	oil	870–915	58–60	150–345	58–54
O1	oil	790–815	63–65	150–425	62–50
O2	oil	770–790	63–65	190–260	62–57
O6	oil	790–815	63–65	150–425	63–50
A2	air	940–970	61–63	205–370	60–57
D2	air	980–1025	61–63	205–370	60–58
D3	oil	955–980	62–64	205–370	62–58
D5	air	980–1025	60–62	205–370	59–57
D7	air	1010–1065	63–65	150–260	65–63
				455–540	62–58
H11	air	995–1025	53–55	540–595	51–43
H12	oil, air	980–1040	53–55	540–595	51–43
H13	air	995–1025	53–55	540–595	51–43
H21	oil, air	1150–1175	50–52	510–620	50–47
T1	oil, air, salt	1260–1300	63–65	550–595	65–63
T2	oil, air, salt	1260–1300	63–65	550–595	65–63
T4	oil, air, salt	1260–1300	63–65	550–595	65–63
T5	oil, air, salt	1260–1315	63–65	565–595	65–63
T8	oil, air, salt	1260–1300	63–65	550–595	65–63
M1	oil, air, salt	1175–1230	63–65	550–565	65–63
M2	oil, air, salt	1190–1230	63–65	550–580	65–63
M3	oil, air, salt	1175–1220	63–65	550–580	65–63
M4	oil, air, salt	1175–1220	63–65	550–580	65–63

The performance of a tool steel is largely dependent upon the condition of its surface, so both decarburization and carburization must be avoided during heat treatment. Neutral salt-bath furnaces are recommended, especially for high-speed steels; controlled-atmosphere furnaces are also satisfactory if properly adjusted.

Preheating is not essential for steels that are austenitized below 1040°C,

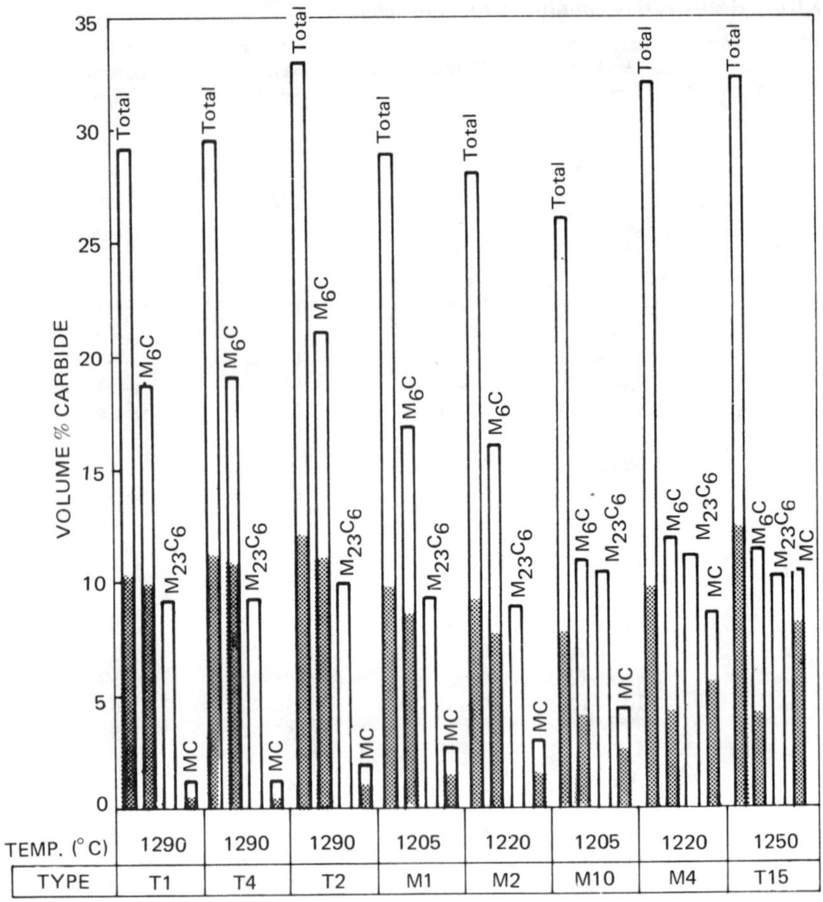

Figure XIII-2. Carbides in high-speed steels in the annealed condition and after being heated to normal hardening temperatures. Open bars represent quantities in annealed steels. Solid bars represent quantities in steels heated to the indicated hardening temperatures. (F.X. Kayser and M. Cohen, *Met. Prog.*, 61:79 (June, 1952), by permission.)

unless the pieces are large or irregularly shaped. High-speed steels are preheated to 815°C before being transferred to the austenitizing salt-bath or furnace. The purpose of preheating, during which heating is fairly slow, is to minimize distortion of the tool.

XIII-3. KINETICS OF TRANSFORMATION OF AUSTENITE

With the exception of the W1 water-hardening type, tool steels have high hardenability, so most of them need not be quenched drastically. In addition, as shown in Fig. XIII-3, the isothermal transformation diagrams of high-speed steels contain a very deep bay, so that cooling can be delayed almost without limit in the temperature range from

600 to 400°C. However, retardation of the transformation of austenite to ferrite plus carbide or to bainite does not make air-cooling of large pieces desirable, for, as shown in Fig. XIII-4, carbides precipitate in austenite in the temperature range from 900 to 750°C. Because of the bay in the isothermal transformation diagram, high-speed steels can be quenched into salt or lead baths at 500 to 600°C and allowed to come to the temperature of the bath before cooling is completed in air. This treatment allows relief of much of the thermal stresses generated during the quench and avoids precipitation of carbides in the austenite. Alternatively, pieces can be quenched in warm oil, held there until they reach a temperature of about 500°C, then cooled in air. Air-cooling from the hardening temperature is suitable for small tools.

With high carbon and alloy content, a considerable volume fraction of austenite, perhaps as much as 20 to 25%, may be retained after the quenching of a tool steel. This has the advantage of allowing straightening of parts immediately after quenching. If they had transformed entirely to martensite, their yield stress would be too high for easy straightening. However, this retained austenite must be transformed to martensite if hardness and wear resistance are to be maximized.

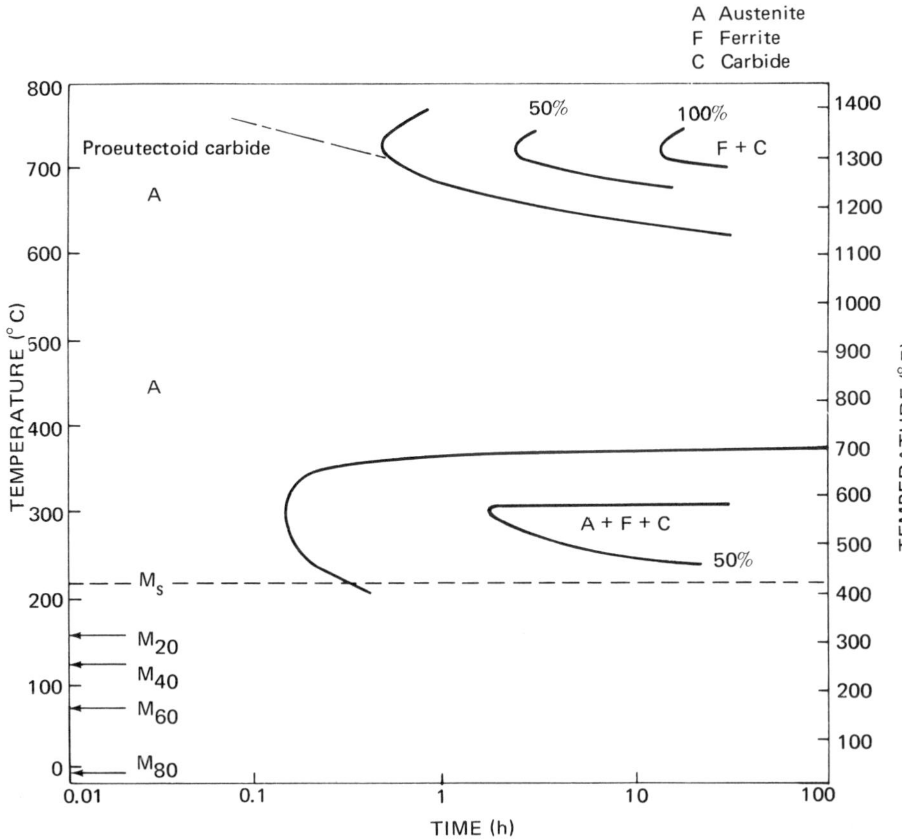

Figure XIII-3. Isothermal transformation diagram for M2 high-speed steel austenitized at 1225°C. (P. Gordon, M. Cohen, and R.S. Rose, *Trans. Am. Soc. Met.*, 31:161 (1943), by permission.)

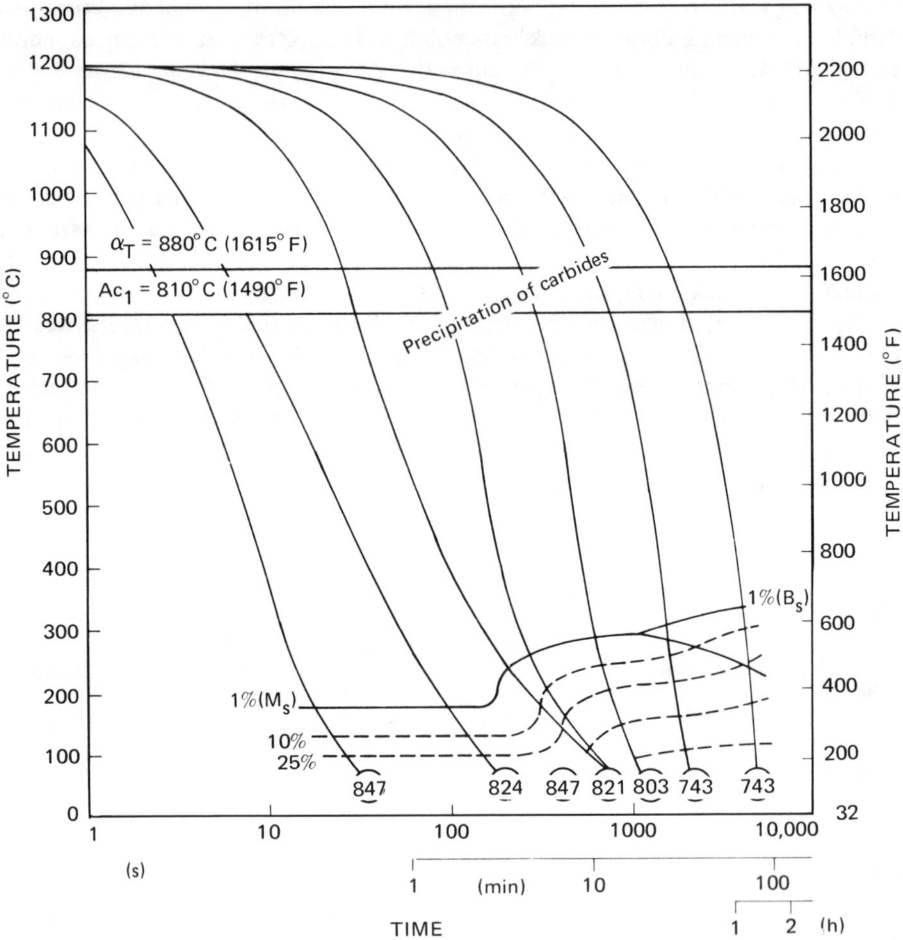

Figure XIII-4. Continuous cooling transformation diagram for M2 high-speed steel austenitized 5 min at 1205°C. Numbers in circles indicate hardness (HV) after cooling to room temperature. (W.W. Cias, Climax Molybdenum Co., Greenwich, Connecticut, 1971, by permission.)

When a tool steel is quenched, it must be allowed to cool to room temperature to decrease the amount of retained austenite, but it should be tempered immediately to reduce the chance of delayed-quench cracking.

XIII-4. TEMPERING, CARBIDE PRECIPITATION, AND SECONDARY HARDENING

As indicated in Table XIII-2, the water-hardening (W), special purpose (L), shock-resistant (S), mold (P), and cold-work (D, O, and A) tool steels are tempered in the usual temperature ranges for steels, with the choice of temperature being dictated by a compromise between strength (hardness) and toughness. The hot-work (H) and high-speed types (T and M) are tempered in the range 500–600°C to promote secondary hardening by alloy carbides, to a hardness equal to or greater than the hardness in the as-quenched condition, with improved toughness and wear resistance. The secondary hardening peak in M2 high-speed steel is shown in Fig. XIII-5, which also indicates the increase of hardening produced by increasing the austenitizing temperature to dissolve more carbides. The curves in Fig. XIII-5 were obtained by triple tempering, which will be discussed further.

The properties of tool steels are critically dependent upon the types and hardness of carbides, whether these are primary carbides formed during solidification or secondary carbides precipitated from solid solution in martensite during tempering. The range of hardness of the carbides that can be present in tool steels is shown in Fig. XIII-6 and compared with the maximum hardness obtainable in martensite. These carbides can vary in the composition of their metallic component, but representative analyses

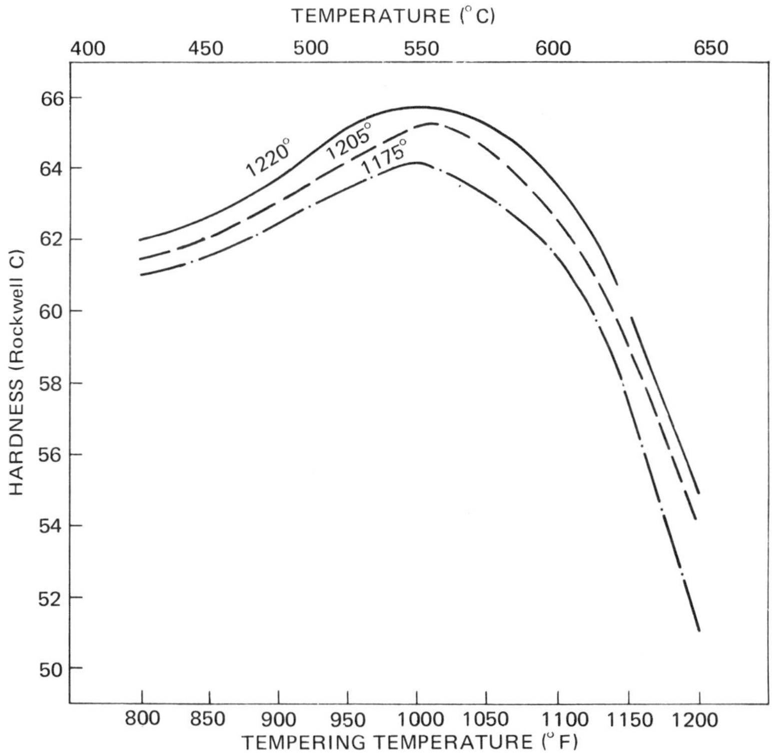

Figure XIII-5. The secondary hardening peak in M2 high-speed steel and the effect of austenitizing temperature on hardness. (*Tool Steels*, Universal-Cyclops Specialty Steel Division, Cyclops Corp., 1972, by permission.)

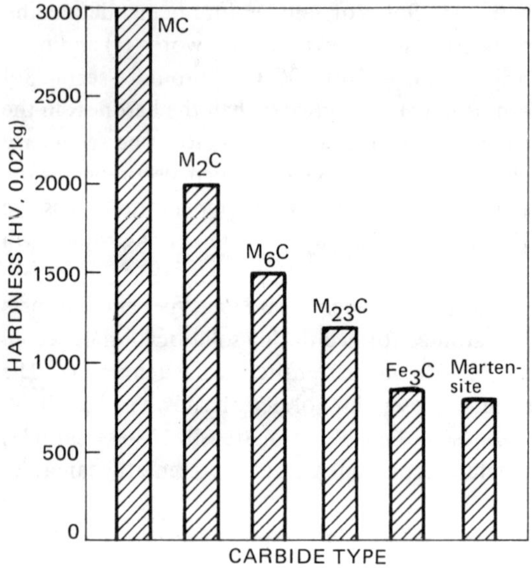

Figure XIII-6. Hardness of martensite and of various carbides: HV, 20-gm load. (E. Haberling, A. Rose, and H.H. Weigand, *Stahl Eisen*, 93:645 (1973), by permission.)

of carbides from an M2 high-speed steel[1] are given in Table XIII-3. From this table it can be seen that MC, the hardest carbide encountered in steels, is predominantly a carbide of vanadium, with tungsten and molybdenum in solution. M_2C is a carbide of tungsten and molybdenum, containing some vanadium, whereas M_3C is principally Fe_3C, with chromium and a smattering of other elements. $M_{23}C_6$ is an iron–tungsten–molybdenum carbide. M_6C, the most voluminous carbide in tool steels (Fig. XIII-2), also contains iron, tungsten, and molybdenum.

During solidification of M2 or M10 high-speed steels, MC forms directly from the liquid and M_6C + MC forms by a eutectic reaction liquid → austenite + carbide.[2] During tempering for secondary hardening the principal hardening agent is MC. Beyond the hardness peak, during overaging, M_2C precipitates; during annealing M_2C disappears and M_6C and $M_{23}C_6$ are formed.[1]

The heat-treatment of tool steels is distinguished by the necessity for multiple tempering operations. When a quenched tool steel containing retained austenite is tempered at 500 to 600°C there is very little tendency for the retained austenite to transform (Fig. XIII-3), but during the tempering this remaining austenite is said to be "conditioned," for at least some of it transforms to martensite during cooling from the tempering temperature. The nature of this "conditioning" process is not completely understood, but the generally accepted explanation[3] is that austenite becomes more susceptible to transformation by losing carbon to martensite. The martensite becomes a sink for carbon because of the precipitation of alloy carbides. It is also possible that microscopic residual stresses opposing the expansion of volume accompanying the transformation to martensite are reduced during tempering. In any event, the martensite formed during cooling from the tempering temperature must be tempered by a second

[1] E. Haberling, A. Rose, and H.H. Weigand, *Stahl Eisen*, 93:645 (1973).
[2] E.J. Galda and R.W. Kraft, *Met. Trans.*, 5:1727 (1974).
[3] R.J. Beltz and R.W. Lindsay, *Trans. Am. Soc. Met.*, 61:790 (1978).

reheating. Double tempering is usually sufficient, but sometimes as many as four tempers may be needed to bring the retained austenite content to an acceptable level.

The hardness of tempered martensite in tool steels is easily measured at room temperature, but the hardness of interest is that existing at the service temperature. Payson[1] has shown a useful correlation between the hardness at room temperature and the hardness at several elevated temperature (Fig. XIII-7).

One persistent problem in the production of tool steels is the nonuniform distribution of carbide particles in the finished product. This condition arises from the large

[1] P. Payson, *Trans. Am. Soc. Met.*, 51:60 (1959).

Table XIII-3 Representative analyses of carbides from a high-speed alloy

Carbide type	Alloying element	Composition (wt %)
MC	C	13.0
	Fe	4.0
	W	23.0
	Mo	14.0
	V	43.0
	Cr	4.5
M_2C	C	6.0
	Fe	7.0
	W	41.0
	Mo	28.0
	V	11.0
	Cr	8.0
M_3C	C	6.7
	Fe	76.0
	W	5.0
	Mo	4.0
	V	2.0
	Cr	8.0
$M_{23}C_6$	C	4.0
	Fe	45.0
	W	25.0
	Mo	18.0
	V	4.0
	Cr	5.0
M_6C	C	3.0
	Fe	35.0
	W	35.0
	Mo	19.0
	V	3.3
	Cr	3.3

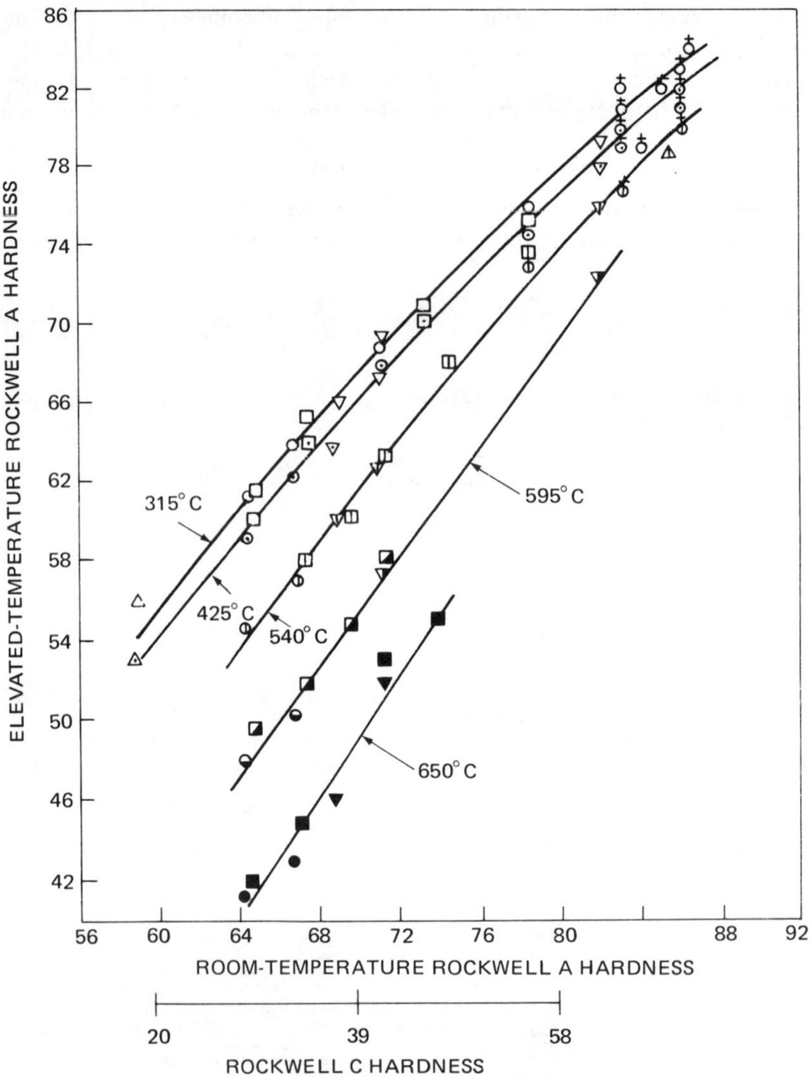

Figure XIII-7. Correlation of room-temperature and elevated-temperature hardness of tempered martensite in tool steels. (P. Payson, *Trans. Am. Soc. Met.*, 51:60 (1959), by permission.)

difference between the liquidus and solidus temperatures, as much as 130°C for the M2 type. The resulting segregation of carbon and alloying elements during freezing leads to segregation of carbides into bands. This is an intractable problem so long as large volumes of liquid metal are being poured into molds, but a highly uniform dispersion of fine carbide particles can be achieved by melting a high-speed steel in the conventional manner, tapping into a closed chamber, atomizing the tapping stream by a gas jet, screening the resulting powder, compacting the powder, then making a solid body by hot isostatic pressing.

APPENDIX

Table A-1 Conversion factors

To convert from	To	Multiply by
Angstrom (Å)	meter (m)	1.000000×10^{-10}
atmosphere (standard)	pascal (Pa)	1.013250×10^{5}
bar	pascal (Pa)	1.000000×10^{5}
calorie/cm·s·°C	watt per meter Kelvin (W/m·K)	4.184000×10^{2}
calorie (thermochemical)	joule (J)	4.184000
day (mean solar)	second (s)	8.640000×10^{4}
dyne	newton (N)	1.000000×10^{-5}
dyne·cm	newton·meter (N·m)	1.000000×10^{-7}
dyne/cm²	pascal (Pa)	1.000000×10^{-1}
electron volt (eV)	joule (J)	1.60219×10^{-19}
electron volt/atom	joule/mole (J/mol)	9.648659×10^{4}
erg	joule (J)	1.000000×10^{-7}
foot	meter (m)	3.048000×10^{-1}
foot·pound (ft·lb) force	joule (J)	1.355818
gauss (G)	tesla (T)	1.000000×10^{-4}
kilogram force	newton (N)	9.806650
kilogram force/mm²	pascal (Pa)	9.806650×10^{6}
kilopond	newton (N)	9.806650
ksi·$\sqrt{in}$	MPa·$\sqrt{m}$	1.098855
oersted	ampere/meter (A/m)	7.957747×10
ohm·centimeter	ohm·meter (Ω·m)	1.000000×10^{-2}
pound avoirdupois	kilogram (kg)	4.535924×10^{-1}
pound/foot³ (lb/ft³)	kilogram/meter³ (kg/m³)	1.601846×10
pound/inch³ (lb/in³)	kilogram/meter³ (kg/m³)	2.767990×10^{4}
pound force/in² (psi)	pascal (Pa)	6.894757×10^{3}
ton force/in² (long ton)	pascal (Pa)	1.544426×10^{7}
tonne	kilogram (kg)	1.000000×10^{3}
torr	pascal (Pa)	1.33322×10^{4}
watt·hour (W·h)	joule (J)	3.600000×10^{7}
year (calendar)	second (s)	3.153600×10^{7}

Table A-2 Physical constants

Avogadro's number, N	6.022169×10^{23} mol^{-1}
Bohr magneton	9.274096×10^{-24} J·T^{-1}
Boltzmann's constant, k	1.380622×10^{-23} J·K^{-1}
Gas constant, R	8.31432 ± 0.00034 J/K·mol
Planck's constant, h	$6.62517 \pm 0.00023 \times 10^{-34}$ J·s

Table A-3 Calculation of ideal diameter[1]

Base hardenability characteristic due to carbon and grain size: Grain size No. 5	No. 6	No. 7	No. 8	Percent of alloys	Mn	Si	Ni	Cr	Mo	V	P	S	Al	Ti
–	–	–	–	0.01	0.014	0.003	0.002	0.009	0.013	0.061	0.011	−0.003	0.006	−0.008
–	–	–	–	0.02	.028	.006	.003	.018	.025	.097	.022	−0.006	.012	−0.018
–	–	–	–	0.03	.041	.009	.005	.027	.037	.137	.033	−0.009	.017	−0.025
–	–	–	–	0.04	.054	.012	.006	.036	.049	.146	.044	−0.011	.022	−0.034
–	–	–	–	0.05	.067	.015	.008	.045	.061	.146	.054	−0.014	.028	−0.043
–	–	–	–	0.06	0.79	.018	.009	.053	.072	.140	.064	−0.018	.033	−0.053
0.021	–	–	–	0.07	.091	.021	.011	.061	.083	.137	.073	−0.020	.039	−0.062
.050	0.012	–	–	0.08	.103	.024	.012	.069	.094	.124	.083	−0.024	.044	−0.072
.076	.038	0.005	–	0.09	.114	.027	.014	.077	.104	.111	.092	−0.027	.049	−0.081
.101	.062	.029	–	0.10	.125	.029	.015	.085	.114	.097	.101	−0.032	.054	−0.092
.120	.084	.052	0.017	0.11	.136	.032	.017	.093	.124	.086	–	–	.059	−0.099
.138	.103	.071	.037	0.12	.146	.035	.019	.101	.134	.072	–	–	.064	−0.112
.155	.121	.088	.056	0.13	.156	.038	.020	.108	.143	.061	–	–	.069	−0.123
.170	.136	.104	.070	0.14	.166	.041	.022	.115	.152	.037	–	–	.074	−0.134
.184	.150	.119	.084	0.15	.176	.043	.023	.122	.161	.025	–	–	.079	−0.146
.198	.164	.133	.097	0.16	.186	.046	.024	.129	.170					
.211	.176	.146	.110	0.17	.195	.049	.026	.136	.179					
.224	.188	.158	.122	0.18	.204	.052	.028	.143	.188					
.236	.199	.169	.134	0.19	.213	.054	.029	.149	.196					
.247	.210	.179	.146	0.20	.222	.057	.030	.156	.204					

Percent of alloys	Hardenability characteristics (continued): Mn	Si	Ni	Cr

EXAMPLE

To calculate the ideal diameter of AISI 8740 Steel having No. 7 grain size and of the following composition:

Analysis	Hardenability characteristics
0.40% Carbon (No. 7 grain size)	0.329
0.85% Manganese	0.584
0.30% Silicon	0.083
0.55% Nickel	0.079
0.50% Chromium	0.318
0.25% Molybdenum	0.244
0.010% Phosphorus	0.011
0.030% Sulphur	−0.009
Sum of hardenability characteristics	1.639

Ideal diameter $D_I = 4.35$ in

.258	.221	.188	.156	0.21	.231	.060	.032	.163	.212	0.91	.606	.214	.124	.472
.268	.231	.198	.166	0.22	.239	.062	.033	.169	.220	0.92	.609	.216	.125	.475
.278	.241	.208	.176	0.23	.247	.064	.035	.175	.228	0.93	.613	.218	.126	.478
.288	.250	.217	.184	0.24	.255	.067	.037	.181	.236	0.94	.616	.220	.128	.481
.297	.260	.225	.193	0.25	.263	.070	.038	.187	.244	0.95	.620	.221	.129	.485
.306	.269	.233	.201	0.26	.271	.073	.040	.193	.251	0.96	.623	.223	.130	.488
.314	.277	.241	.209	0.27	.279	.075	.041	.199	.258	0.97	.627	.225	.131	.491
.322	.285	.250	.216	0.28	.287	.078	.042	.205	.265	0.98	.630	.227	.132	.494
.330	.292	.259	.223	0.29	.294	.080	.044	.211	.272	0.99	.633	.229	.134	.497
.337	.299	.267	.230	0.30	.301	.083	.045	.217	.279	1.00	.637	.230	.135	.500
.343	.306	.274	.238	0.31	.308	.085	.047	.222	.286	1.02	.643	.234	.137	.506
.350	.313	.281	.246	0.32	.315	.088	.048	.228	.293	1.04	.650	.238	.139	.511
.356	.320	.288	.253	0.33	.322	.090	.049	.234	.299	1.06	.656	.241	.142	.517
.362	.327	.295	.260	0.34	.329	.093	.051	.239	.306	1.08	.662	.245	.144	.522
.368	.333	.301	.266	0.35	.336	.095	.052	.244	.312	1.10	.669	.248	.146	.528
.374	.339	.306	.272	0.36	.343	.098	.053	.249	.318	1.12	.675	.251	.148	.534
.380	.345	.312	.278	0.37	.349	.100	.055	.255	.324	1.14	.681	.255	.150	.539
.386	.351	.318	.284	0.38	.355	.102	.057	.260	.330	1.16	.687	.258	.153	.545
.392	.357	.324	.290	0.39	.362	.105	.058	.265	.336	1.18	.694	.262	.155	.550
.398	.362	.329	.296	0.40	.368	.107	.059	.270	.342	1.20	.702	.265	.157	.555
.403	.368	.334	.301	0.41	.374	.110	.061	.275	.348	1.22	.710	.268	.159	.561
.408	.373	.339	.306	0.42	.380	.112	.062	.280	.354	1.24	.718	.271	.161	.566
.413	.378	.344	.310	0.43	.386	.114	.063	.285	.360	1.26	.725	.275	.164	.571
.418	.383	.349	.315	0.44	.392	.117	.064	.290	.365	1.28	.733	.278	.166	.576
.423	.387	.354	.320	0.45	.398	.119	.066	.295	.371	1.30	.741	.281	.168	.581

Conversion of the sum of hardenability characteristics into the ideal critical diameter D_I

Sum	D_I (in)	Sum	D_I (in)
0.740	0.55	1.550	3.55
0.778	0.60	1.556	3.60
0.813	0.65	1.562	3.65
0.845	0.70	1.568	3.70
0.875	0.75	1.574	3.75
0.903	0.80	1.580	3.80
0.929	0.85	1.585	3.85
0.954	0.90	1.591	3.90
0.978	0.95	1.597	3.95
1.000	1.00	1.602	4.00
1.021	1.05	1.607	4.05
1.041	1.10	1.613	4.10
1.060	1.15	1.618	4.15
1.079	1.20	1.623	4.20
1.097	1.25	1.628	4.25

[1] *Republic Alloy Steels*, Republic Steel Corp., 1968, Table 4-1, pp. 78–79

Table A-3 Calculation of ideal diameter *(Continued)*

Base hardenability characteristic due to carbon and grain size				Hardenability characteristics (continued)						Hardenability characteristics (continued)					Conversion of the sum of hardenability characteristics into the ideal critical diameter D_I			
Grain size				Percent of						Percent of								
No. 5	No. 6	No. 7	No. 8	alloys	Mn	Si	Ni	Cr	Mo	alloys	Mn	Si	Ni	Cr	Sum	D_I (in)	Sum	D_I (in)
0.428	0.392	0.358	0.325	0.46	0.404	0.121	0.067	0.300	0.377	1.32	0.749	0.284	0.170	0.586	1.114	1.30	1.633	4.30
.433	.397	.362	.330	0.47	.409	.124	.069	.304	.382	1.34	.757	.287	.172	.590	1.130	1.35	1.638	4.35
.438	.402	.366	.334	0.48	.415	.126	.070	.309	.387	1.36	.765	.290	.175	.595	1.146	1.40	1.643	4.40
.443	.407	.372	.338	0.49	.420	.128	.072	.313	.393	1.38	.772	.294	.177	.600	1.161	1.45	1.648	4.45
.448	.412	.377	.343	0.50	.426	.130	.073	.318	.398	1.40	.780	.297	.179	.605	1.176	1.50	1.653	4.50
.452	.417	.382	.348	0.51	.431	.133	.074	.323	.403	1.42	.787	.300	.181	.609	1.190	1.55	1.658	4.55
.456	.422	.387	.352	0.52	.437	.135	.076	.327	.408	1.44	.794	.303	.183	.614	1.204	1.60	1.663	4.60
.461	.427	.391	.356	0.53	.442	.137	.077	.331	.413	1.46	.801	.306	.185	.618	1.217	1.65	1.667	4.65
.465	.431	.396	.360	0.54	.447	.139	.078	.336	.418	1.48	.808	.309	.187	.623	1.230	1.70	1.672	4.70
.469	.435	.400	.364	0.55	.452	.141	.079	.340	.423	1.50	.815	.312	.190	.627	1.243	1.75	1.677	4.75
.473	.439	.404	.367	0.56	.457	.144	.081	.344	.428	1.52	.822	.315	.193	.632	1.255	1.80	1.681	4.80
.477	.443	.408	.371	0.57	.462	.146	.082	.349	.433	1.54	.828	.318	.196	.636	1.267	1.85	1.686	4.85
.481	.447	.412	.375	0.58	.467	.148	.084	.353	.438	1.56	.835	.321	.198	.641	1.279	1.90	1.690	4.90
.485	.450	.416	.378	0.59	.472	.150	.085	.357	.442	1.58	.841	.323	.200	.645	1.290	1.95	1.695	4.95
.489	.454	.419	.382	0.60	.477	.152	.086	.361	.447	1.60	.848	.326	.203	.650	1.301	2.00	1.699	5.00
.493	.458	.423	.386	0.61	.482	.154	.087	.365	.452	1.62	.854	.329	.205	.653	1.312	2.05	1.703	5.05
.497	.461	.427	.389	0.62	.487	.157	.088	.369	.456	1.64	.860	.332	.208	.657	1.322	2.10	1.708	5.10
.500	.464	.430	.393	0.63	.492	.159	.090	.373	.461	1.66	.866	.335	.210	.661	1.332	2.15	1.712	5.15
.504	.467	.433	.396	0.64	.496	.161	.091	.377	.465	1.68	.872	.338	.212	.665	1.342	2.20	1.716	5.20
.507	.470	.436	.400	0.65	.501	.163	.092	.381	.470	1.70	.878	.340	.215	.670	1.352	2.25	1.720	5.25

.510	.473	.439	.403	0.66	.505	.165	.094	.385	.474	1.72	.884	.343	.217	.673	1.362	2.30	1.724	5.30
.513	.476	.442	.407	0.67	.510	.167	.095	.389	.479	1.74	.890	.346	.219	–	1.371	2.35	1.728	5.35
.517	.479	.446	.410	0.68	.514	.169	.096	.393	.483	1.76	.896	.349	.222	–	1.380	2.40	1.732	5.40
.520	.482	.449	.413	0.69	.519	.171	.097	.396	.487	1.78	.902	.351	.225	–	1.389	2.45	1.736	5.45
.523	.485	.452	.415	0.70	.523	.173	.099	.400	.491	1.80	.908	.354	.228	–	1.398	2.50	1.740	5.50
.526	.488	.455	.418	0.71	.527	.175	.100	.404	.496	1.82	.914	.357	.231	–	1.407	2.55	1.744	5.55
.530	.491	.458	.422	0.72	.531	.177	.101	.407	.500	1.84	.920	.359	.234	–	1.415	2.50	1.748	5.60
.533	.494	.461	.425	0.73	.536	.179	.102	.411	.504	1.86	.925	.362	.237	–	1.423	2.65	1.752	5.65
.536	.497	.464	.428	0.74	.540	.181	.104	.415	.508	1.88	.930	.365	.240	–	1.431	2.70	1.756	5.70
.539	.500	.467	.431	0.75	.544	.183	.105	.418	.512	1.90	.936	.367	.243	–	1.439	2.75	1.760	5.75
.542	.502	.470	.433	0.76	.548	.185	.106	.422	.516	1.92	.941	.370	.245	–	1.447	2.80	1.763	5.80
.544	.505	.473	.436	0.77	.552	.187	.107	.425	.520	1.94	.946	.373	.246	–	1.455	2.85	1.767	5.85
.547	.508	.476	.439	0.78	.556	.189	.109	.429	.524	1.96	.951	.375	.247	–	1.462	2.90	1.771	5.90
.549	.511	.479	.441	0.79	.560	.191	.110	.432	.528	1.98	.955	.378	.249	–	1.470	2.95	1.775	5.95
.551	.513	.481	.444	0.80	.564	.193	.111	.436	.531	2.00	.960	.381	.250	–	1.477	3.00	1.778	6.00
.554	.516	.484	.447	0.81	.568	.195	.112	.439	.535	2.10	–	–	.262	–	1.484	3.05	1.785	6.10
.556	.519	.487	.450	0.82	.572	.197	.113	.443	.539	2.20	–	–	.275	–	1.491	3.10	1.792	6.20
.559	.521	.490	.453	0.83	.576	.199	.114	.446	.543	2.30	–	–	.288	–	1.498	3.15	1.799	6.30
.561	.524	.492	.456	0.84	.580	.201	.116	.449	.547	2.40	–	–	.303	–	1.505	3.20	1.806	6.40
.563	.526	.494	.458	0.85	.584	.203	.117	.453	.550	2.50	–	–	.318	–	1.512	3.25	1.813	6.50
.566	.529	.497	.461	0.86	.588	.205	.118	.456	.554	2.60	–	–	.333	–	1.519	3.30	1.820	6.60
.568	.531	.500	.464	0.87	.592	.207	.120	.459	.558	2.70	–	–	.351	–	1.525	3.35	1.826	6.70
.571	.534	.502	.467	0.88	.596	.208	.121	.462	.561	2.80	–	–	.369	–	1.531	3.40	1.833	6.80
.573	.537	.504	.469	0.89	.599	.210	.122	.466	.565	2.90	–	–	.387	–	1.538	3.45	1.839	6.90
.574	.539	.507	.471	0.90	.602	.212	.123	.469	.568	3.00	–	–	.405	–	1.544	3.50	1.845	7.00

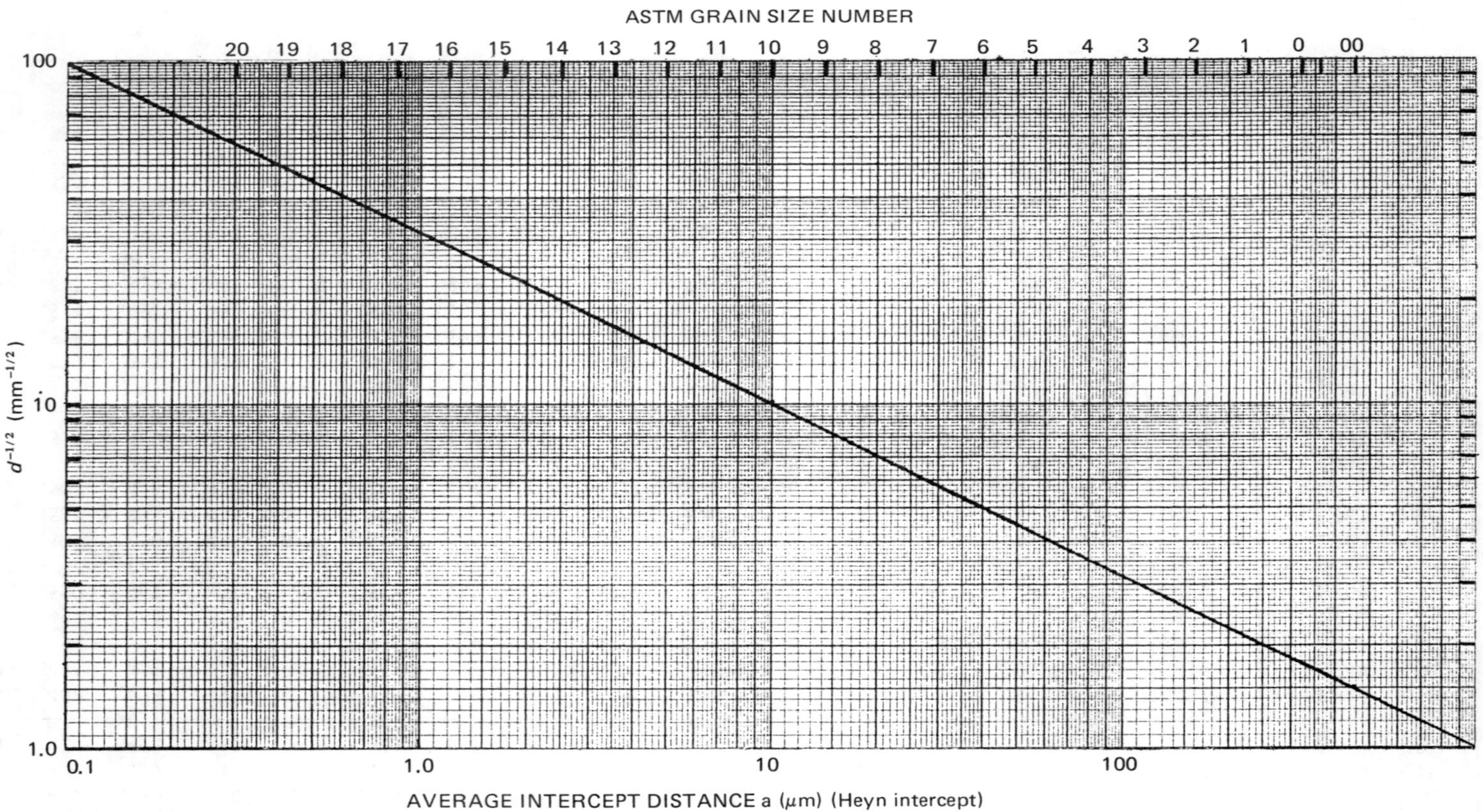

Figure A-1. Grain size relations.

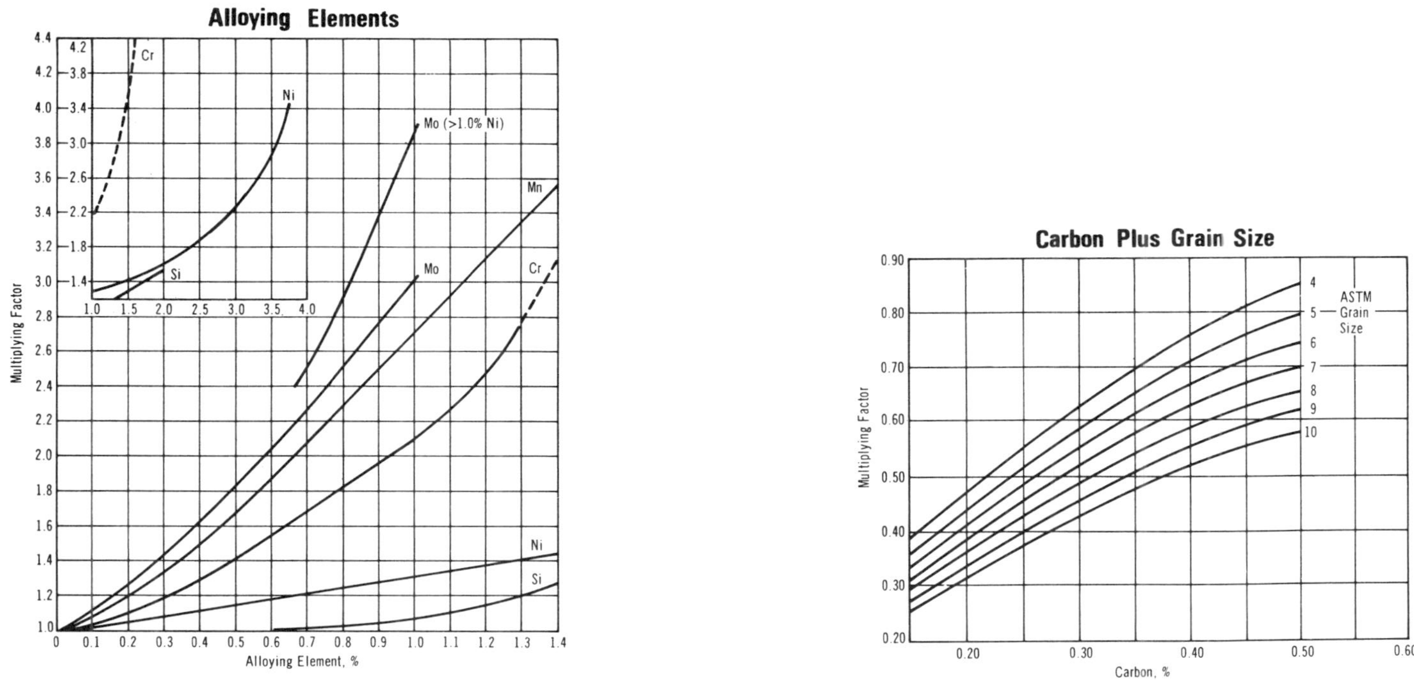

Figure A-2. Factors for predicting hardenability of carburizing steels. To determine the hardenability of a given steel in terms of its ideal diameter D_I, ascertain the multiplying factor corresponding to the amount of each element in the composition. Then, to obtain the D_I, multiply these factors with the carbon-plus-grain-size factor also determined for the composition. Example: An 8822 steel contains 0.24%C, 0.95%Mn, 0.28%Si, 0.56%Ni, 0.44%Cr, and 0.31%Mo. Its grain size is 7.5. Corresponding factors, taken from the graphs, are 0.43 (carbon plus grain size), 2.60 (Mn), 1.00 (Si), 1.18 (Ni), 1.33 (Cr), and 1.44 (Mo). Multiply these factors to get the D_I, 2.53. The alloying multiplying factors were empirically derived from a large number of single- and multialloyed steels with carbon contents in the range 0.15 to 0.30%, and are especially useful in this carbon range. (A.F. de Retana and D.V. Doane, *Met. Prog.*, 100:65 (Sept. 1971), by permission.)

NAME INDEX

SUBJECT INDEX